The Bride of Science

BENJAMIN WOOLLEY is an author and broadcaster whose
work covers subjects ranging from the origins of virtual reality to
the history of colonial America. His books have been translated into
German, Italian, Spanish, Japanese and Chinese, and his
documentaries broadcast across the world.
He lives in London.

BENJAMIN WOOLLEY

The Bride of Science

Romance, Reason and Byron's Daughter

PAN BOOKS

First published 1999 by Macmillan

First published in paperback 2000 by Pan Books

This edition published 2015 by Pan Books
an imprint of Pan Macmillan
20 New Wharf Road, London N1 9RR
Associated companies throughout the world
www.panmacmillan.com

ISBN 978-1-4472-7254-0

3 5 7 9 8 6 4 2

A CIP catalogue record for this book is available from the British Library.

Typeset by SetSystems Ltd, Saffron Walden, Essex
Printed and bound by CPI Group (UK) Ltd, Croydon CR0 4YY

Contents

Acknowledgements

The main sources of primary material I have used in the writing of this book are the Lovelace Byron and Somerville papers, respectively catalogued by Mary Clapinson and the late Elizabeth Patterson. These collections are held at the Bodleian Library, and I spent many hours consulting them in the reading room for Modern Papers, known only (in rather Orwellian fashion) as 'Room 132'. Thanks to Mrs Clapinson, now Keeper of Western Manuscripts at the Bodleian, to the staff there, and to Somerville College, Oxford for making the Somerville papers available.

Thanks also to the staff at the following archival services and repositories: the National Register of Archives, the Carl Pforzheimer Collection of Shelley and his Circle, The New York Public Library Astor, Lennox and Tilden Foundations; Cambridge University Library; the John Rylands University Library at the University of Manchester; the Church of England Record Centre; the county record offices for Hertfordshire, Essex and the North Riding. I am especially indebted to Duncan Mirylees at the Surrey Local Studies Library (now the Surrey History Centre) for helping me track down the elusive John Crosse, and to Tom Mayberry at the Somerset Archive and Record Service in Taunton, not only for the free copy of his excellent book *Coleridge & Wordsworth in the West Country*, but for pointing me in the direction of a variety of important documents relating to Ada's home at Ashley Combe. North Somerset emerges as one of Britain's great literary landscapes, and Mr Mayberry's work provides a perfect introduction to its glories. Another wonderful literary landscape is Newstead Abbey, Nottinghamshire. My enthusiasm for the place is only matched by my gratitude to Haidee Jackson, the Abbey's Keeper of Collections, for her welcome and time.

For their local knowledge, I must also thank Barbara Milne, author of the booklet 'The Rise and Fall of Ashley Combe Lodge near Porlock Weir', Audrey Meade at Fyne Court, and Ann Norbury of the Ealing Local History Library.

Thanks to Yale University Press for permission to quote from *From Mesmer to Freud* by Adam Crabtree, and to John Murray (Publishers) Ltd for permission to quote from *Ada, Countess of Lovelace* and *The Late Lord Byron* by Doris Langley Moore, and *Lord Byron's Family* by Malcolm Elwin.

The descendants of the main characters who appear in the story that follows have been generous both in sharing their family histories and allowing access to papers. The Earl of Lytton, who inherited the Lovelace estates, showed me around the site of Ashley Combe, and gave me permission to consult and quote from the Lovelace Byron papers at the Bodleian; Lord Shuttleworth allowed me access to the fascinating journal kept by his ancestor Dr John Kay Shuttleworth; Dr Neville Babbage provided invaluable suggestions for sources of information about Charles Babbage, one of which was the Powerhouse Museum in Sydney, Australia, where computing curator Matthew Connell found interesting new material for me; and Andrew and Peter Hamilton, Andrew Crosse's relatives, talked to me about their family's fascinating and, at least during the period covered in the following pages, mysterious past.

I have depended heavily on the help of staff at a number of excellent general and specialist libraries: the British Library, the National Library of Scotland, the Wellcome Institute History of Medicine Library, the Science Museum Library and Thomas Carlyle's splendid London Library.

For specialist knowledge, thanks go to Betty Alexander Toole, who devoted so much effort to collating and editing Ada's correspondence (published in her volume *The Enchantress of Numbers*), to Doron Swade, curator of computing at London's Science Museum, and to Howard Nicholson, Leslie Cowan and Douglas Lister for their advice on archaic forms of shorthand. The latter three all expended considerable effort in their attempts to decipher a shorthand letter sent to Ada in 1833. It was Mr Lister who finally cracked the code, and I am grateful to him for revealing its contents, which inevitably turned out to be less sensational than I had hoped.

Special thanks go to Dorothy Stein, for suggesting the title of this book and for helping me in so many ways during its writing; to Clare Alexander for commissioning it, and to Georgina Morley and Nicholas Blake at Macmillan for nursing it into publication. Dorothy Stein, Duncan Wu and Peter Cochran read the manuscript, and I have them all to thank for correcting numerous errors.

Personal thanks go to Anthony Sheil, David Stewart, Asha Joseph, Joy Woolley and Matthew Woolley for support and inspiration.

List of Illustrations

Chapter One: Portrait of Lord Byron in Albanian dress by Thomas Phillips, 1835. *(National Portrait Gallery)*

Chapter Two: 'Fare Thee Well', Cruickshank's vision of Byron departing from England. *(British Museum)*

Chapter Three: 'Ada! Sole daughter of my house and heart.' Ada Byron, aged about five, in an engraving after a painting by F. Stone. *(Murray Collection)*

Chapter Four: View of the London and Croydon railway near New Cross, circa 1840. *(Mary Evans Picture Library)*

Chapter Five: Culbone Church, drawn from nature by Miss Sweeting, on stone by R. Pockock. *(Somerset Studies Library, Taunton)*

Chapter Six: 'A full discovery of the strange practices of Dr Elliotson on the bodies of his female patients!' Front cover of pamphlet. *(Wellcome Institute Library, London)*

Chapter Seven: Drawing of the side view of Babbage's Difference Engine No. 2. *(Science Museum/ Science and Society Picture Library)*

Chapter Eight: Sections of the brains of adult organisms compared. From *Rudiments of Physiology*, John Fletcher, 1835–7.

Chapter Nine: Newstead Abbey. From *The Mirror of Literature*, 24 January 1824.

Chapter Ten: Ada on her deathbed, sketched by her mother. *(Bodleian Library)*

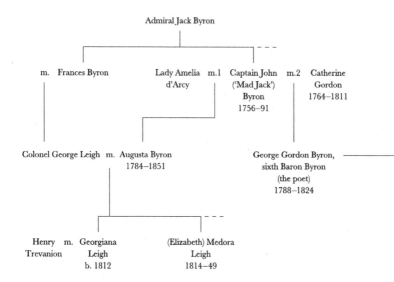

Ada's Family Tree

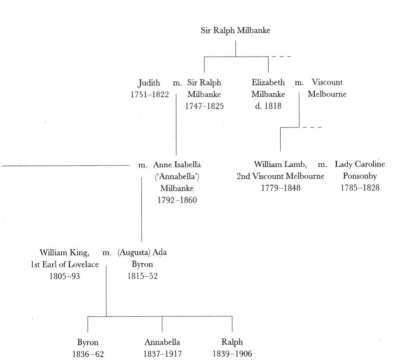

Sir Ralph Milbanke

Judith m. Sir Ralph   Elizabeth m. Viscount
1751–1822   Milbanke   Milbanke   Melbourne
             1747–1825   d. 1818

m. Anne Isabella   William Lamb,   m. Lady Caroline
   ('Annabella')   2nd Viscount Melbourne   Ponsonby
   Milbanke   1779–1848   1785–1828
   1792–1860

William King,   m. (Augusta) Ada
1st Earl of Lovelace   Byron
1805–93   1815–52

Byron   Annabella   Ralph
1836–62   1837–1917   1839–1906

Introduction

'Tis strange, but true, for truth is always strange,
Stranger than fiction.

Don Juan, Lord Byron

Ada Lovelace was the daughter of one of the world's first true celebrities, a man whose poetry was read and likeness seen by everyone – except by her, for she was not allowed to see his likeness in its full romantic glory until she reached her twentieth year. She was brought up by her clever but embittered mother, Annabella, Lady Byron, whose aim, Byron protested in lines addressed to his daughter, was to 'drain my blood from out thy being', to make her everything he was not – mathematical, methodical, moral, scientific. And Annabella's efforts apparently succeeded. Ada rejected poetry in favour of mathematics, art in favour of science. She worked with some of the most interesting and important scientists of the day, figures like Andrew Crosse, a researcher into electrical power who was said to be a model for Mary Shelley's Dr Frankenstein, and Charles Babbage, the inventor of calculating engines. It was the latter collaboration that provided the connection with computers, as in 1843 she wrote a paper about Babbage's most ambitious invention, the Analytical Engine. This paper contains the first published example of what could be called a computer program – written over a century before the emergence of the technology needed to run it.

However, as we shall see, Ada's scientific career took her far beyond an interest in mechanical computation. She began to dabble with dangerous new ideas – mesmerism, phrenology, materialism. She experimented with social and sexual conventions, too. She was flirtatious, outspoken and often shocking; she consorted with people at the margins of society as well as those at the heart of it; she rebelled against the impositions of her gentility and gender; she yearned for financial independence, for fame, to escape the expectations of her peers, to become a 'completely professional person'.

She tried for all this at a time when so much was becoming possible. Her life spanned the era that began with the Battle of Waterloo and ended with the Great Exhibition – a period of barely forty years that saw the world utterly transformed. This was the age when social, intellectual and technological developments opened up deep fissures in culture, when romance began to split away from reason, instinct from intellect, art from science. Ada came to embody these new polarities. She struggled to reconcile them, and they tore her apart.

*

Just as the formation of the earth and the evolution of life have passed through various eras and aeons – the Devonian, the Permian, the Triassic, the Jurassic – so has human history and culture: the Medieval, the Renaissance, the Enlightenment, the Romantic, the Victorian, the Modern. Every historian uses such categories, every historian adapts or questions them. Any work on the Romantic Movement considers whether there was ever any such thing, any analysis of the Victorian Age discusses its obstinate refusal to coincide with the reign of Queen Victoria. We even have 'long' and 'short' centuries. Some have argued, for example, that for historical purposes the nineteenth century began not in 1800 but in 1789, and did not end until 1914.

It is right to be suspicious of these slices through the historical record. The divisions are not obvious and are sometimes arbitrary. They are not like rings in a tree trunk or layers in sedimentary rock. Nevertheless, despite overlappings and interminglings, patterns seem to be discernible. Lines can be drawn, or at least lightly pencilled in. In the final decades of the eighteenth century and the beginning of the nineteenth, poems, paintings, journals and novels – the fossil record of cultural history – show a distinct change in character. Poets begin to write about new themes, thinkers to think about new ideas. These changes do not amount to a coordinated artistic or conceptual movement, more a 'mood', as one observer put it. But what a mood – a real, raging passion, which we now know as Romanticism.

Romanticism was a scream of protest and cry of anguish against a world emptied of spirituality – the legacy of that lower stratum in the rock face of human civilization, the Enlightenment. Religion in some quarters seemed to be in headlong retreat – the hand of God cuffed by the discoveries of Copernicus, Galileo and Newton. The Romantics were on the lookout for something to take His place, and alighted upon themselves. They discovered that, given free reign, their own imagination, a faculty previously dismissed as a mere species of 'fancy', could give them the creative power of any deity. They could invent new ideas, behaviours, moralities, worlds! Just as Prometheus, the mythological hero of Romanticism, brought down from the gods the divine gift of fire, so the Romantic poets brought humanity the divine gift of imagination.

Artists were not the only ones to surge with this new sense of power. Scientists, too, began to believe that they could manage without God. Though centuries of repression at the hands of the ecclesiastical authorities meant that they were often reluctant to declare so openly, many were now convinced that the cosmos, the natural world, life itself were not divine miracles, but the mere flatus of physical mechanics. And, using the scientific method, they were ready to prove it.

Thus we find mixed together in the sediment of the later eighteenth and early nineteenth century signs of new, nimble species of artist and scientist ready to challenge the dominance of the great dinosaurs of divinity that had ruled the world since the Middle Ages.

We also find something else: a sudden and catastrophic environmental change. A new force comes into play, shaping the world in which these creatures live, accelerating their evolution. That force is technology, and the era that erupts out of it is the Victorian Age.

We think of the period in which we now live, already dubbed the 'Information Age', as one of unprecedented technological change. However, the first decades of the Victorian Age saw transformations just as dramatic, if not more so – of time, travel, wealth, health, power and speed. This was when the first, fast, far-reaching public transport and communications systems emerged: the railway network, the postal service, the electric telegraph. This was when a person could for the first time travel from London to Edinburgh in less than a day rather than two weeks, and news could travel the same distance in an instant; when the urban population overtook the rural one; when London was transformed into the world's first modern metropolis; when steam power enabled the mass production of everything from pins to newspapers; when state education, democratic government and public sanitation made their first tentative appearances. The effects of industrialization and mechanization were felt across all levels of society, each new development feeding another in a chain reaction that shook the world, and shakes it still.

Most women of Ada's class were insulated from these seismic shifts, but she felt every tremor. Even her body seemed to resonate. Her health was assailed by the diseases that defined the age, nervous disorders such as hysteria, infectious illnesses such as cholera. But it was in her response to the cultural climate of the time that she particularly captured what had come to be called the 'spirit of the age' – a phrase freshly minted for an era that seemed to be haunted with a sense of anticipation and uncertainty.

*

The 'two cultures' divide between the arts and sciences is now generally treated almost as a genteel dispute, a matter of a slight misunderstanding. If only scientists were a little clearer with their language, if only artists were a little broader in their interests, one could be happily united with the other. But in Ada's time, this division was more ominous, marking the boundary between two world-views that would soon be at war.

To begin with, most of the Romantics rather liked science. They saw it as a promising source of exciting new ideas for their work. Poets should follow in the steps of scientists, Wordsworth wrote, 'carrying sensation into the midst of the objects of science itself'. 'The remotest discoveries of the chemist, the botanist, or mineralogist, will be as proper objects of the poet's art as any upon which it can be employed,' he announced.

Samuel Taylor Coleridge may have thought one Shakespeare or

Milton was worth five hundred Isaac Newtons, but he too was drawn to scientific ideas, which seeped even into the supernaturalism of his masterpiece 'The Rime of the Ancient Mariner'. Science had paved 'the common road to all departments of knowledge,' he once wrote, 'and, to this moment, it has been pursued with an eagerness and almost epidemic enthusiasm which, scarcely less than its political revolutions, characterise the spirit of the age.'

However, just as these artists appeared to succumb to this 'epidemic enthusiasm', many started to rebel against it. In his *Prelude*, Wordsworth, having told his fellow poets to follow in the footsteps of scientists, now began to wonder where they might be taken. These 'natural philosophers', as scientists were then known, should be condemned as

> Sages who in their prescience would control
> All accidents, and to the very road
> Which they have fashion'd would confine us down
> Like engines . . .

Coleridge's 'common road' was now reduced to a railway track, to be traversed without any hope of deviation or diversion.

For their part, the scientists were generally surprisingly respectful of their artistic counterparts. However, they would not defer to them. No aspect of nature and life was allowed to lie beyond their reach. By the mid-nineteenth century, the physician and mesmerist John Elliotson was using Coleridge's drug-induced poetical effusions as examples of the pathology of 'diseased sleep', proving that, to the scientist, the eccentric behaviour of poets no more transcended the scientific order of things than the eccentric orbit of planets.

Such hubris was not new. In 1754 the *abbé* Nicolas Trublet, an influential critic and philosopher, confidently declared that poetry was about to become obsolete. 'As reason is perfected, judgement will more and more be preferred to imagination, and, consequently, poets will be less appreciated,' he wrote. 'The first writers, it is said, were poets. I can well believe it: they could hardly be anything else. The last writers will be philosophers.'

And a century before that the poets were already becoming concerned, Milton himself attacking the astronomers for their attempts to 'gird the sphere'. By the early nineteenth century the girds had become tighter than ever before. Technology meant that science was not just interpreting the world, but changing it. From now on, all a poet could apparently do was stand and watch as the laws of mechanics and mathematics, enacted by engineers and technicians, caused the scientific realm to materialize relentlessly before their very eyes. Subjective experience was to be subordinated to objective measurement, fictional imaginings to factual observations.

The enduring attraction of the Romantics was that at this crucial moment they did not just stand and watch. They challenged the very idea of uniformity, of laws of nature as well as God, and threatened to see if they could break them. In so doing they kept art – indeed, some would argue our very humanity – alive.

Ada's life reveals the flow of this great tide of history just at the point it reached its spring. This book attempts to chart her struggles with the powerful eddies and currents it set in motion, one woman's heroic swim across the Hellespont of history, through turbulent waters that in years to come would engulf us all.

CHAPTER ONE

A Thing of Dark Imaginings

ADA WAS PROBABLY HARD AT WORK in the library studying Dionysius Lardner's hefty *Analytical Treatise on Plane and Spherical Trigonometry* when the carriage took a tangent off the London road and followed the gentle arc leading to Ockham Park. Its cargo was a large flat package – a Christmas present from her mother Annabella, Lady Byron. To innocent eyes, the gift sealed inside would have seemed a generous and fitting one: a precious heirloom that celebrated the bloodline which Ada, just married and expecting her first child, was carrying into future generations. Those familiar with Ada's life and susceptibilities would have thought differently. To them it was a potential parcel bomb. If handled incorrectly, it could devastate its recipient's life the moment it was opened.

Someone was on hand to try to prevent this from happening: Ada's mentor, Dr William King. He was currently resident at her new marital home to keep an eye on her. A physician, lunatic asylum manager, devout evangelical Christian and enthusiastic promoter of the Co-operative Movement, Dr King knew a lot about moral incontinence, the disease it was feared Ada had inherited from her father Lord Byron. It was he who had saved her from ruination by prescribing a course of trigonometry and sums, the study of which he promised would stop her having the 'objectionable thoughts' that constantly assailed her, and would cultivate the sense of discipline she so badly needed.

The medicine had apparently worked. Barely a year after she had nearly plunged to her destruction and dragged her mother down with her, she had married the very respectable William, Lord King (no relation of his namesake, the good doctor), become pregnant, acquired a reputation as an intellectual bright light in the scientific firmament of London, and for the first time shown herself to be properly submissive to her mother's will. There was now not a trace of her father's romanticism or old Regency licentiousness in her. She was a thoroughly modern young woman, an example of what her gender could become in the Victorian age that would be inaugurated in eighteen months' time.

Thus it was decided that she was ready to receive the present that even now approached her husband's elegant, Italianate country seat, trundling past the intricate geometry of ice crystals that had formed on the frosted hedgerows and frozen puddles, towards the line of servants standing in perpendicular attention next to the entrance to receive it.

Lord King himself would have supervised the package being carried into the house and prepared for its presentation. He liked to organize, one of many merits that had won him such adulation from his mother-in-law. For the actual unveiling, however, Dr King would have been required too. Ada's mother Annabella had no doubt insisted on it, to monitor her daughter's response.

Possibly even the clipped wings of Dr King's emotions experienced a little flutter when the moment arrived for the present to be opened. What would Ada do? After so much mathematical training, after so much exposure to the latest ideas about natural philosophy and physical laws, would her reaction be a scientific one – an act of detached observation focused by a certain amount of curiosity? Or would it be of a more dangerous sort, an unreasoning, instinctive, passionate engagement with what she saw?

The sealed casket was opened, the packaging removed, and there it was. For the first time in her life Ada, the daughter of Lord Byron, beheld in its full romantic glory the life-size face of her father.

The portrait of Byron that Annabella had given her was by Thomas Phillips. It created the image by which its fabulous subject would be for ever recognized. It depicted him dressed in Albanian costume, clasping a ceremonial sword, his high forehead wrapped in a billowing turban with a braided silk tassel that cascaded over his shoulder like a tress of hair, his long, smooth jawline containing just a hint of the shape of Ada's own, his cupid's-bow lips framed by a pencil-thin moustache above and a prominent dimpled chin below, his eyes . . . those eyes – years ago, a girl visiting Byron in his bachelor rooms just days before his marriage to Ada's mother stared into them and declared them the finest in the world, large, grey, with long dark lashes making them black: 'I never did behold such eyes before or since.'

It was a picture Ada had passed many times. When as a little girl she had stayed with her grandparents at their Leicestershire seat, Kirkby Mallory, it had hung over the mantelpiece. When her mother had taken refuge at Kirkby just a year after her marriage to Byron and a month after her daughter's birth, a green curtain was drawn over it.

The little girl acquainted at so early an age with this strangely explicit act of concealment may not have questioned it, nor the censorship of her father's poetry, nor the distance that was kept between her and her affectionate aunt Augusta – until Byron's death, the only connection the girl had with her exiled father. But she would have known that her father represented a threat. Everyone from the servants to the local parson's daughter had been told that he might snatch her away at any time, and drag her down like Persephone to the Hades of his dissolute existence. While she was at Kirkby she was to be kept under the constant surveillance of servants to protect her from 'agents' assumed to be lurking in the grounds, waiting to abduct her. Her protective grandmother became convinced one of the maids was in league with these kidnappers and scheming to let them in at night while the rest of the household slept, so she kept loaded pistols next to her bed to fight them off.

Now, just turned twenty, armed with the reflective shield of science and Victorian morality, Ada could safely behold this moral Medusa, and apparently did so without flinching. A satisfied Dr King could reassure himself that his remedy had worked, that she had passed the ultimate test, confronted the man whose countenance was so dangerous that even a glance at his face was once considered potentially fatal. As Lady Lovell, a friend of Ada's mother, had warned her own daughter when she spotted Byron during a visit to Rome: '*Don't look at him, he's dangerous to look at.*'

*

On 25 March 1812 another young twenty-year-old woman dared look upon that dangerous face. Annabella Milbanke was watching Byron attentively from the other side of the drawing room at Melbourne House, where they were attending a morning party given by Lady Caroline Lamb, Annabella's flamboyant London cousin. Annabella observed the young poet as a zoologist would an unfamiliar species of exotic creature in its sultry jungle habitat, noting minutely the way he kept covering his mouth with his hand, how his lips would curl ('thicken', as she put it) with disdain, his eyes roll with impatience. She saw Byron in his full plumage, in other words, which he himself poetically captured two years later in *Lara*:

> There was in him a vital scorn of all:
> As if the worst had fall'n which could befall
> He stood a stranger in this breathing world,
> An erring spirit from another hurled;
> A thing of dark imaginings . . .

Most women were unable to resist this thing of dark imaginings, but Annabella could. She was too clever, too self-possessed to fall for him. She only noted how disgustingly the panting gaggle of girls surrounding him fawned, and passed on to observe with equal condescension the other specimens parading themselves around the room.

That evening, though, she felt obliged to find out a little more about the rare creature she had seen. She learned from one acquaintance, the mother of a Cambridge friend, that Byron's feelings were 'dreadfully perverted'; another told her that he was – she hardly dared breathe the word – an 'infidel'.

Annabella already knew that he was a celebrity, a species that was still extremely rare. As Elizabeth, Duchess of Devonshire, observed, not even the Napoleonic Wars then raging across the Channel could rival him for attention: 'The subject of conversation, of curiosity, of enthusiasm almost, one might say, of the moment, is not Spain or Portugal, Warriors or Patriots, but Lord Byron!' His books sold in quantities that would rival any published in the twentieth century. Ten thousand copies of *The Corsair* were bought on the day of publication. It appeared when the population of Britain was a fifth of its current size and when less than half that population could read or write, and that rate of sale puts Byron's work on a par with *Diana: Her True Story in Her Own Words* – the best-selling book in British publishing history.

His popularity became so great traffic jams would form outside his house in St James's Street as cabs and couriers jostled to deliver invitations to his door. Samuel Rogers, a poet then (as now) better known for knowing Byron than writing poems, observed caustically that all requests for the pleasure of his company carried the imploring postscript, 'Pray, could you not contrive to bring Lord Byron with you?'

The cause of Byron's fame was *Childe Harold's Pilgrimage*, which was published on 10 March of that miraculous 1812, and sold out within three days. In this great poetic travelogue he created the archetype of the romantic hero, first tragic, then Satanic, then melancholic. And this hero, this Childe Harold was – everyone knew it, though he then denied it – himself. It was him who had really run through Sin's long labyrinth, who had loved one who alas! could ne'er be his (plenty of speculation as to who that might be). It was him who, drugg'd with pleasure, almost longed for woe, and from his native land had resolved to go . . .

> Ah, happy she! to 'scape from him whose kiss
> Had been pollution unto aught so chaste;
> Who soon had left her charms for vulgar bliss,
> And spoil'd her goodly lands to gild his waste . . .

There was a hint of something dreadful and dark in the background, unspecified, but signs of which 'in his maddest mirthful mood' would flash along his brow. What could it be? What had he done, this young lord who wrote such potent poems by day, and by night would frolic with friends in the ancient abbey of Newstead, his ancestral seat in the middle of Sherwood Forest, where they would dress up as monks and quaff claret out of skulls plundered from the old monastic burial ground? Whatever it was he had done, it had left him, like the age he so brilliantly captured in his poetry, ruined and ready for redemption, and the idea began slowly to form in Annabella's solemn mind that her destiny was to be his redeemer.

It is an indication of Annabella's self-confidence that she considered it possible that she would even be noticed by such a man. She was, after all, just a country girl, the only child of a humble provincial baronet and Whig MP from the far north of England. She was by no means beautiful, though she had a good figure and, as Byron himself would later observe, the sweetest pippin cheeks.

Nevertheless, there was something about her. She was 'quite the fashion' during the 1812 season, she told her mother. 'Mankind bow before me, and womankind think me *somebody*.' She was barely exaggerating. Suitors tried to burrow their way into her affections like looters into a rich pharaoh's tomb, and all of them were repulsed by the impenetrable wall of her reserve.

One of her attractions was that, through multiple accidents of birth, she had become the heiress presumptive to the venerable and substantial estates of her uncle Lord Wentworth, who had no children of his own and had recently fallen into ill health. She also had some high-powered family connections. Her father, Sir Ralph, was brother to Elizabeth, wife of Viscount Melbourne. Lady Melbourne was a powerful, clever, beautiful Regency socialite and courtesan. She was a matriarch of the louche London world that had emerged in the eighteenth century and would soon disappear from the nineteenth. She represented the life that would so disgust her Victorian successors, the life of late breakfasts, languid luncheons,

> Then dress, then dinner, then awakes the world!
> Then glare the lamps, then whirl the wheels, then roar
> Through street and square fast flashing chariots, hurled
> Like harnessed meteors . . .

It was her ladyship who was to push Annabella into this whirl, and bring her closer to the man then at its hub, Byron.

However, Annabella had something beyond rank and connections that allowed her to think she might attract the attention of such a celebrated man. She, like him, was considered to exist on a plane that floated above the fripperies of society, he elevated by his poetical genius, she by her intellect and piety. He was the Prince of Passion, and she, as he later dubbed her, the Princess of Parallelograms.

She acquired her reputation as a precocious intellectual not just from her manner, which could be decidedly humourless and haughty, but from her knowledge of the latest ideas and philosophies. While others used the London season to indulge the body, she saw it as an opportunity to improve her mind. She attended lectures and read books on the latest ideas, seizing every opportunity to track significant new developments in science and religion, art and literature. She was as interested in geology as poetry, in rock formations as Romanticism. She even dabbled in scientific esoterica – mesmerism, phrenology and mnemonics – and unorthodox theologies, such as Socinianism, which upheld the existence of God but denied the divinity of Jesus and the Christian sacrament.

She could do all this without any risk of disturbing the bedrock of her principles because of her extraordinary powers of self-control. Since the age of thirteen she had resolved to expunge any shred of romantic attachment to the heroes and heroines she read about in fiction and devote herself wherever possible to improving texts by great thinkers, poets and preachers. From then on, even the most inflaming texts were to be considered in the same cold light of her austere rationality and rigid piety. *Childe Harold*, for example, interested her but apparently did not excite her. It was the work of a mannerist, she pronounced, but allowed that its author demonstrated some understanding of deep feeling and human nature.

Thus she could coolly tell her mother that even when she saw Byron that day at Melbourne House, she felt no need to make an 'offering at the shrine of Childe Harold'.

For the next few weeks she was quite content to continue with making her observations of the rituals of sexual selection then being enacted in the dazzling parties and grand balls held across London. With her trained zoological eye, she noted how the 'Super-fine Ladies' had developed a habit of suggestively sweeping up their gowns from around their ankles before sitting down, and the way their dresses 'horribly caricatured' nature. She observed the various calls of the different species of socialite, how one group (which included Lady Caroline Lamb) affected a sheep-like 'baa' sound in its speech, while another (the 'Greville' set) adopted 'a languid, listless, languishing, lifeless *sha'* and yet another (the *'Blues'*) a characteristic bray.

She also entertained herself with a little poetry writing. She composed a short verse on what she dubbed 'Byromania', a not entirely convincing satire on the hero-worship from which she assumed herself to be immune:

> Reforming Byron with his magic sway
> Compels all hearts to love him and obey –
> Commands our wounded vanity to sleep,
> Bids us forget the *Truths* that cut so deep . . .

*

On 13 April Annabella went to a party given by her friend Lady Gosford. Byron – 'the comet of the year', Annabella now labelled him – was there, and he 'shone with his customary glory'.

On this occasion, she was no longer content to observe the object of her interest from afar. She wanted an introduction, and hoped the instrument for achieving one might be the suitor currently forced to dangle humiliatingly from the upper reaches of her lofty pride, William John Bankes. Bankes had been a friend of Byron's at Cambridge. It was his presentation copy of *Childe Harold* received from the author that Annabella had originally read.

Bankes, unfortunately, had other ideas. Sensing he was in the presence of a rival, he started to obstruct Annabella's efforts, holding her in conversation while she darted hungry glances over his shoulder at her quarry.

Perhaps the least forgivable aspect of Bankes's behaviour, which doomed him to eventual rejection, was that he forced Annabella's hand. As the evening, and the opportunities to achieve her objective, were drawing to a close, she was compelled to shake him off and contrive an introduction with Byron on her own.

The great poet was friendly, and flatteringly indicated that he knew of her, in particular her connection with Joseph Blacket. This was not a diplomatic choice of subject. Blacket was Annabella's protégé, a humble cobbler and aspiring poet whom she and her mother Judith had nurtured into publication. Judith had put him up in a cottage on the Milbanke family estate and sustained him with little parcels of cash conveyed by servants. In return for this he wrote touching poems on themes suggested by his patrons, read and complimented Annabella's own poetical efforts, and expired romantically at the age of twenty-three.

Byron had treated his fellow poet very differently. He snootily caricatured him in his collection *English Bards and Scotch Reviewers* as little more than a poetic peasant, 'the tenant of a stall' with 'a pen less pointed than his awl', who had given up his store of shoes and now 'cobbles for the Muse'.

Confronted by the poor man's patron, Byron was more charitable. She was kind, he told Annabella, to support him, and gravely cautioned her against putting too much faith in a certain Samuel Pratt, the man who had been appointed Blacket's literary executor. In 1809 Pratt had published a collection of Blacket's work, which had been dedicated to the Milbankes. The proceeds from its publication were supposed to go to the child Blacket had left behind, and Byron suspected Pratt of pocketing them.

Annabella was touched . . . and, after this first meeting, eager to be touched again.

She sought him out on two further occasions that week, and with each encounter more powerful feelings seeped into the foundations of her self-assurance. They discussed questions concerning poetry and literature. Must a poet have experienced deep emotions in order to make his readers feel them? Byron argued not. Which was the best English novel? Byron cited *Caleb Williams* (a book Annabella would hear him mention again in very different, much darker circumstances).

As they conversed, their views of each other began to change. He found her to be clever as well as confident. She did not flatter him nor respond to everything he said with fluttering eyelashes or inappropriately intense bursts of giggling.

She, too, began to see something different. Indeed, what had begun as an act of observation became transformed into revelation. She suddenly noticed how truly handsome he was, how modest and well mannered, how much one of Nature's gentlemen. He repented his sins

– a close friend of Byron's swore to her he did; he was alone, she had beheld impulses of sublime goodness that would burst through. 'Do you think there is one person here who dares look into himself?' he suddenly asked her in the middle of a crowded room. Then she overheard him saying, 'I have not a friend in the world.' Yes, he had – there and then, he had her. She would be his friend for ever!

And so in a matter of a few weeks the woman who had arrived in London so sure of herself and the control she exercised over her emotions, who had been described by the mother of one of her spurned suitors as an icicle . . . melted. The moment had come for her to reveal who she really was, submit her true, naked self to Byron's gaze – she would show him her poetry.

*

Among the verses she sent were these 'Lines supposed to be spoken at the Grace of Dermody'. They were written when she was seventeen, about the poet Thomas Dermody who had died young from drink and disease:

> Degraded genius! o'er the untimely grave
> In which the tumults of thy breast were still'd,
> The rank weeds wave, and every flower that springs
> Withers, or ere it bloom. Thy dwelling here
> Is desolate, and speaks thee as thou wert,
> An outcast from mankind . . .

The courier appointed by Annabella to deliver these lines about a dead degraded genius to a living one was Lady Caroline Lamb, the daughter-in-law of Annabella's aunt, Lady Melbourne.

Lady Caroline was, at the time, infatuated with Byron. Ever since she had read an advance copy of *Childe Harold*, she was determined she must know its author. She was warned that he bit his nails and had a club foot, but she did not mind if 'he was as ugly as Æsop'.

She quickly discovered that despite the foot and the nails he was by no means as ugly as Æsop, and, as soon she was able, embarked on an obsessive pursuit of him, of the man she famously dubbed 'mad – bad – and dangerous to know'.

Byron had suffered the uninvited attentions of several fans. There was Christina, for example, whose husband, Lord Falkland, had been killed in a duel while she was pregnant. She soon after produced a child, and Byron dropped £500 into the christening cup to help her provide for it. She interpreted this typically extravagant gesture as a betrothal, and proceeded to bombard him with declarations of love which Byron, in desperation, handed over to his lawyers.

But with admirers like Lady Falkland he did not make the mistake he made with Lady Caroline: he did not have an affair with them.

He must have calculated that it would be safe to dally with Caroline, as she was married and presumably unlikely to jeopardize the union by allowing their relationship to develop beyond the casual. At first he was fascinated by her – her thin, androgynous looks (Hobhouse dubbed her the 'mad skeleton'), her unorthodox manners. Sometimes he felt quite affectionate towards her. But he soon went off her, and even became frightened of her. As his interest dwindled, hers multiplied, and following a pattern of behaviour that to contemporary eyes looks remarkably like celebrity stalking, she responded by pursuing him yet more obsessively, alternating gestures of adoration with acts of self-destruction and malice. When adoring, she stood outside houses he was visiting so she could see him; she pledged to give him her jewellery; she sent him locks of her pubic hair, demanding that the loss of blood she had suffered when she had cut 'too close' be reciprocated. When malicious, she threatened to kill either him or herself, sent forged letters signed in his name, set fire to his correspondence.

While the pendulum of her emotions swung from one extreme to another, the mechanisms of her mind remained perfectly regulated. She developed ever more elaborate schemes designed both to pleasure Byron and to punish him, and Annabella now unwittingly became entangled in them.

The poems Annabella had given Lady Caroline were interesting ammunition in their deliverer's hands. They were sentimental, romantic, very nearly passionate, things that the haughty little Miss Milbanke herself tried so hard not to be. Caroline happily passed them on to Byron for his opinion, presumably in the expectation that Annabella would get the same treatment as the poor cobbler Blacket.

Byron did not oblige. His reply to Lady Caroline, sent on 1 May 1812, was unexpectedly generous and candid. He thought there were glimpses of real talent, particularly in the lines on one of her favourite themes, the shoreline of Seaham, which were better than anything produced by Blacket. He was unable to get through the entire letter without a little satirical seasoning, and noted that he liked the Dermody lines so much he rather wished they had rhymed. Nevertheless, he wondered if there were any more samples of her work. Annabella was certainly an 'extraordinary girl', he observed – 'who would imagine so much strength & variety of thought under that placid countenance?' 'I say all this very sincerely,' he concluded. 'I have no desire to be better acquainted with Miss Milbanke, she is too good for a fallen spirit to know or wish to know, & I should like her more if she was less perfect.'

Byron invited Caroline to pass on to Annabella whatever of his opinions she thought proper.

Caroline did exactly that, sending on the entire letter except for the last page, which she retained. On the face of it, if Caroline wanted to dispense with a rival for Byron's affections, this would seem to be a self-defeating strategy. Byron's warm reception for Annabella's treasured poetic effusions, stripped of the final judgement that they were penned by a woman who was 'too perfect' for him, would surely be taken as encouragement. Perhaps, though, that was Caroline's intention. Maybe she was being mischievous, using the letter to draw Byron and Annabella into a relationship that she must have guessed would ruin Annabella. It may seem impossible to imagine anyone being devious enough to concoct such a scheme, but, as subsequent events would demonstrate, a desperate Lady Caroline was very devious indeed.

A few weeks later, Caroline sent Annabella a drunken, disjointed letter, bubbling with biblical warnings: Annabella was in danger, surrounded by worms, vulnerable to being snatched up by 'fallen angels' – Byron, of course – 'who are ever too happy to twine themselves round the young Saplings they can reach'. Everything that comes into contact with London is tainted, not by 'routs, Coxcombs & Gossips' but by those who come in the guise of geniuses and heroes. If

Annabella chose to live like Caroline, 'playing ever on the verge of the precipice & every one almost wishing you in it', then, for all the pleasures, she would endure 'bitter bitter pain'.

Annabella left no record of what she made of this strange epistle. At the time she received it, her feelings for Byron had momentarily cooled. She had seen him at several events and found him increasingly irritable and sarcastic. So what was Caroline playing at? Was this another example of her reverse psychology, another attempt to pique Annabella's interest, rouse her redemptive instincts into action?

If that was not the intention, it was certainly the effect. Ignoring Lady Caroline's portents, Annabella began to allow her feelings for Byron to deepen again, to the extent that she sought reassurance that they were not in some manner diseased. She consulted the Milbankes' family physician, Dr Fenwick of Durham, to help her sort them out. Dr Fenwick had treated her for chickenpox as small child, and was one of many with a deep, paternal attachment to her. He was supportive. It was impossible not to be interested in Byron, he wrote, and his poems demonstrated that he should surely be an object more of compassion than censure.

Annabella spent the remaining weeks of the 1812 season in an uncharacteristically relaxed mood, her journal dotted with exclamations about attending parties and balls at which she 'stayed till sunrise!'. As she prepared to leave London, she drew up a list of men she had got to know during the previous months: it ran to some twenty names, none of which seemed to excite her interest. Byron was not listed.

And so he might have remained but for Lady Caroline Lamb. She had been dragged off to Ireland by her mother in an attempt to keep her out of trouble, but from there she continued to shower Byron with deranged letters. He made matters worse by replying to them, apparently in the hope of keeping her from doing something frantic (like returning to London). Feeling trapped and hounded, he put his ancestral home, the beautiful, romantic ruined abbey at Newstead, up for auction and made plans to leave the country.

The Abbey did not even reach its reserve price. In desperation, Byron turned to Lady Melbourne, Caroline's mother-in-law and

Annabella's aunt, for advice. She had become a close confidante, and, though several decades his senior, perhaps his lover. He wrote a letter begging her to help him get rid of Caroline.

She needed evidence that he meant it – he had, after all, encouraged her by accepting, even inviting, her erotic overtures. So he provided evidence: he was, he blurted, 'attached to another' – Annabella Milbanke.

He knew it was unconvincing, and the harder he tried to explain it, the less convincing it sounded: true, he had mentioned in his last two letters that he had no interest in any other woman, but he had been deceiving himself; yes, he had been writing to Caroline despite being told not to, but that was to pacify her; no, he knew nothing of Annabella's financial situation (i.e., he was not after her money). Having exhausted all avenues of justification, he finally fell to supplication: Lady Melbourne must help set him free, 'or' – a hint here of his more familiar satirical tone breaking through – 'at least change my fetters'.

Lady Melbourne thought he was being ridiculous, and told him so, but he persisted. A few days later he came up with some new reasons for being interested in Annabella: she was of 'high blood', and he still harboured prejudices which made this an attraction; she was clever, amiable and pretty – but not so pretty as to attract the attentions of rivals. 'As to *Love*, that is done in a week . . . Besides, marriage goes on better with esteem & confidence than romance.' Lady Melbourne was still unconvinced, so he finally came clean. '. . . Nothing but marriage and a *speedy one* can save me,' he wailed. And if he could not marry Annabella, then it would have to be 'the very first woman who does not look as if she would spit in my face'.

*

Byron had sent his imploring letters to Lady Melbourne at the end of September 1812. In the early days of October, Annabella said farewell to yet another suitor, the poor William Bankes, who back in the spring had so irritated her by frustrating her efforts to talk to Byron for the first time. After he had taken his final leave of her, she noted down her opinion of him in her diary: he was feeble, incapable of ever expressing what he really thought because he was so anxious to say what he thought people wanted to hear.

That was when she received a proposal of marriage from Byron via her aunt, together with extracts from the letters he had written to justify the match.

She knew she had to reject it and, as with William Bankes, worked out exactly why by sitting down and composing a 'character', a considered analysis of the person's qualities and weaknesses. Whereas Bankes's 'character' was an assassination executed in just three sentences, Byron's was dwelt upon over a period of days and ran to several pages. The resulting document clearly identified the reason why they should never unite, together with all the excuses she could think of to ignore it.

Translating from her almost technical language, she described a man with a very superior intellect oppressed by the tyrannical rule of the passions. He lived for the moment, since he had no faith nor any belief in an afterlife, yet his 'love of goodness' and hatred of human weaknesses showed that he was fundamentally moral. He was impulsive, easily provoked into malevolence, but then recovered and repented. He opened his heart to those he regarded as good, and was humble towards those he respected.

Annabella clearly numbered herself among those he regarded as good and those he respected – encouraged by the proposal she had received from Lady Melbourne, she even thought she was the only one he saw in such a way. So that provided at least one reason why he should want to marry her. Another was a chance to escape the misery resulting from his 'restless inconsistency'. He lamented the lack of 'tranquillity' in his life, and wanted to concentrate on his writing. Annabella, with her devout principles, her seriousness, her maturity, her disdain of flippancy and fads and her circle of close friends – each one of them prototypes of bourgeois Victorian probity – was an obvious candidate to provide this tranquillity.

In a subsequent letter to a friend, she continued her exhaustive exploration for reasons to accept him. She was not, she decided, afraid of his passions, because they were an aspect of his goodness, which she regarded as more sincere than the '*systematic* goodness of colder characters' (she did not include herself in this category). And it was all a part of his 'genius'. The meaning of the word has changed over the years. In the period she was writing it was not, as it is now, reserved to describe people with transcendental intellect and imagination (though, in Byron's case, he was regarded as a genius in that sense too). Rather, she was referring to what he was born with, his innate wit and imagination. Such genius was, she concluded, a requisite of any successful marriage, because it enabled one spouse to understand the other. Indeed, she went further: marriages built on the expectation of contentment were often unhappy ones.

Thus Annabella laboured to justify what her heart so evidently wanted, but her mind knew to be impossible. But even she could not out-argue her own remorseless rationality: a union of passion and intellect would, she knew, never work.

She sent her rejection via Lady Melbourne, together with the character sketch she had laboured over – clear evidence of her interest in him. And Lady Melbourne added a covering note, asking him to treat the rejection gently. He happily did so, pledging never to raise the subject again, and hoping they would become good friends as a result. His heart had never been much concerned with the business, except that it provided an opportunity for Lady Melbourne to become his relation.

In subsequent letters to Lady Melbourne he revisited the subject, more in amusement than rancour. Having read Annabella's character of him, he now seemed to have formulated his own character of her. He concluded she was a creature more at home with numbers and lines than words and imagery. He dubbed her the 'fair Philosopher', the 'amiable Mathematician', a 'Princess of Parallelograms'. 'Her proceedings are quite rectangular, or rather we are two parallel lines prolonged to infinity side by side but never to meet.' He was right, but little did he realize that, for him, Annabella would soon be prepared to try to break the laws of geometry.

Silence followed. Both Annabella and Byron made their rounds of the following London season, seeing each other, but not talking. In her journal, there are occasional jottings: 'Lord Byron makes no profit by his publications' . . . 'Lord Byron never suffers the slightest hint in disrespect to Religion to pass at his table' . . . 'went to Lady Spencer's where I saw Lord Byron at a distance for the first time this year' . . . 'again saw Lord Byron, but without renewing my acquaintance with him'. Two parallel lines . . .

Spring passed into summer, there was another rejection, this time of Frederick Douglas, Lord Glenbervie's heir. Then, at a party given by the spurned Douglas's mother, Annabella saw Byron for the first time with Augusta, and her feelings for him seemed to be instantly and irreversibly transformed.

*

Augusta was Byron's half-sister by Captain John Byron. 'Mad Jack', as he was known, was a ne'er-do-well of the first order, a Guardsman who gave up the army to live a life of dissipation in London, where he had an affair with the lovely Amelia, Lady Carmarthen. She was divorced by her husband the Marquis when the affair became public, and she married Mad Jack. They lived extravagantly in London on the £4,000 per year allowance she received from her father until they ran out of money and were forced by creditors to flee to France. There, in 1784, she gave birth to Augusta, the only one of three children she bore him to survive infancy.

After Amelia's death a year later, Mad Jack returned to England and prowled the ballrooms of Bath in search of a replacement heiress. He soon alighted upon the dumpy but rich Catherine Gordon of Gight, whom he promptly married. Two years later he had squandered her fortune and estates, impregnated her, and was fleeing back to France. Despite being heavily pregnant, she chased after him. No sooner had she found him than she was handed his sickly daughter, Augusta, to care for. Catherine nursed the girl back to health and returned to England, taking Augusta with her. On Tuesday, 22 January 1788, living on her own in a rented flat behind Oxford Street in London, she gave birth to a son with a club foot. Though her husband had temporarily returned to London to harass her for handouts, he failed to turn up to his son's christening, so she named him George Gordon after her father, who had committed suicide by throwing himself into the Bath Canal in 1779.

Annabella would have known a great deal of this history when she saw Byron and Augusta together. What she would not have known was the nature of the relationship developing between them – but she immediately saw that it was something special. Augusta was, in the perceptive assessment of one of Annabella's grandchildren, extremely loveable – an unusual quality among an elite of such unlovely people. With Byron, she was cheerful, childish and playful, and he, as Annabella now saw, responded by being comical and affectionate, revealing a gentle side that she had never seen before, and to which she felt strongly drawn.

Annabella decided she wanted to renew her acquaintance with Byron, and searched with uncharacteristic desperation for a pretext. The one she found was almost pitifully transparent. She sent a letter to her aunt, Lady Melbourne, reporting a rumour that Byron had behaved badly towards the 'imprudent youth' who had just bought Newstead. She had decided to write, she tortuously explained, because she was not going to be seeing Byron again, and wanted it made plain to him that she disbelieved such gossip and was sure he had acted correctly.

In fact, the 'imprudent youth' was a Lancashire lawyer called Thomas Claughton. He had agreed on a price of £140,000 for the Abbey, but finding he could not raise the money was now resorting to various legal devices to delay having to pay the £25,000 deposit.

Lady Melbourne dutifully passed on Annabella's awkward overture to Byron, but he was too preoccupied to respond and left it up to Lady Melbourne to do the job, which she did in her most diplomatic manner.

It was not much, but it was enough. Annabella used the slender excuse of Lady Melbourne's letter to send off one of her own directly to Byron. It was long, very long; Annabella's authorized biographer Ethel Colburn Mayne described it as 'one of the longest letters in the world, containing some of the longest words in the English language', and she was being kind.

The letter begins with a desperate attempt to correct Byron's mistaken impression of her. She may appear to him serene, she wrote, but that did not mean she was a 'stranger to care', or that her future was secure. She too could nurture strong feelings, 'deeply, & secretly'.

If she had admonished his behaviour, it was not the result of 'cold calculation', because *she* had suffered too, suffered as he had suffered.

She recalled their early encounters, 'when I was far from supposing myself preferred by you'. She had studied his character, and had observed that he was in a 'desolate situation'. She could see he was surrounded by worthless cronies and feckless friends. He was flattered and persecuted. How she had felt for him, felt *with* him. And how she had yearned to be able to ignore social manners and express her true feelings.

She cared about his welfare not because she was blind to his errors, but because of the strength and generosity of his feelings, his pure sense of moral rectitude, which could not be perverted by the 'practice of Vice'. He had told Lady Melbourne in the past that, despite her rejection of him as a husband, he was still prepared to conform to her wishes, and she was now ready to hold him to that promise, which she was entitled to do because her only concern was his happiness. She begged him to stop being a 'slave of the moment', to do good, to love mankind, to tolerate his infirmities, to feel benevolence towards others . . . the list goes on and on, as will the steps he will have to climb to reach his salvation.

Finally, she asks him to keep what she has just written secret, particularly from Lady Melbourne, with whom she shared little sympathy, and who might therefore read designs into her words.

Byron's reply was quick and, as he pointed out (twice), brief. After a few words in defence of Lady Melbourne, he announced that he could not trust himself to be her friend – 'I doubt whether I could help loving you'. 'I cannot yet profess indifference,' he concluded, but 'I fear that must be the first step'.

Annabella's response was briefer yet, and spoke volumes: 'I will trouble you no more.'

There, of course, it could have, and should have ended: all square, their lives carrying on along parallel – more likely, diverging lines, never to meet. But it was not to be. The lines were jumped and the result was calamity.

*

Earlier, Annabella had sent Lady Melbourne an outline of her idea of a model husband, which Lady Melbourne had shown to Byron. It provoked him to compare her to Clarissa, the eponymous heroine of Samuel Richardson's novel. Clarissa was a well-bred young lady 'of great Delicacy, mistress of all the Accomplishments, natural and acquired, that adorn the Sex', whose sense of infallibility led her into a ruinous elopement with an unscrupulous man of fashion. Despite all the miserable omens he should have read in this comparison – perhaps, mischievously, because of them – Byron decided to write back to Annabella, giving her what he knew she really desired. He told her he wanted to be friends, that he would obey her, and invited her to 'mark out the limits of our future correspondence & intercourse'. She replied encouraging him to treat her in whatever way he wanted, offering to serve him in the hope that her calming influence might relieve his despondency. And she provided an example of the sort of influence she would exercise. She recommended that he get to know the elderly dramatist Joanna Baillie, a model of the sort of woman with whom he ought to associate. Baillie had all the virtues Annabella valued, and would become a familiar figure in the support-group of respectable, middle-class female friends that was to become such a feature of Annabella's life. Joanna had a simple and truthful countenance, a modest cheerfulness and lack of vanity – and she would be a perfect corrective for the man who had experienced none of these.

Thus began a correspondence in which both enjoyed playing their roles – he the incorrigible creature of passion and levity, she the indulgent but determined therapist of sin. He protested against her calling him despondent: 'On the contrary – with the exception of an occasional spasm – I look upon myself as a very facetious personage ... Nobody laughs more ... The great object of life is Sensation – to feel that we exist, even though in pain ...' She replied to this provocative manifesto to hedonism by saying her tolerance for his faults was inexhaustible, and that she knew his levity was really a disguise for feelings of self-dissatisfaction.

The agenda was set: he argued for indulgence, she for restraint, he for passion, she for reason, he for flippancy, she for seriousness. In the exchange of letters that followed, they ranged far and wide in their efforts to argue their various cases. They discussed religion, he refusing to accept it, she trying to persuade him into embracing it. This provoked him to come out with a powerful, even moving proclamation of scientific alienation, the view that the world meant nothing, that there was no future, that man was but a powerless particle in the great, indifferent mechanical universe revealed by science: he was a mere atom and 'in the midst of myriads of the living & the dead-worlds – stars, systems, infinity – why should I be anxious about an atom?' Being a Unitarian, Annabella had an answer for this that embraced both science and religious faith. Of course we are just atoms in the universal scale, she replied, but even an atom means something to the 'Supreme Being'.

Unitarianism was the Nonconformist religion of choice for the scientifically and industrially minded – a 'featherbed to catch a falling Christian', as Erasmus Darwin, the grandfather of Charles, put it. In the hands of believers like Annabella, it developed into an austere form of faith; God was revealed through reason and conscience rather than miracles and worship. Benjamin Disraeli, the novelist and Tory politician who would himself fictionalize the story of Annabella and Byron's relationship in his 1837 novel *Venetia*, thought that Unitarians were the religious equivalent of Utilitarians, who believed that usefulness, as opposed to beauty or pleasure, was all that mattered in the world. Both doctrines, in Disraeli's view, 'omit Imagination in their system, and Imagination governs mankind'.

Annabella omitted imagination when it came to other topics of discussion, too. In the letters that were now falling like confetti on Byron's doormat, she quoted the great philosophers of science and empiricism – names like Locke and Bacon – to demonstrate how reason provided the route to salvation. 'I have not as high an opinion of your powers of Reasoning as your powers of Imagination. They are rarely united,' she scolded in as playful a manner as she could manage. But perhaps they *could* be united. Perhaps she and he could unite them. What a union that would be!

Byron readily accepted her assessment of him. He claimed to have no head for logic or arithmetic, and took every opportunity to prove it. Being told that two and two made four simply provoked him to find a way of making them make five, he wrote. He was a poet, and poetry had nothing to do with reason. It was 'the lava of the imagination whose eruption prevents an earth-quake'.

When he wrote this, Annabella had no conception of just how explosive Byron's position was. The eruptions to which he was referring were a revised version of *The Giaour* and a completely new poem, *The Bride of Abydos*. And the subterranean tensions they relieved had been caused by events which, had she known of them, would have shaken even Annabella's belief in herself as a judge of character.

*

At the same time as he had was corresponding with Annabella he had been spending more time with his half-sister Augusta, and found his brotherly love intensifying into something more sinister. In this he was following family tradition. Not only had generations of Byrons inter-married, but his father, Mad Jack, had developed incestuous feelings for his sister, Frances Leigh (the mother of Augusta's cousin and husband, the irascible gambler Colonel George Leigh).

And he was following family tradition in another sense, one of which he was acutely self-conscious. He revived it each time he quaffed from the human skull he used as a drinking cup at Newstead, and re-enacted it each time he organized routs in its grounds. In the eighteenth century, the Byrons had become a byword for depravity. Just as the great thinkers and artists of the Enlightenment had mapped the heights of human capability, so they dived into its depths. Just as the French Revolution had shown how the dawn of pure humanism produced the reign of terror, so the Byrons showed how individual genius was accompanied by selfish depravity.

It was not just Mad Jack. There was the 'Wicked Lord', Byron's great-uncle, the fifth baron, who killed a cousin in an argument over the hanging of grouse, re-enacted naval battles in the lake at Newstead, built model castles along its shores in which orgies reputedly took place and out of rage at his son and heir's elopement with a cousin spent the last decades of his life despoiling the estate the poor boy was due to inherit. When he died in 1798 he left behind only pillaged fields, razed woods, stagnant ponds, dilapidated buildings and some pet crickets that he would lure into the house to chirrup him to sleep.

Fifty years later, Byron's daughter and the Wicked Lord's great-grandniece would visit Newstead and, surveying the 'mausoleum of my race', spontaneously proclaim how much she loved her 'wicked forefathers'. The Byron blood, distilled by generations of intermarriage, seemed to compel those who had it running through their veins to dabble with the inner workings of human nature, to test it to its extremes, to run their lives without regulation.

Augusta – in combination with fame, notoriety, vulnerability, loneliness, debts, lust, age and bachelorhood – provided Byron with a chance to dabble to the limit, and he appeared to seize it. In the summer of 1813, while Augusta's husband was away at various horse-racing fixtures, he spent several weeks with her at Six Mile Bottom, the quaintly named hamlet near Newmarket where the Leighs lived. During that time he asked her to go abroad with him, and despite the fact that she had three children and debts almost as catastrophic as his she accepted. Back in London, in a turmoil of excitement and fear, refusing to confide to Lady Melbourne what had happened between them, he lost his resolve. What provoked this is unclear, but he returned to Six Mile Bottom to discuss what to do, and it was decided not to pursue the planned escapade further. He left a few days later, determined to find something, anything, that would distract him from his now uncontrollable incestuous impulses.

First he went to Cambridge to indulge in a six-bottle claret-quaffing binge with his friend Scrope Davies. That didn't work, so he headed on for Aston Hall, the Rotherham home of James Wedderburn Webster. He had gone there, he confided to Lady Melbourne in one of a series of intimate letters, to 'vanquish my demon' – his love of Augusta – by 'transferring my regards to another', that other being the pert if rather fragile twenty-year-old wife of Webster, Lady Frances. Unfortunately, his plan of seduction was going wrong, because he found his passions were not aroused by their object. So he vented his frustration on the hapless Webster, who was trying to wheedle an invitation to Newstead for the purpose – as Byron well knew – of having his way with one of the maids. Teased mercilessly by Byron about his true motives, the poor man fell to defending his honour and that of his wife. She was as good as Christ, the traduced host proclaimed, a comparison that reduced Byron to fits of laughter which in turn reduced Webster to fits of rage. The following morning Byron had to leave for London.

Within a few weeks they had patched up their quarrel and Webster secured his invitation to Newstead, where he spent most of his time emptying Byron's wine cellar and boasting of his carnal conquests. It was this provocation, rather than passion, that persuaded Byron to have another go at Lady Frances. He arranged it so that they would be left alone in the Abbey for one day. The day arrived and he embarked on his mission with the determination of a burglar picking a lock. According to the account he gave to Lady Melbourne, the breakthrough did not come until two in the morning, when suddenly, taking him completely by surprise, she submitted. 'I give myself up to you,' she cried. She could not bear to think of the consequences, yet she was entirely at his mercy – 'Do as you will!' She made no scene, she put up no struggle, yet . . . he could not bring himself to complete his task. She was, as he put it to Lady Melbourne, mixing callousness with genuine feeling, 'spared'. There was something in her voice – hers was not 'decorous reluctance', it was real.

He returned to London in a state of yet greater confusion, his passions for Lady Frances now aroused, and entangled with his unresolved feelings for Augusta.

The lava welled up and poetry erupted: new lines for later editions of *The Giaour* and *The Bride of Abydos*, written within a week to wring his thoughts 'from reality to imaginings'. Both poems are set in Turkey, and both set a new standard of oriental romanticism, opening up to their readers a rich imaginary terrain 'of the cedar and vine, Where the flowers ever blossom, the beams ever shine'.

For all their foreign exoticism, the works toyed with issues much closer to home. In stanzas added to *The Giaour*, there are references to a dangerous love that . . .

> . . . will find its way
> Through paths where wolves would fear to prey . . .

And the subject of *The Bride of Abydos* was incest, the erotic love of Zuleika for her brother Selim, the 'companion of her bower, the partner of her infancy'. The poetical voice is thick with passion:

> He lived – he breathed – he moved – he felt;
> He raised the maid from where she knelt;
> His trance was gone – his keen eye shone
> With thoughts that long in darkness dwelt;
> With thoughts that burn – in rays that melt . . .

Before publication, Byron took fright and turned the two main characters from siblings into cousins. But the poem still scandalized, attracting both censure and sales. Zuleika's erotic language, in particular, outraged the critics:

> Thy cheek, thine eyes, thy lips to kiss,
> Like this – and this – no more then this:
> For, Alla! sure thy lips are flame . . .

No woman should express herself so. The *Antijacobin Review* of March 1814 declared the language 'indecent even in the mouth of a lover' and considered Zuleika's forthright manner unnatural. Even sympathetic liberal journals were shocked. The December 1813 edition of *Drakard's Paper* (which later became the *Champion*, a journal that took a key role in the media frenzy surrounding Byron's later life) accused the author of using 'fine writing' to 'obtain mastery over a story which is in itself positively objectionable. So far from sharing Zuleika's passion . . . our feelings revolt from its contemplation'.

For Byron, there was no question of mastery. The subterranean magma of the imagination welled up inside him whether he liked it or not. What was within him boiled so turbulently it had to find its release, and poetry was, in a sense, just one weakness in a character that he himself admitted to be riven with faults.

*

Annabella was very taken by this image of molten passions forcing his poetry to the surface, and in her next letter to Byron was anxious to reassure him that she was not as mathematically minded as he had implied – indeed, she did not like people of 'methodised feelings', she claimed. She was also concerned to correct the impression she may have given of being superior in giving him religious advice. She was not really like that. She could be poetical, too. And to prove it, she asked to see his new outpourings, as she got more pleasure from his poetry 'than from all the QEDs in Euclid'.

These were unconvincing claims, reserved only for Byron. To her friends, she remained as methodized in her feelings as ever. Her interest in Byron was purely spiritual, she now claimed. Beholding his 'Heaven-born genius' with its lack of 'Heavenly grace' made her Christian heart 'clasp the blessing with greater reverence and love'.

All her efforts to neuter her passion by turning it into piety were soon to be undermined. Byron wrote back to her, and just as she had tried to conform to him in her last letter he now attempted to conform to her, pointing out that, though she might think him capricious, 'I am not quite a slave to impulse'. The proof? The fact that he had managed to live to the ripe age of twenty-six without marrying foolishly. He had searched for an ideal partner, and found only two candidates. One was too young at the time he knew her, and subsequently turned out not to be so ideal after all. The second, the only one he had seriously considered as a wife, had 'disposed of her heart already, and I think it too late to look for a third'.

Annabella probably did not then know the identity of the first candidate. It was Mary Chaworth, his first love and a descendant of the man the Wicked Lord Byron had shot. When Byron's lawyer-cum-guardian, John Hanson, had jokingly suggested at the time that Byron should marry Mary, the young lord had replied: 'What, Mr Hanson? *The Capulets and Montagues intermarry?*' As to the second candidate, the reference was obvious: Byron meant Annabella herself. For it was her who had told him her heart was disposed to another.

Unfortunately for her, her heart was not disposed, but still in her possession, and now palpitating. So why had she told him that she was already attached when she was not? And why, when she had several chances to set the record straight later, did she not do so? She rehearsed one possible explanation in a letter to her friend Lady Gosford. She was being – how could she put it – 'prudential'. In fact she was being positively considerate: she had wanted to avoid hurting Byron, so had allowed this little . . . 'deception' to persist so that he would not develop feelings for her that could never be reciprocated.

Deception, prudence – being forced to apply even these pallid euphemisms to herself was too much for Annabella, and the machinery of her logic started to break down. She repeated to Lady Gosford his words, that she had *disposed of her heart to another* – and added nonsensically that she was 'fully convinced' that she had never done any such thing, as though the idea came from him and not from her. Then she declared that she wanted to put an end to her deception as 'a continuation of silence is an acquiescence in untruth'.

She drafted a reply to Byron, and tried it out on Lady Gosford. She had unintentionally misled him, she would explain, and now felt compelled to reveal the truth 'for the sake of that Sincerity which I have invariably desired to practise'. She added woefully to Lady G. that she would be unhappy until she managed to set the record straight, and blamed the whole episode on the 'effects of Imagination', that mercurial faculty which she came to so despise and would do everything to expunge from her life and her daughter.

The battle between her sense of righteousness and her arousal to an unexpected passion was, at this moment, pulling Annabella apart. She had lashed herself to the mast of propriety with her little 'deception', but now tugged desperately at the ropes, even though she threatened to tear apart the very substance of her moral and rational being in the process. The strain was too much, and she fell seriously ill, her condition made worse by the news that Byron had left the country.

The news was false. Byron was enjoying his continuing part in the self-made melodrama of his relationship with Lady Frances Webster. Despite their abortive liaison at Newstead, she had started to correspond surreptitiously with him, and he had sent her a brooding portrait of himself, just to keep her going. He had already imagined a detailed scenario, appropriately theatrical: he would be found out by the hot-tempered Webster, challenged to a duel, killed, all womanhood would then fall in love with his memory, all wits would have their jest, all moralists their sermon, and, best of all, Lady Caroline, who was still pestering him, would go 'wild with *grief* that – *it did not happen about her*'.

Lady Frances was not his only correspondent. He also received a lock of hair from Augusta tied in white ribbon and enclosed in paper signed with her name. He wrote on the paper that it was the hair 'of the *one* whom I most loved'. And he received a letter from Mary Chaworth, the love of his boyhood, the woman he had identified to Annabella as his 'early idol'. Her husband had just left her and she had taken refuge with a friend and she wanted to see him.

Byron felt overwhelmed. He no longer wanted anything to do with the problems that had so recently piqued his interest, so a policy of mischievous engagement was substituted by a mood of languor and inertia. The 7 December 1813 entry in a journal he was haphazardly keeping at the time read as follows:

Went to bed, and slept dreamlessly, but not refreshingly. Awoke, and up an hour before being called; but dawdled three hours in dressing. When one subtracts from life infancy (which is vegetation), – sleep, eating and swilling – buttoning and unbuttoning – how much remains of downright existence? The summer of a dormouse.

It was more like the winter of a literary lion, that December. He was working on another oriental tale, the one that would prove the most successful and notorious yet: *The Corsair*. It outraged convention not only by making a hero of its villain, but by making a saviour of its heroine, a murderess. It also contained the most complete portrait of the Byronic hero, Conrad, 'That man of loneliness and mystery', of 'one virtue and the thousand crimes', whose

> . . . features' deepening lines and varying hue
> At times attracted, yet perplex'd the view,
> As if within that murkiness of mind
> Work'd feelings fearful, and yet undefined . . .
>
> Slight are the outward signs of evil thought,
> Within – within – 'twas there the spirit wrought!
> Love shows all changes – Hate, Ambition, Guile,
> Betray no further than the bitter smile;
> The lip's least curl, the lightest paleness thrown
> Along the govern'd aspect, speak alone
> Of deeper passions . . .

The Corsair was not the only portrait of the Byronic hero to emerge that spring. Byron also sat for the artist Thomas Phillips, in the full Albanian dress that placed his image in the exotic setting of his poems.

There he was, in the pose that twenty-one years later his daughter, the girl he was only to see for a single moment in her infancy, now beheld as she stood on the threshold of womanhood.

CHAPTER TWO

Wanting One Sweet Weakness

AGAINST ALL EXPECTATIONS, logic, emotion, experience, advice, and on the sofa, Byron 'had' Annabella Milbanke. It was his derisive term, used in his memoirs, according to his friend Thomas Moore, to describe the consummation of his marriage. And never had he chosen a word less wisely. Annabella, it turned out, would be had by no one.

Following Byron's letter identifying her as one of only two women he had ever considered marrying, it took them nearly a year of further correspondence to achieve the union he had assumed was impossible. First, Annabella had to find a way of telling him she was available, without actually admitting either to herself or to him that when she had told him otherwise she was lying. She managed this to her satisfaction by burying her confession in a sermon so incomprehensible that Byron had to refer back to her to explain what she meant (sample: 'I have found that Wisdom (often the most difficult Wisdom, Self-knowledge) is not less necessary than Will, for an absolute adherence to Veracity. How I may in a degree have forsaken *that* – and under an ardent zeal for Sincerity – is an explanation that cannot benefit either of us . . .').

Meanwhile, Augusta had given birth to a daughter, Elizabeth Medora (the name 'Medora' came from a character in *The Corsair*, and is one that will arise again with terrible consequences in our story). Byron wrote a strange letter to his confidante Lady Melbourne on the birth, making an allusion to it being his fault if the baby was an 'ape', arousing the suspicion that he was referring to the medieval belief that the offspring of an incestuous union would be a freak. Whether the suspicion was true – something possibly even he did not know – he now saw marriage as all that could save him from his feelings for Augusta, which were, as he put it, a mixture of 'good & diabolical'.

Annabella apparently offered salvation, as well as sermons on virtue and reading lists of improving texts like Locke's treatise on *The Reasonableness of Christianity*. Unfortunately, her own virtue was being severely strained by her feelings for Byron. She had been reading the new, enlarged edition of *The Giaour*, which contained the lines about love finding its way through paths where wolves would fear to prey, and she was being tempted to stray. The ardent language made her yearn for its author's 'acquaintance'. Indeed, she was now so desperate that even she, the Princess of Primness as well as Parallelograms, would be prepared to risk being called a flirt to achieve it, she admitted.

So she invited Byron to Seaham to meet her parents. While she waited, she took long walks along the seaside, trying to control passions that pounded her sense of self-control as relentlessly as the waves pounded the rocky shore.

Byron was now living in Albany, a famous suite of bachelor's chambers off Piccadilly, and enjoying the so-called 'summer of the sovereigns', when the heads of states of the victorious alliance against Napoleon were in London being entertained by the Prince Regent. He went to the theatre at Covent Garden and Drury Lane, was captivated by Edmund Kean's Othello, attended 'balls and fooleries', 'masquerades and routs', and with his old, faithful friend Hobhouse, back from abroad, talked and drank through to the morning hours.

He also continued to write, starting the sequel to *The Corsair* that became *Lara*, and dashing off two of his most lyrical, romantic poems, 'Stanzas for Music' and 'She Walks in Beauty'. 'Stanzas' was clearly about Augusta, and its beauty only served to prove the intensity of his feelings for her:

> I speak not, I trace not, I breathe not thy name,
> There is grief in the sound, there is guilt in the fame:
> But the tear which now burns on my cheek may impart
> The deep thoughts that dwell in that silence of heart . . .

Since March 1814, Byron had been considering Lady Charlotte Leveson-Gower as a possible match, if only because of her close friendship with Augusta. This resulted in evasive responses to Annabella's repeated invitations to join her in Seaham. In his letters to her, he made teasing references to the union that might have been, that her feelings and his character made impossible – references to which she felt compelled to acquiesce to maintain her image of virtuous piety, but which were winding her up to higher and higher pitches of excitement and frustration.

In early September, she attempted another discourse on religion, trying to correct his pessimism by asking rhetorically: 'Is it proof of reverence for the Governor of all worlds to mock his government of this?' – but then, in a postscript, the pressure valve that had so far done such a good job of regulating her feelings gave way: she wanted to burn the letter, she wrote; her 'apparent inconsistencies' would disappear if they could only meet.

At around the same time as Annabella sent her imploring postscript, Byron's prospects of a match with Lady Charlotte Leveson-Gower came to an abrupt end when he learned that the girl's family were trying to marry her off to someone else. He immediately sat down and wrote to Annabella, asking for her hand, and she as immediately replied, accepting.

There followed a flurry of letters in which they struggled to reassure each other and everyone else that their engagement made sense. Byron tried to explain his 'former conduct' as a matter of being sinned against as much as sinning, Annabella tried to explain her former opinion of his character as being based on 'false accounts', Byron wrote sensible and practical letters to her about his having to stay in London to sort out his affairs, she admired him for being 'mathematical', meaning rational.

Annabella's friends attempted to look on the bright side, most expressing confidence that she had done the right thing, a few, such as Joanna Baillie, only managing to express a hope that she had done so. Lady Granville pithily summed up a more private view: 'How wonderful of the sensible, cautious Prig of a girl to venture upon such a Heap of Poems, Crimes & Rivals.'

How wonderful indeed, and London society was breathless to see the outcome. They had to wait awhile for satisfaction. The journey to the altar was proving hazardous and slow. Byron continued to make excuses for not coming up to Seaham, forcing Annabella to wait out each morning in the blacksmith's cottage, where the post arrived from London.

It took him nearly two months to find the time to make the journey up north, and having finally fixed a date, he still managed to arrive two days late, by which time Annabella's mother Judith was frantic with anticipation.

The first encounter was inauspicious. Annabella was – where else? – in her reading room when she heard his carriage arrive. She put out the candles, and sat for a while in the dark pondering what to do. She decided that the two of them should meet alone. He was ushered into the drawing room, and she found him standing next to the fireplace when she walked in. Upon her entrance, he did not move or greet her. She walked up to him, and extended her hand, which he kissed. She then took up her position on the opposite side of the mantelpiece, and they both stood in awkward silence, as still as a pair of cast-iron firedogs. Finally he spoke, observing how long it had been since they last saw each other, whereupon Annabella blurted out an unintelligible reply and rushed out of the room.

She eventually returned, now in the custody of her parents. She sat next to him, probably in silence, certainly in a state of nervous tension, as Byron and the Milbankes conversed, Byron holding forth on the acting brilliance of Edmund Kean. He was tense, too. He had not remembered her being so quiet; it made him uncomfortable ('I like them to talk, because they *think* less,' he facetiously observed to Lady Melbourne in his report to her on the proceedings). They all parted for bed, Byron asking Annabella as he left what time she rose. She told him ten.

She awoke early, and waited for him. He had not appeared by noon, and she went for a lonely walk among the craggy rocks of her beloved Seaham shoreline. She felt insecure – perhaps for the first time in her life – about what she had done. When they did finally get to talk, she started to probe everything he said, trying to make it fit with the reasoning that had led her to suppose him to be her ideal mate. Nothing did fit.

Things improved when Annabella's father was around. Byron took a shine to old Sir Ralph, who spoke as he found, and remained straightforward and uncomplicated in the presence of this famous poet and prospective son-in-law. Judith, on the other hand, remained in a state of quivering agitation, and got on Byron's nerves.

As their first week together progressed into the next, things began to deteriorate. Byron was becoming aware that Annabella did not add up to quite the mathematical creature he had assumed her to be. To begin with, she seemed to be a hypochondriac, falling ill every few days without the warning of any symptoms (a pattern that was to continue throughout her life, until her death a day before her sixty-eighth birthday). Worse, there had been a 'scene' – one worthy of Caroline Lamb, he told Lady Melbourne.

*

The prelude to the 'scene' was Annabella noticing certain 'strange words and ways' emanating from her husband-to-be. At first she assumed that it was another symptom of his overexcited imagination. But as the days went by, and she continued to dwell on the matter, she became convinced that there must be some other, more concrete cause. As she retold the event years later, she painted herself as having approached the subject in an almost medical manner: she systematically observed a pattern of symptoms, and set about determining a diagnosis. In fact, she was in a state close to panic – anxious and self-tormenting, as Byron described it. She was discovering things about herself as well as him that she had not expected. For example, the Princess of Parallelograms turned out to have very powerful passions. In private moments, they had experimented with caresses, and she was unexpectedly responsive. For Byron, this proved that her behaviour was caused by sexual frustration – she was hysterical and just needed a dose of what he coyly called the 'calming process'.

He was trying to reassure himself, or boasting, or joking. Whatever the cause of Annabella's growing anxiety, it was more than a roll among the seaside rocks could solve. She had created an intricate structure of rationalizations to justify her choice of him above so many as her life partner, and, exposed to just a few days of direct contact, it was already looking shaky. He no longer seemed so keen on the idea of reforming, of letting her choose which religious texts he should read, of making her his 'Guide, Philosopher and friend', of making his heart worthy of hers.

Thus the 'scene' was set.

One evening, when they were alone together, she asked him – 'gently and affectionately', she later claimed – if there was anything that had changed his feelings about their marriage, because if there was, she, acting as a 'true friend', would be prepared to break off the engagement. She portrayed it as an innocent enough enquiry, and was astonished to see it send Byron into a paroxysm. He became 'livid' and fainted on the sofa. When he came round again, he murmured 'indistinct words of anger & reproach' – 'You don't know what you've done,' he said. She pleaded with him to forgive her. He was implacable. From that moment on he would never confide in her again, and his past became a foreign country to which they could never safely return.

Annabella had a habit of veiling accusations as innocent questions. They would be delivered in her most humble, pious manner, and she would measure guilt by the violence of Byron's response. Needless to say, it was as likely to be that insufferable manner as guilt that was responsible for the violence. Elizabeth Barrett Browning knew just how insufferable it could be. She observed many years later of Annabella, 'the pretension to the calmness of absolute justice is – horrible'.

Annabella, beginning to see that she had stronger feelings for Byron than he had for her, was trying to restore the balance, regain control – and control was, to Annabella, the key to her existence. Since Byron had arrived, it had been leaking away. She could not even get him to follow her timetable, any timetable – what chance that he would 'conform to her wishes', as he once put it? Where the malleability or contrition that were so essential to her project of salvation? Putty in her hands? He was more like flint.

Where Annabella was driven by control, Byron was steered by impulse. This was, indeed, one of his greatest attractions to friends like John Cam Hobhouse, Francis Hodgson, and Scrope Davies. He expected nothing from them, and nothing he gave them was expected. He was a master of the extravagant gesture, the gesture that was so

impulsive he could not possibly have calculated the risk or reward it
might bring. Everyone could provide an example of this. In late 1814,
for instance, Hodgson found himself denied the hand of the girl he
loved by the girl's mother on account of his debts. Byron had just
returned from his abortive tryst with Lady Frances Webster at Newstead and learned of his old Cambridge friend's quandary. Setting all
his other concerns aside, he rode all night to meet the mother and
persuade her that Hodgson's debts would be cleared. He then returned
to London, and took his friend straight to Hammersley's Bank, where
he ordered that £1,000 be handed over from his own account,
depriving himself of the funds he needed for his own longed-for trip
abroad.

The obverse of this generosity was an offhand attitude to debts. He
happily incurred them to finance his lavish lifestyle, but never thought
to modify his spending habits to pay them off, driving many a
tradesman to the edge of bankruptcy in the process.

Contrast this with Annabella's attitude to giving money. She was
not stingy, and donated willingly from the considerable sums she would
inherit via the Wentworth estate to anyone she thought would benefit.
But she only gave on a long-term, contractual basis, often using lawyers
as intermediaries, and after careful consideration of the circumstances.
This was of far more practical use to the recipients than Byron's
spontaneous discharges of cash, but it also gave them the leisure to
contemplate their obligations to the donor.

Byron's spontaneity was not, as far as he was concerned, a weakness;
it was a philosophy, and it extended to his relationship to women and
attitude to wedlock. He wanted an open marriage. Any wife of his, he
confided in Lady Melbourne just before he made his second, successful
proposal to Annabella, would be free to do as she pleased just as long
as she left him 'in the same liberty of conscience'. Annabella, of course,
had no such conception of marriage. For her, it was based on
obligations, obligations not just to her as a wife, but as a fellow human
being, an equal.

In subsequent decades, radical writers like Harriet Beecher Stowe
and Harriet Martineau adopted Annabella as a sort of icon for equal
rights (or at the very least, equal treatment) for women. They detected

in the way she applied her radical attitudes the first, faint pulse of what would much later be called feminism. It can be seen in her correspondence with Byron, when he, contemplating with despair the meaninglessness of existence in a mechanical universe, compared humans to aimless atoms. To her, those aimless atoms were all made of the same substance, regardless of gender. Indeed, later in life she would argue that they were the same substance regardless even of class.

Byron, on the other hand, saw the sexes as made of fundamentally different matter. There is ample evidence of this – much of it in his correspondence with a woman, Lady Melbourne. Byron excused her the deficiencies of her sex by electing her an honorary man. In one letter, he applauded Annabella's 'abilities and excellent qualities', but saw these manly qualities undermined by the surreptitious way she tried to make him admire her, which proved that, despite everything, there remained 'something of the *woman* about her'. He acknowledged that he had a poor view of women, and blamed it on their attitude to him. He was disgusted by their shallowness, the way they used him as a substitute for the Byron they really wanted, the romanticized Byron, the Childe Harold/Conrad/Lara/Byron.

*

To his alarm, the Seaham visit was confirming that there really *was* something of the woman about Annabella – something manipulative, scheming, possessive. She only served to confirm this in her behaviour after their 'scene'. Byron cut short his visit, voicing 'grave doubts' about the marriage, and made his way back to London pursued by Annabella's pleas for reconciliation. He had been forced to leave because she was so tense and edgy in his presence. Now she now wanted him back! There were professions of love, promises, supplications – and the ultimate declaration of her desperation: 'I would rather share distress *with* you than escape it without you.'

Breaking his journey at Boroughbridge, he wrote one of his most acidic letters to her. He was, she would presumably be satisfied to learn, 'as comfortless as a pilgrim with peas in his shoes'. He had some business letters which he would pass on to Annabella's mother, 'who having a passion for business will be glad to see any thing that looks like it'.

Within a few days, however, he was softening. Still en route to London, he had called in at Cambridge, where the students, against explicit orders, had roared him to the rafters when he visited the Senate to take part in an election. Buoyed by the experience, surrounded by male company, he wrote gaily to Annabella, 'opposite to me at this moment is a friend of mine – I believe in the very act of writing to *his* spouse-elect . . . "My last will have made you anxious to hear again – and indeed I am so myself." this is a sentence which I have borrowed by permission from my neighbouring suitor's epistle to his Ladye.' (The suitor was Francis Hodgson, the man who only had a Ladye to write to because Byron had paid off his debts.) Annabella should stop scolding herself, Byron continued; she had not offended him. He was 'as happy as Hope can make me, and as gay as Love will allow me to be till we meet'.

As he settled himself back in London, their correspondence continued in the same reassuring vein, he pretending there was no problem, she that the problem could be solved. 'I certainly was not myself during your stay,' she told him – her true self being much more than the 'grave, didactic, deplorable' person he encountered in Seaham. He had called her 'the most silent woman in the world', but at least he now knew how to interpret her silences. Everything would be fine.

Once more, the sound of wedding bells could be heard, not just by the couple, but an increasingly curious public and press. On 4 December 1814 a Sunderland newspaper reported that the marriage had already taken place – setting off premature peals in neighbouring village churches, much to Annabella's amusement. A day later, Byron sent Annabella a clipping from the *Morning Herald*, which reported the marriage of 'two such interesting persons' and commented on Annabella's 'poetical spark' which she was offended to note was described as 'borrowed'.

The real event did not take place for another month. There were problems with the sale of Newstead, Byron wrote, responding to

Annabella's pleas to hasten their nuptials. No excuse, she replied. Young lovers do not need money. They could manage on a small income, with just one house and one carriage, receiving only 'that quiet society which I think we both prefer' (this just three days after Byron had written to her reporting that his friend Thomas Moore was in town and that they had drunk so much he had awoken the following morning with his head in a 'whirlwind'). She would be happy, provided they lived within their means, as she had a particular horror of debt.

Byron, of course, was horrendously in debt. He was also heavily involved with another affair arising out of impetuosity. Just before he had made his first visit to Seaham he had been doorstepped by an admirer called Eliza Francis. She was an aspiring poet, and, much taken by her demeanour, he had sent her home in his own carriage with a £50 cheque to support her work. She was now to be found dawdling on the doorstep of his Albany apartment again, and was welcomed in. She had come to say goodbye to Byron, she said, on the eve of his wedding. She was later to describe what then happened in a narrative that shows how one of the world's great romantic poets became the model hero of cheap romantic fiction.

They talked for a while, and she was about to leave when she felt overcome and reeled. Byron caught her in his arms.

> As I stood with my head bent down, he lightly put aside some curls which had escaped from under my cap . . . and kissed my neck – this completely roused me and I struggled to free my hand, but then he clasped me to his bosom with an ardour which terrified me . . . 'Don't tremble, don't be frightened, you are safe – with *me* you are safe,' he said impetuously, and throwing himself in an armchair he drew me towards him – for a moment I clung to him – I loved for the first time and this must be my final parting with this transcendent Being . . . he had drawn me down upon his knee, his arms were round my waist, and I could not escape . . .

She made to leave, but he 'clasped me to his bosom with frenzied violence while his passionate looks showed what his feelings were'. She spoke gently to him, and he released her, telling her she should go, adding 'not for the world would I distress you'.

On 24 December 1814 Byron finally set off for Seaham, his last journey as a bachelor. 'Never was lover less in haste,' remarked Hobhouse, his best man and companion during the desultory journey. They dawdled up the Great North Road, managing to take three days to reach Newark, another two to get to Thirsk. They finally arrived at Seaham late in the evening of Friday, 30 December. They were so late, Judith, Annabella's mother, was once again overcome with agitation and had been forced to confine herself to her room.

Hobhouse had not met Annabella before, and he was not much impressed. She was wearing a 'long and high dress' – a fashion which rejected the Regency taste for low necklines. He admired her ankles, but not her face, though it improved on acquaintance. When she and Byron were alone, she threw her arms around him and burst into tears. In Hobhouse's company, she was once again the most silent woman in the world, but gazed on her bridegroom dotingly.

The following day, the atmosphere was more relaxed, and Hobhouse even began to see Annabella's attractions, principal among them her unaffected manner. Sir Ralph was on good form, regaling his guests with more of his stories, and the day ended in a jolly mood with a mock marriage, in which Hobhouse took the part of the bride. On that cold New Year's Eve, everyone was happy, everything was fine.

On New Year's Day, 1815, the eve of the wedding, tension returned. Dinner was an altogether less convivial affair, and afterwards Byron bade Hobhouse farewell, as though it was the last time they would ever be together.

The accounts of the actual wedding day and all that followed it conflict. According to the son of Sir Ralph's steward, writing many years later, Byron went missing just before the ceremony, eventually being found in the grounds with a pistol, taking potshots at his glove, which he was using as a target. Hobhouse makes no mention of this incident (but then, being Byron's most loyal friend, he probably would not have done so), remarking only that having awoken and dressed he found Byron already prepared in his room.

The ceremony was to take place in the drawing room. Two kneeling mats were arranged on the floor, and the party assembled to await the bride's arrival. Annabella walked in accompanied by her childhood nanny, Mrs Clermont. She was dressed in a simple white muslin jacket and gown trimmed with minimal lace – her sackcloth for the supplication. During the ceremony, Annabella remained composed, looking steadily at Byron throughout, repeating her vows clearly. Byron was more hesitant.

By eleven o'clock they were married. A six-musket salute sounded out from the front of the house, and the bells of Seaham church were rung. Annabella, now Lady Byron, left the room for a moment, then returned to sign the register. Having completed the task, she headed once more for the doors, glancing at her parents as she went, her eyes filled with tears.

Within the hour she was dressed in her travelling clothes – a slate-grey cape – and was escorted to a carriage waiting outside. It was snowing. She climbed in to find a copy of Byron's poems bound in yellow morocco leather lying on the seat, a wedding gift from Hobhouse. Byron then climbed in. He put his hand out of the carriage window to grasp Hobhouse's. The carriage set off, and he would not let go of his friend, who ran alongside the carriage as it started to make its way up the drive. Finally, their hands were pulled apart and the married couple set off alone on their forty-mile journey to Halnaby, the hall in Yorkshire lent to them for their honeymoon by Sir Ralph.

*

During the journey, they reverted to the behaviour they had manifested during their first days together at Seaham: Byron was temperamental and caustic, Annabella quiet and apprehensive. He began singing in a 'wild manner', she later recalled, possibly chanting one of his favourite Albanian folk songs.

They first arrived at Durham, just a few miles down the road, and the great bells of the magnificent Romanesque cathedral rang out to greet their arrival. 'Ringing for our happiness, I suppose?' Byron asked sarcastically – the first words, Annabella later claimed, he had uttered to her since they set off.

As falling snow softened the landscape outside, the mood froze inside. Recalling the journey years later in a mood of recrimination that makes at least the tone of her account dubious, Annabella claimed that he made a series of bitter remarks designed to unsettle, even frighten her: that he had only married her to humiliate the others who had sought her hand, that he had managed to capture her by outwitting her, that she would suffer for holding out so long before consenting to marry him, that he had been wearing a mask of affection that he now intended to discard, that he hated her mother, who had been the subject of certain unpleasant remarks made by Lady Melbourne.

Thus began the opening act of a melodrama that was to become one of the great sensations of Victorian times. Restaged in the numerous fictionalizations (such as Disraeli's novel *Venetia*), revelations (such as an account of the wedding night in the rabid proto-tabloid *John Bull*), and polemics (such as Harriet Beecher Stowe's *Lady Byron Vindicated*), it came to dramatize the anxieties of the age. Both parties had their roles to play, and both performed them brilliantly. Annabella came to be identified with the modern world: cold and calculating, pious and rational, capable of exercising 'absolute control' over 'the baser kind of spirit'. Byron was the Gothic romantic, 'chain'd to excess, the slave of each extreme'.

According to a report published by Harriet Martineau, a servant recalled that as soon as the carriage arrived at Halnaby Byron jumped out and walked away, leaving Annabella to make her way to the entrance on her own. Her expression, the servant claimed, was one of horror and despair. In another account, given by another servant to a newspaper in the 1860s, Annabella arrived in buoyant and cheerful mood, and gratefully received the welcome extended to her by the staff and tenants. It is Thomas Moore's biography of Byron – a controversial account based on what Moore could recall of his subject's memoirs, which he had read soon before they were destroyed – that tells us of Byron's brusque consummation of the marriage on the drawing room sofa. Those same memoirs were claimed as the source for a theatrical account of the wedding night reproduced in 1824 by *John Bull*:

> It was now near two o'clock in the morning, and I was jaded to the soul by the delay. I had left the company and retired to a private apartment. Will those, who think that a bridegroom on his bridal night should be so thoroughly saturated with love, as to render it

impossible to yield to any other feeling, pardon me when I say, that I had almost fallen asleep on a sofa when a giggling, tittering, half-blushing face popped itself in at the door, and popped as fast back again, after having whispered as audibly as a *suivante* whispers upon the stage, that Anne was in bed. It was one of her bridesmaids.

Yet such is the case. I was actually dozing. Matrimony soon begins to operate narcotically – had it been a mistress – had it been an assignation with any animal covered with a petticoat – any thing but a wife – why, perhaps the case would have been different.

I found my way, however, at once into the bedroom, and tore off my garments. Your pious zeal will, I am sure, be quite shocked when I tell you I did not say my prayers that evening – morning I mean. It was, I own, very wrong in me, who had been educated in the pious, praying kingdom of Scotland, and must confess myself – you need not smile – at least half a Presbyterian.

The account was widely believed at the time of its publication, despite two very obvious errors that reveal it to be a forgery (Annabella was never called Anne by Byron or anyone else; there were no bridesmaids at either the wedding or the honeymoon). A further revelation in the same account was also believed, this time based on a story that came via Samuel Rogers, the poet and gossip who had been so piqued at being used by society women to wheedle introductions to Byron during his *annus mirabilis*. It concerned the following dawn:

It was a clear January morning, and the dim grey light streamed in murkily through the glowing red curtains of our bed. It represented just the gloomy furnace light with which our imaginations have illuminated hell. On the pillow reclined my wife . . . She slept, but there was a troubled air upon her countenance. Altogether that light – that cavern bed – that pale, melancholy visage – that disordered and dark hair so completely agreed with the objects I had just seen in my slumbers that I started . . . 'Hail Proserpine' was again upon my lips, but reason soon returned.

The details may have been wrong, but the tone was right. Both felt their Honeymoon was turning into a private Hell. Whatever the truth of the days that followed, which can never be fully known, all the accounts of them that remain (principally Annabella's own, written for the benefit of her lawyers) portray a couple sinking into a mixture of Gothic romance and horror equal to any work of fiction – indeed the model for many, embracing passion, crime, love, superstition, and madness.

*

On the morning of the first full day of their marriage they awoke to a glacial world, with snow covering Halnaby's grounds and ice its ornamental lake. No doubt partly because her fingers were so cold, Annabella used a piece of black ribbon to secure her wedding ring to her finger. When Byron, who unlike his wife was superstitious, saw what she had done, he became terrified that it was some sort of omen, and told her to remove the ribbon. She did as he commanded, and as they warmed themselves by the dressing room fire the ring slipped off her finger and into the grate, which aroused him to a yet more agitated state of foreboding.

His temper worsened when a letter was delivered. It was from his sister and now Annabella's friend, Augusta. He read sections from it in which she expressed her feelings for him, and asked Annabella's opinion of them. He openly lamented his sister's absence and craved her love.

The mood continued, with Byron becoming preoccupied with his past, the madness in his family which had pushed his mother's father to suicide and a cousin to arson. He kept going on about his past crimes, about being a villain, about a curse that lay upon their marriage, about some act he had performed for which he could never forgive himself, which he might have avoided if only she had married him the first time he had proposed to her.

Annabella began to suspect that these ravings, particularly the ones about his sister, were more than the emanations of an overactive imagination. Yet she claimed there was no ulterior motive to her casual remark, made a few days later, that an incestuous union between brother and sister might arise if they were ignorant of their parentage. Byron flew into a rage, demanding to know what had provoked such a comment. Why, her reading of Dryden's tragicomic play *Don Sebastian*, she said, in which just such an episode takes place – what else? At this point, Annabella later claimed, Byron picked up his dagger, which lay next to his pistols on the table, and left the room. He walked along the gallery to his own room, and slammed the door.

He reappeared later that evening, wearing an expression that recalled for her one described in his poem *Lara*:

> For Lara's brow upon the moment grew
> Almost to blackness in its demon hue . . .

During the long nights, which he so reluctantly shared with her, he would get up and restlessly pace the long, dark gallery outside his room, giving every indication that he might yet fulfil a threat to commit suicide. When he lay with her in their hotbed of hellfire, he was in torment, like the child locked in an oven – the Victorian description of damnation, dwelling on the poor little creature's agony as it writhed in the inescapable furnace.

He told her he had children, and another wife. She lay her head upon his chest, and he said, 'You should have a softer pillow than my heart' – and it moved her to despair.

He spoke of a seduction – not of Lady Frances Webster, to which he had already confessed, but another, far worse. Annabella asked if Augusta knew about it; don't ask *her*, he said, and went to sleep. The following morning Annabella raised the issue again, appealing to him to let her help, to let her share the burden of his conscience. He refused, then, staring into the fire, said that perhaps one day she would know his secret – but warned that she should not forget *Caleb Williams*. It was the book that he had mentioned to her so admiringly during one of their first encounters. Its author was William Godwin, husband of the radical feminist Mary Wollstonecraft, and father of the future creator of *Frankenstein*, Mary Shelley. Godwin's aim in writing the novel was to reveal 'the modes of domestic and unrecorded despotism'.

The story concerns the murder of a country squire. Falkland, a proud, high-minded man described as the 'fool of fame', is suspected of committing it. He appoints the humble Caleb Williams as his secretary, and Caleb becomes convinced of his employer's guilt. The rest of the story concerns the complicity and suspicion that binds them together and tears them apart – Falkland's relentless persecution of the servant that suspects him, Williams' desperate desire to serve his master loyally and thereby save him. The similarities between fiction and fact did not need explaining to Annabella, though she began to wonder how far the parallel extended when it came to the crime her own Falkland had committed. Was it incest? Or was it . . . ? Hints and comments drew her irresistibly to the suspicion of a murder. But he would tell her nothing. She was stranded in what she described as a 'wilderness of doubt'.

'Now – sit down and convert me,' Byron challenged her one day. She refused, telling him that 'his own dispassionate reflection' would do the job better. And there was no point in him trying to unconvert her either, she warned. Her beliefs – those 'fixed rules and principles squared mathematically', as Byron would later describe them – were unaffected by his attacks. But she did rankle at his charge that she was a Methodist (a charge she knew came from his drinking friend Hobhouse). She had too much feeling for such an austere philosophy, she protested – had demonstrated as much in her forgiving attitude, her desire to bestow 'peace and cheerfulness . . . in an atmosphere of darkness & doubt'.

She was desperate to find some firmer ground. She began to study books like *Free Enquiry* by Conyers Middleton, a militant eighteenth-century theologian who attacked biblical mysticism – just the sort of bracing rationalism that she needed at a time when she was immersed in so much emotion. She wrote a few notes about what she had read, celebrating her delight in the open-minded contemplation of religious ideas, her belief in the importance of each individual to confront their faith.

As she ascended higher into the abstractions of divinity, Byron dabbled with damnation. He was a fallen angel, he told her, dragged down by his love for mortal women – and by his deformed foot, a curse from Heaven which he would avenge through his wickedness. He spilled over with Old Testament rage, which emerged in a poem he was writing at the time, one of the *Hebrew Melodies* entitled 'The Destruction of Sennacherib':

> For the Angel of Death spread his wings on the blast,
> And breathed in the face of the foe as he pass'd;
> And the eyes of the sleepers wax'd deadly and chill,
> And their hearts but once heaved, and for ever grew still!
>
> And there lay the steed with his nostril all wide,
> But through it there roll'd not the breath of his pride;
> And the foam of his gasping lay white on the turf,
> And cold as the spray of the rock-beating surf . . .

Poetry. It was the one thing that could unite them, a higher dimension where different laws of geometry applied, where two parallel lines could meet. In her own poetry she had struggled to find that other realm, with Byron at Halnaby she perhaps came closer to it than ever before. They would spend time in the library together, he working on the later portions of *Hebrew Melodies*, she making neat copies of the earlier verses, verses such as:

> She walks in beauty, like the night
>   Of cloudless climes and starry skies;
> And all that's best of dark and bright
>   Meet in her aspect and her eyes:
> Thus mellow'd to that tender light
>   Which heaven to gaudy day denies . . .

Those moments together she later described as 'springs in the deserts'.

*

The honeymoon came to an end three weeks after it had started, and in much better humour. When they arrived at Seaham, Nanny Clermont noted a change in her beloved Annabella's appearance – the woman who returned was not the flower she had seen depart. But the couple seemed happy, even playful. They frolicked during parlour games – 'Duck & Devil' 'Ramble-Scramble' – Byron going perhaps a step too far when he snatched Judith's wig from her head. In a more tender act of coiffure, Annabella cut a lock of Byron's curly hair, which she preserved to her death. Their physical feelings intensified, with Annabella discovering a yearning for moments of private 'mischief'. Defying his lameness, he would scramble across the shoreline rocks, challenging her to follow, until they both reached a favourite spot, a crag dubbed the 'Featherbed', where he stood. '. . . That form is before me now,' Annabella later lyrically recalled,

> That eye is beholding the waters roll,
> It seems to give them a living soul;
> That arm by mine is tremblingly prest,
> I cherish the dream – he shall be blest!
> Oh yes – tho' the phantoms fade in air,
> The heart's devotion may not despair!

Then, one night, he said to her, 'I think I love you – better even than Thyrza . . .'

It was the most meaningful declaration that Byron could bestow. The poem *To Thyrza* had originally appeared as part of *Childe Harold* in 1812, but had been written in response to a very different impulse. It was an elegy about the death of a loved one, identified only by the woman's name Thyrza. Following the publication of the poem, Byron was besieged with letters from admirers wanting to identify themselves with the subject of the poem and the object of his heart-rending endearments. Lady Falkland even imagined he had meant the lines for her, that it was she who gave

> . . . the glance none saw beside;
>> The smile none else might understand;
> The whisper'd thought of hearts allied,
>> The pressure of the thrilling hand;
>
> The kiss, so guiltless and refined
>> That Love each warmer wish forebore;
> Those eyes proclaim'd so pure a mind,
>> Even passion blush'd to plead for more . . .

Those lines were not for Lady Falkland, nor for any woman; they were addressed to a youth, John Edleston, for whom Byron had developed a violent passion at Cambridge – a passion that for all its declared purity has been the source of speculation ever since. Edleston died soon after Byron had returned to England from the travels that inspired *Childe Harold*, and the effect had been devastating, changing both him and his first masterwork, casting a deep shadow across both.

Annabella had no idea who the true Thyrza was, and assumed that it was another woman, perhaps the one he had written about in his first love letter to her, the only other he had ever considered marrying. But now it was *her*. That glance, that smile, that whispered thought of hearts allied – they were *hers*. At last there was, she thought, a gleam of hope.

He had no hope; only fear, fear about the past he could not escape, and the future it would contaminate. One midnight, after a few brandies, he sat down and wrote

> There's not a joy the world can give like that it takes away,
> When the glow of early thought declines in Feeling's full decay . . .

Byron wanted to return to London, and at first suggested he do so alone, so he could sort out his financial affairs. They were getting worse, because of the problems he was having with the sale of Newstead. Annabella insisted on accompanying him, and the moment she stepped into the carriage noted a change in his manner which chillingly recalled the trip to Halnaby. As they approached Six Mile Bottom, where they were to spend a few days visiting Augusta, Byron's mood became even more unpredictable. Within a few miles of their destination, he suddenly began to caress her, and asked her to kiss him, even though they shared the carriage with a maid.

Their stay at Six Mile Bottom was a disaster, at least as Annabella portrayed it in her later recollections. All his madness seemed to return, his cruelty, his brandy drinking, his violent mood swings, his dark allusions to unspeakable crimes. When the three of them were together, he would talk openly to Augusta about what he had been getting up to while he was engaged to Annabella. On one occasion, he lay down on the sofa and asked Annabella and Augusta to take turns in kissing him, a ritual which left Annabella with the impression that he loved his sister more than his wife. Having tired of these games, he would wave Annabella away so that he could be alone with his sister, she then being confined to her room until he came to bed, usually drunk and bad-tempered.

The conversation turned one evening to the subject of portraits, and Annabella suddenly suggested that Byron should be painted while he was looking at Augusta's youngest daughter, Medora, then not a year old, because he would gaze at the little infant with such adoration. This innocent suggestion (but was Annabella really being so innocent?) produced a livid reaction from Byron, which Annabella then claimed not to understand. Nearly forty years later, she confided to a friend that he had then referred to little Medora as his child. She had assumed he meant godchild, but wondered if the ambiguity was deliberate, an apparition of the terrible truth about the little girl that was to haunt Annabella and her own child, Ada, years later.

In her account of the final days of their stay at Six Mile Bottom, Annabella resorted to a shorthand system to disguise several allusions to sexual issues. The first of these was Byron's boast that he knew that Augusta wore drawers, thus proving to Annabella his intimate knowledge of his sister's undergarments. The next coded observation was altogether less prim. Byron said that Augusta had been looking particularly tired since his arrival – and he left Annabella to draw her own conclusion as to why this should be, rightly assuming it would be the worst conclusion imaginable. After his intimate caresses in the carriage as they arrived at Six Mile Bottom, he had not touched Annabella, who began to suspect he was satisfying his urges elsewhere. But then, as the visit was drawing to a close, he suddenly resumed sexual relations. Annabella subsequently discovered that this renewed vigour coincided with Augusta's period.

The newly weds left Six Mile Bottom after two weeks and headed for the house found for them by Lady Melbourne at 13 Piccadilly Terrace, one of London's smartest streets. As the carriage carried them away, and Byron waved his handkerchief to his sister, Annabella decided to suppress her suspicions and content herself with the thought that in Augusta she had not a rival but a friend, a collaborator who could help her in the management of her fractious husband. She would need such help in the coming months, as she had just discovered she was pregnant.

Meanwhile, beyond the bumpy ride they were enduring in their carriage and their marriage, the world they were about to re-enter was itself suffering from a series of shocks. Old Sir Ralph had written to Annabella about Napoleon's escape from Elba, describing 'Nap' as 'a kind of Raw head & bloody bones, to keep the world in continual alarm', and lamenting the prospect of more war when at his age he craved only peace. Two days later, Judith wrote about riots in Sunderland, which she anxiously observed imitated those performed by the 'Mob of the Capital', referring to agitation against political repression and fluctuating food prices in London. Over the coming months, these two events were respectively to culminate in the Battle of Waterloo and the passing of the 1815 Corn Law, the last great victories of one generation about to hand on power to the next. And over the coming months, Byron's marriage was to reach its own, equally dramatic culmination.

*

Annabella and Byron moved into their new home, an imposing terraced house overlooking Green Park, during the spring of 1815. Byron was in excellent humour, looking forward to resuming metropolitan life. Annabella's pregnancy must have added to the sense of a new beginning and the hope of a happy future. Byron plunged back into the publishing world, particularly enjoying an encounter with Sir Walter Scott, his only rival at the time in terms of literary celebrity. Though they were politically and philosophically at odds, they got on well, Scott mischievously suggesting that Byron's views would change as he got older (Byron was then twenty-seven, Scott forty-three – an age Byron would never achieve). Byron took this to mean that he would turn Methodist, but Scott cannily recognized that a character as baroque as Byron could only ever be Catholic.

Annabella was probably not having as much fun dealing with the domestic side of their lives – she was never a natural home-maker – and was relieved of her duties by the news that her uncle, Lord Wentworth, had been taken seriously ill. Byron, who stood to inherit the Wentworth estates through his wife, had been anticipating this development for some time, and not all that patiently. The viscount was 'immortal', he had complained to Hobhouse in January, growing healthier by the day and 'at this moment cutting a fresh set of teeth'. Not any longer. Annabella went to his London home, and found him close to death. She installed herself at his bedside and remained there for several days while her mother Judith (Wentworth's sister) made arrangements to come down to London. He died on 17 April.

If Byron had hopes of immediate financial relief, they were to be dashed when it was discovered that the venerable viscount's beneficiary was not Annabella but the detested Judith, Annabella's mother. Under the terms of the will, the Wentworth family name, Noel, the ancestral seat at Kirkby Mallory in Leicestershire, and an income of £7,000 a year were all to go to the Milbankes.

Despite the news, Byron, who always tried to rise above monetary matters, seemed unaffected by such developments. If anything, his mood improved, and as spring passed into warm summer, so apparently did their marriage. A new alchemy had somehow managed to unite the pious and the profane, the romantic and the mathematical. Byron's friends commented on how he doted on Annabella, how at parties he would hang over the back of her chair, talking to her, proudly introducing her to passing acquaintants. An American traveller who paid the couple several visits noted how affectionate Byron was towards her, and Murray, Byron's publisher, admired her temper and good sense.

Then, sabotage – not by Byron, but, astonishingly, by Annabella. Despite Byron's explicit warnings, despite all the evidence from Six Mile Bottom, despite all the good sense Murray had so correctly observed in her, she decided to invite her sister-in-law to stay with them in Piccadilly.

Annabella provides no rational explanation for this decision, which leaves plenty of room to speculate about some irrational ones.

In his book *The Worm in the Bud* Ronald Pearsall observed of Victorians that 'when sex semaphored its presence, reason retired in confusion'. In this, as in so much, Annabella seems to have anticipated the age about to begin. It is a mistake to think of her as repressed – she was quite open, almost in a Regency fashion, about her erotic feelings for Byron. But when not directed towards their legitimate object, her urges seemed to cause her more problems.

It is, of course, expected of all modern biographers that they question the sexuality of their subjects. In Annabella's case, the evidence is slender. But it is still suggestive. She had a number of infatuations for older women during her early years. They would be called schoolgirl crushes if she had ever gone to school, but the way she remembered them in later life clearly demonstrated that they had left a deep mark. She recalled experiencing an 'indwelling fervid feeling' towards Sophia Curzon, her cousin. Sophia had been adopted by her mother after her aunt's death, and would later become Ada's godmother in favour of Augusta. Following Sophia's marriage to Lord Tamworth, Annabella was so miserable that she 'shrank into an existence of solitary reverie'. She felt a similar way after the departure of the next object of her devotion, the 'beautiful, lonely' Harriet Bland.

Augusta also aroused strong feelings. When Annabella first saw her, sitting next to Byron at a party, she was captivated. She later confessed to Augusta that she stared at her as intently as she could 'decorously' do so. From then on, she seemed almost as enraptured by Augusta as she was by Byron, and it seems possible that among the passions thus aroused were some illicit ones, desires that would soon fuel overwhelming feelings of jealousy. This would explain her positively Byronic mood swings towards Augusta, from loving to murderous. And the toll of all that repression might explain the vindictive attitude she held towards her sister-in-law in later years . . .

Another, less exotic explanation for her inviting Augusta to Piccadilly is attention-seeking. Perhaps she saw it as her only means of rousing Byron to a response, of regaining his attention – cruel, if it could not be kind – when she saw it being diverted by the distractions of metropolitan life. If that was her aim, she only partially succeeded. As soon as Augusta arrived, Byron went berserk, treating his wife as savagely as ever. But soon the threat (she claimed) of a miscarriage confined her to her room, and there she had to lie, neglected, listening to Byron and Augusta laughing as they passed through the house.

The pressure lifted fleetingly when Augusta returned to Six Mile

Bottom, and Byron followed to escape London for a few days. Absence made their hearts grow infinitely fonder, and the letters they sent each other are full of the infantile endearments and insignificant personal details that smitten lovers like to exchange. He addressed her as his 'Dearest Pip' (as in pippin, an allusion to her apple cheeks), and signed himself as 'not Frac.' – i.e., not fractious. She replied from the midst of a house being spring-cleaned in his absence (not by her, of course – by the servants, who, she complained, had woken her up early with the din), missing him desperately – 'better a *nau*[ghty] Byron . . . than no Byron'. He replied that he had hurt a toe in a mousetrap left by 'Goose' (his pet name for Augusta) in his room.

When they were together again, relations began to collapse. Byron's financial affairs, which had remained ominously unresolved throughout their marriage, now intruded with a new force. In a last, desperate attempt to raise some cash, he put Newstead Abbey and some other property up for auction, but they both failed to reach the reserve price. This left him with debts amounting to £30,000, rent on their Piccadilly Terrace house, a large establishment of servants to keep up, a coach to pay for, and all to be financed by the £700 a year he was then getting from the Milbankes as part of the wedding settlement. It was nowhere near enough, and soon there were bailiffs in the house, one taking up residence to ensure that Byron did not abscond with anything of value.

Their new, very unwelcome guest boasted to his hosts that he had just spent a year in the home of the playwright Richard Sheridan, who had also suffered some temporary embarrassment. Sheridan had treated him very civilly, which helped him keep himself inconspicuous from the neighbours – if they took his meaning.

Byron's response to the crisis was the familiar one of denial. Rather than make any attempt to sort out the mess, he began to devote his attentions to the Drury Lane Theatre, which had elected him to its management committee, and in particular to one of its actresses, Susan Boyce. She had apparently become his mistress two months after Annabella became pregnant, and his by now wretched wife would be regaled with reviews of Boyce's carnal performances.

Byron also turned to the company of drinking companions, relating with zest the antics of drunken parties, such as one shared with his fellow debtor Sheridan – 'Sherry', who was later found insensible in the middle of the street by a night watchman (this being a decade and a half prior to Robert Peel's Metropolitan Police Force). Byron takes up the story in a letter to Thomas Moore: ' "Who are you, sir?" – no answer. "What's your name?" – a hiccup. "What's your name?" – Answer, in a slow, deliberate, and impassive tone – "Wilberforce!!!" ' William Wilberforce was well known as a teetotaller, as well as the slave trade abolitionist, and Sherry's ability to satirically cite the name while so inebriated deeply impressed Byron.

Annabella tells the story of his return from another riotous night out, when he found her waiting for him in a mood of bitter indignation. He was all contrition, chastising himself for being such a monster, throwing himself at her feet. She pledged that she would never forget the injuries he had inflicted. He had lost her for ever. He then looked up at her with an expression of such desperate supplication that she was instantly reduced her to tears which flowed over his face. She promised to forgive him and never mention the matter again, whereupon he leapt to his feet, folded his arms and burst into laughter. Science! It had all been an experiment, he told her mockingly, designed to test just how unforgiving she could be.

She could be very unforgiving, as men more resilient than Byron would discover to their cost. But she wanted to forgive Byron, and, true to what would become the motto of the social sciences, the only way she could think to forgive all was to understand all. In one of her 'characters', she attempted to isolate the cause of his complexities, and fingered a familiar culprit, his imagination. It was, she wrote, 'too exalted', always wandering off into the past or the future, never applied to the present. In a letter to Augusta sent around the same time, she remarked on his 'habitual *passion for Excitement*' driven by the 'Ennui of a monotonous existence'. She speculated that his excessive drinking and eating might be aggravating this condition, and optimistically wondered whether a spot of exercise and breath of fresh air might not fix the problem.

It would take more than air and exercise. A darkness had descended on Byron in those months that spilled out into his poetry.

It seeped through *The Siege of Corinth*, in which the hardened renegade Alp, a familiar Byronic figure, wanders alone along the beach towards Corinth, the enemy's besieged and battered stronghold, where he sees . . .

> . . . the lean dogs beneath the wall
> Hold o'er the dead their carnival,
> Gorging and growling o'er carcass and limb;
> They were too busy to bark at him!
> From a Tartar's skull they had stripp'd the flesh,
> As ye peel the fig when its fruit is fresh . . .

In the final lines, which Annabella herself helped to transcribe, he describes a great explosion that finally destroyed the city, its inhabitants and its conquerors . . .

> All that of living or dead remain,
> Hurl'd on high with the shiver'd fane,
>   In one wild roar expired!
> The shatter'd town – the walls thrown down –
> The waves a moment backward bent –
> The hills that shake, although unrent,
>   As if an earthquake pass'd –
> The thousand shapeless things all driven
> In cloud and flame athwart the heaven,
> By that tremendous blast . . .
>
> . . . That one moment left no trace
> More of human form or face
> Save a scatter'd scalp or bone:
> And down came blazing rafters, strown
> Around, and many a falling stone,
> Deeply dinted in the clay,
> All blacken'd there and reeking lay.
> All the living things that heard
> That deadly earth-shock disappear'd:
> The wild birds flew; the wild dogs fled,
> And howling left the unburied dead . . .

As 1815 drew to its close, the crisis point approached. Byron admitted to his friend Hobhouse that he had been treating Annabella badly, and blamed his liver and the live-in bailiffs. Annabella wrote a series of letters to Augusta – one desperate because he had been out drinking again, this time until four thirty in the morning, the next relieved because of some tiny token of kindness from him, the next full of fears and written with an aching heart. Piccadilly seemed full enough of passion in those days to blow apart as violently as Corinth.

*

Augusta returned to Piccadilly, and she and Annabella spent many evenings alone together, discussing the man now out and indulging in the 'depravities' which, according to Annabella, he was in the habit of minutely describing to her when he returned. During one of these evenings, Augusta became overcome with her own feelings of remorse, and cried, 'You don't know what a what a fool I have been about him.' 'The bitterness of her look as she threw her hair back from her forehead with a trembling hand, wrung my heart,' Annabella later recalled. 'I kissed that forehead where I saw penitential anguish, and left the room. Will you condemn me, reader?'

Fletcher, Byron's faithful valet, who would stay by his master's side to his master's death and accompany the body to its final resting place, would apparently during those dreadful days watch over Byron after he returned from one of his binges, to make sure he did not attack Annabella.

Annabella spent most of 9 December feeling ill. She saw an old friend of her parents', a Sergeant Heywood, and discussed with him whether she should leave the house and her husband for the imminent birth. It seems Byron overheard her. In the evening, she suddenly realized she was in labour, and went down to the drawing room to tell her husband. He apparently ignored the announcement, and demanded to know if she wanted to continue living with him. She fled the room in tears. Mrs Clermont met her running up the stairs, sobbing about her treatment, and she was no doubt conducted back to her room in her old nanny's arms. A slam of the front door would have marked Byron's departure for the theatre. When he returned much

later, he asked Augusta how Annabella was feeling. Having been told
in no uncertain terms that she was not feeling very well, thanks to his
careless remark, he ruefully agreed that it was a tendency of his to
choose his moments badly.

That night, Annabella complained of being kept awake by a
thumping sound coming up from the room below. It was Byron hurling
bottles of soda water at the ceiling to keep her awake, she assumed,
though Byron's friend Hobhouse later inspected the room and could
find no evidence for this, preferring the explanation that he had been
indulging in his odd habit of knocking off bottle-tops with a poker.
Whatever the cause, the racket was sufficient to awake Augusta too,
whose room was two flights up.

A baby daughter was born the following day.

Everyone had their own story as to Byron's reaction to the event:
he asked Annabella in the midst of the delivery if the child was dead;
he suggested that if it looked like Augusta, it must be because it was
conceived during their stay at Six Mile Bottom; he gazed with an
exultant smile at his newborn daughter lying beside her mother, and
exclaimed, 'Oh what an implement of torture have I acquired in you!';
on the following day he asked Mrs Clermont anxiously about the
baby's feet, and was reassured to learn that she had not inherited his
own deformity.

Annabella was confined to her room for the following two weeks,
and it was during that period that they decided upon the name for
their child: Augusta – a dangerous choice, in the circumstances, though
perhaps at the time seen as safe since the adult Augusta was, as they
obviously hoped their daughter would be, a common object of affection. The second name, Ada, had been found among the deepest
thickets of Byron's family tree.

The following evening, Hobhouse called and spent a few convivial
hours with the new father. He paid his respects to mother and child
the next day, and perhaps it was then, as Annabella watched him look
at the little baby, that a dreadful thought first struck her. Since
Halnaby, she had believed that the key to Byron's bouts of madness
was a person, a certain someone who reminded him of something
terrible he had done in the past. Whenever he encountered this person,

it aroused feelings of guilt and remorse that drove him to distraction. Until then, she seemed to suspect that someone to be Augusta and that something to be incest. But she began to wonder if it was Hobhouse and the crime alluded to in their violent honeymoon row about *Caleb Williams* was murder. Hobhouse had accompanied Byron on the foreign trip that inspired *Childe Harold*, his appearances in Byron's life seemed to set off bouts of her husband's madness, and she distrusted him – indeed, was disgusted by the way he would flatter her. Could he, she wondered, have witnessed Byron commit some terrible act while they were away? Had he helped the real Childe Harold escape 'Sin's long labyrinth', but then, under the 'mask of friendship', threatened to reveal everything if Byron did not continue to see him? Was that the bond Byron could not escape, that was dragging him screaming down to his damnation?

Dreamed up at the other end of the nineteenth century, such a theory would have been worthy of one of the veiled ladies ushered into Sherlock Holmes's study. So it is perhaps appropriate that in a pioneering act of forensic medicine and psychological profiling Annabella should turn to science to find an answer to her predicament.

She consulted Dr Matthew Baillie, brother of the inestimable Joanna Baillie, whose acquaintance Annabella had so heartily recommended to Byron during their courtship. He was a distinguished physician, famous for his lectures on morbid anatomy and diseases of the nervous system, the latter having become an increasingly fashionable field since the 1760s, when William Cullen began to argue the importance of its role in health and disease.

Annabella had discovered an article in a journal about hydrocephalus, water on the brain. Though no record remains of which journal she consulted, the article appears to have been one by a Dr Reed published that year in the *Edinburgh Medical and Surgical Journal*, which attributed mental derangement to the final, fatal stages of the disease. She discussed with Dr Baillie whether this might explain Byron's behaviour. Baillie was unable to give a diagnosis, but advised that she should leave her husband, at least on an experimental basis, while other medical opinions were sought.

Annabella was not yet ready to take such a course of action. She still needed to find some external, measurable, perhaps curable cause for her husband's madness.

In a last desperate attempt to find one, she decided to creep up to his room and go through his papers and possessions. There she found a vial of laudanum, together with a copy of the Marquis de Sade's *Justine*. What she would have made of the latter is hard to imagine. Perhaps it was like an earthquake – watching and feeling what she had assumed to be, *knew* to be the immutable bedrock of morality and faith heaving and fracturing before her eyes. She would have understood de Sade – a woman with her breadth of reading and intellectual accomplishments must have done. She would have understood that he showed where the principles of the Enlightenment led if not regulated by religion: to probing the limits of human nature, to an exploration of the uncharted terrain of experience that would 'extend unbelievably . . . the sphere of one's sensations', as de Sade put it in the very volume she now held in her hands. Was that not Byron's manifesto, Childe Harold's mission, the artistic impulse?

She found herself tottering at the edge of a chasm of scepticism and profanity, as she watched Byron clamber down into its darkest recesses, into places beyond even her redemptive reach. She resolved to follow Dr Baillie's advice and move out immediately.

*

According to Annabella, on the evening of 14 January 1816 she announced to Byron that she would be leaving the following day, and would take the child. 'When shall we three meet again?' he asked. She replied: 'In heaven.' She left the room sobbing, her self-control having deserted her. That night, as boxes and cases were packed and carried downstairs, she slept soundly 'as I believe is often surprisingly the case in such cases of deep sorrow'. The following morning she arose feeling exhausted, went downstairs and past the door to Byron's room. There was a mat on the floor outside, where his beloved Newfoundland dog used to lie, and she contemplated throwing herself on it and there 'wait at all hazards'. But she passed on to the carriage that was ready for her outside to take her to Kirkby Mallory.

She stopped the first night in Woburn, and wrote a letter to her husband that would become notorious for its affectionate tone – proof, for Byron's supporters, that her later protestations about his treatment of her were exaggerated. Anticipating this very reaction, she wrote a letter to her friend Selina Doyle explaining that she was following Dr Baillie's advice, which was to write to Byron only on light and soothing topics. And she made a copy of this explanation so it would be available to show to anyone disbelieving her true motives. In other words, the letter became not proof of exaggeration about Byron, but an example of what her future enemies came to regard as her most revealing and repellent characteristic: the ability of this Princess of Parallelograms to be calculating.

The association between Annabella's scheming nature and her mathematical bent became a continuing motif that persists to this day. In a 'fantasy' newspaper interview with Byron by the historian and newspaper columnist Paul Johnson, Byron fulminates about 'that monster, Miss Milbanke'. 'All she cares about is mathematics. I call her the Princess of Parallelograms. She spends all her time with an old fossil called Charles Babbage, who had invented a calculating machine he calls a computer, which will solve all the mysteries of the universe. Be damned to them!' Some of the details about Babbage are wrong, but the caricature is absolutely right, and in those affectionate letters she sent to Byron she contributed a great deal to its formation.

For Byron's allies, the Prince of Passions was, of course, the very opposite of calculating: spontaneous, sincere, generous. But Byron was not the one in need of defending. It was Annabella. She had embarked on what for a woman of her time was the most perilous, one might also say courageous, journey imaginable: away from her husband. If Annabella was to survive her planned separation without risk of losing her child, her property or, as she at times claimed would happen if she was forced back to the marital home, her life, she had to calculate with all the genius of Euclid.

For the first few days with her parents at Kirkby, as everyone settled in and calmed down, Annabella still hoped that such extreme remedies would not be necessary. The separation was to be temporary, giving Byron a chance to recover his sanity under a strict regime of abstinence to be applied by Augusta with the help of George Anson, Byron's cousin. It also provided an opportunity for Annabella to proceed with her forensic examination, as she managed to persuade Byron to be seen by her doctor, Francis Le Mann, who would surreptitiously assess the man's mental state.

Meanwhile, at the risk of over-egging the pudding, Annabella sent Byron another affectionate letter, in which she addressed him as her 'Dearest Duck', signed herself 'Ever thy most loving Pippin', and filled the lines in between with playful allusions to the lavatory at Kirkby being the ideal '*sitting*-room or *sulking*-room all to yourself' should he choose to visit.

Then the situation changed irreversibly.

Annabella learned of Le Mann's report on Byron's health: there was nothing wrong with him other than a 'liver complaint', the doctor announced, for which he had prescribed a course of calomel pills. This meant that whatever the cause of his behaviour, it was moral not physical, within the domain of personal responsibility, not medical science.

Annabella also received a letter from her friend Selina Doyle, in reply to Annabella's sent on the day of her departure from Piccadilly. Selina, a key character in the future lives of Annabella and the little baby she now nursed, advised her friend to take the 'final step', and leave Byron permanently. She had consulted a 'friend' (who turned out to be her brother Francis, a stiff-necked colonel), who agreed that the cause of Byron's malady must be a 'morbid passion . . . acted upon by a distempered imagination which has by degrees led to the outrages committed one after another with a view to make you Love the Man as it were for his Essence in spite of his qualities & of your Sense of virtue'. These were the sorts of capital letters Annabella could understand, and they had the intended effect. The moment, she now realized, had come, the die was cast: this was to be a permanent separation.

Only now did Judith seem to learn what had really been going on during the marriage of her beloved only child, and the outraged mother immediately set off to London to consult lawyers. For her part, Annabella began to gather evidence to support her case.

Then, fortuitously, the most damning and demonstrative evidence imaginable dropped straight into her lap. She received a note from

Lady Caroline Lamb, proposing a secret assignation, and promising to reveal secrets that would make Byron 'tremble'. Annabella kept the appointment to find Lady Caroline in a state of apparently genuine agitation. Byron, she explained, had confessed to her his incest with Augusta. But there was worse. He had also owned up to indulging in homosexual acts while at Harrow – acts which were, at that time, regarded with a particular horror reflected in the law making sodomy a capital crime.

Annabella regarded this evidence as decisive, and wrote it down to support a series of detailed 'narratives' about the marriage she had been compiling for her lawyers. These notes, collected together by her grandson fifty years later to vindicate the permanent estrangement from her husband she now sought, were to form the foundation for an edifice of self-justification that it became her life's mission to build. Though they were not to be published in any form until much later in the century, they helped set the agenda for the great Separation debate that was to follow, revealing it to be a matter not of mere personal antagonisms, but of moral and philosophical principle. The press and public realized this, indeed treated it from the beginning as a symbolic event that captured the spirit of the age. *Blackwood's*, the Scottish Tory journal widely quoted in newspapers at the time, even went so far as to proclaim that 'had [Byron's] marriage been a happy one, the course of events of the present century might have been materially changed'.

*

By current standards, it took a while for news of the Separation to appear in the papers, but when it did, it hit with the same impact as any modern-day scandal. Indeed, in many respects it was the prototype celebrity sensation. Here we see in nascent form the same foaming outrage, the same competition between editors to have higher double standards than each other, the same prurience, the same posturing, the same fake prudery.

The *Champion*, a pugnaciously patriotic Sunday paper which published under the slogan 'Let not England forget her precedence of teaching nation's how to live', broke the story on 14 April 1816. The paper had managed to secure copies of two poems, 'Fare Thee Well' and 'A Sketch from Private Life', written by Byron about the Separation for private circulation. His publisher Murray had somehow allowed the poems to leak to John Scott, the *Champion*'s ambitious editor. Scott splashed them over the centre spread – then the most prominent position in the paper, as the front page was covered with classified advertising – under the heading LORD BYRON'S POEMS ON HIS OWN DOMESTIC CIRCUMSTANCES.

'Fare Thee Well' transferred Byron's relationship with Annabella into the stuff of romance, appealing to a bond between them that went beyond the conventions of marriage or the strictures of morality, to some unattainable yet undeniable ideal:

> Though my many faults defaced me,
>   Could no other arm be found,
> Than the one which once embraced me,
>   To inflict a cureless wound?
>
> Yet, oh yet, thyself deceive not;
>   Love may sink by slow decay,
> But by sudden wrench, believe not
>   Hearts can thus be torn away:
>
> Still thine own its life retaineth –
>   Still must mine, though bleeding, beat;
> And the undying thought which paineth
>   Is – that we no more may meet.

And then, proving to his critics that the blackguard would not stop even at using his own child to win sympathy, an appeal to Annabella's pity for the sake of their daughter:

> And when thou would solace gather,
>   When our child's first accents flow,
> Wilt thou teach her to say 'Father!'
>   Though his care she must forego?
>
> When her little hands shall press thee,
>   When her lip to thine is press'd,
> Think of him whose prayer shall bless thee,
>   Think of him thy love had bless'd!

> Should her lineaments resemble
>   Those thou never more may'st see,
> Then thy heart will softly tremble
>   With a pulse yet true to me . . .

A few days later, one newspaper, speculating on the effect of these lines on the heartstrings of the female population, reported that 'one fair [i.e., female] correspondent' had said that if her husband had bade her such a farewell, she could not have stopped herself 'running into his arms and being reconciled immediately – *Je n'aurais pu m'y tenir un instant.*' – 'I could not hold myself back for a moment'.

The effect, however, was rather reduced by the less gallant verses that followed. Like 'Fare Thee Well', 'A Sketch from Private Life' was about a woman, Annabella's old nanny Mrs Clermont. Byron blamed her for turning his wife against him (not, as other correspondence reveals, a fair charge). The poem contained some of the nastiest lines ever to be published as poetry, made all the nastier by his mastery of the satirical form:

> Born in the garret, in the kitchen bred,
> Promoted thence to deck her mistress' head . . .
>
> Quick with the tale, and ready with the lie,
> The genial confidante, and general spy . . .

'A Sketch' also contained a portrait of the Annabella Mrs Clermont had helped to raise:

> Foiled was perversion by that youthful mind,
> Which flattery fooled not, Baseness could not blind,
> Deceit infect not, nor Contagion soil,
> Indulgence weaken, nor Example spoil,
> Nor mastered Science tempt her to look down
> On humbler talents with a pitying frown,
> Nor Genius swell, nor Beauty render vain,
> Nor Envy ruffle to retaliate pain,
> Nor Fortune change, Pride raise, nor Passion bow,
> Nor Virtue teach austerity – till now.
> Serenely purest of her sex that live,
> But wanting one sweet weakness – to forgive . . .

Though dressed up as a paean, these few lines were, if anything, even more damaging than his poisonous depiction of the 'hag of hatred' Mrs Clermont. It is instantly recognizable as a portrait of a prig, of a creature held aloof by her scientific detachment and religious piety. And from this moment on, such would be Annabella's public profile.

This image is the one that persists to this day. Nearly every biographer of Byron and Annabella (and there have been many) has taken Byron's own caricature of his wife as a 'mathematician', a scientist, the 'Princess of Parallelograms' at face value. She was none of these. She had an interest in maths and science, as she did in literature and art. She did not have any special expertise. When, many years later, she first witnessed Babbage's experimental Difference Engine, her response to it was that of the well-informed but baffled layperson.

Annabella was cast as a mathematician not because she was one, but because it was demanded of her role in what came to be called the 'Separation drama'. She was the embodiment of the modern, almost mechanical breed of human needed in an industrialized world; he represented the spirit of Romanticism railing against the tide of progress. These two were thus engaged not in a mere marital dispute, but in the heroic struggle between the two sides of a culture cleaving in two, and the public that so uncomfortably straddled that divide was desperate to see who would win.

*

The effect of the publication of Byron's two poems on his 'domestic circumstances' served *The Champion*'s purpose perfectly. They showed this libertine to be not only treacherous (Byron was infamous for his sympathetic view of Napoleon) but hypocritical.

In a commentary appended to the poems, the paper laid about him as brutally as any modern-day tabloid. After the usual preamble condemning what would now be called media intrusion, it proceeded to attack Byron with relish. It pronounced him 'the advocate of drunkenness, the calumniator of women', who had acted in 'the lowest degree of human baseness' in writing the poems.

'Lord Byron will not pretend that these poems were not designed as

an appeal, to throw the blame of his early separation from Lady Byron
on the weak and defenceless party,' it went on, steam now rising from
the page.

> The most artful language, and elaborate combinations of touching
> reproaches, are employed in the piece, entitled 'Fare Thee Well',
> to prove Lady B. utterly deficient in a wife's first duty and a
> woman's best grace. Lord Byron takes advantage of his ground as
> a popular Poet to attempt to turn the whole current of public
> reproach and displeasure against his wife, and of its sympathy and
> admiration towards his injuries and tender sorrow!

The *Champion*'s intervention had a decisive effect on the outcome of
the Separation in a number of ways. Firstly, it brought the poems into
the public domain, and most of the other London papers and many
provincial ones (such as the *Durham County Advertiser*, the local paper for
Seaham) now felt able (or 'compelled', as they preferred to put it) to
join in. In the following two weeks *The Times*, the *Morning Chronicle*, the
Morning Post, the *Morning Sketch*, the *Examiner* and the *Courier* all carried
the story prominently, over several editions.

Sir Ralph succeeded magnificently in keeping the bandwagon
careering along by lashing out at the *Morning Chronicle*. The *Chronicle*
had launched an attack on the editor of the *Champion*, suggesting that
the paper was part of a conspiracy against Byron. Sir Ralph assumed
he and his family were somehow being implicated in this conspiracy,
and stormed into the offices of the *Chronicle*'s editor, James Perry,
demanding a retraction. Since Perry had never intended to imply such
a thing, he published a story the following day in which he asserted
that

> we have authority from Sir Ralph Milbanke for saying, that *he knows*
> of no conspiracy against the domestic peace of Lord Byron. We
> cheerfully yield to the Honourable Baronet's desire to insert his
> declaration, of the truth of which no man, who is acquainted with
> him, can doubt . . . Sir Ralph assures us that the insinuations have
> been published not only without his knowledge but also much to
> his disquiet and condemnation.

Apparently driven to distraction by the whole affair, this was not enough for Sir Ralph. He sent Perry a letter for publication claiming that the statement was 'utterly unsatisfactory'. When Perry refused to print the letter, Sir Ralph took it to the *Courier*, the *Morning Chronicle*'s main rival, which was only too delighted to publish it, provoking another story the following day which included Perry's response. At that point, the *Courier* was one of the few papers which had refrained from publishing the infamous poems. Now, as Sir Ralph himself had recruited it to his cause, it too felt 'compelled' to publish them, 'following demands from our readers'.

Another effect of what within the family came to be known as the Newspaper War was that the only child of the marriage became publicly embroiled in its collapse. She, too, was now a celebrity – so much so that, still an infant, she would draw crowds when Annabella took her out. The lines in 'Fare Thee Well' referring to her established what was to be her lifelong role – reinforced by lines Byron had yet to write – as the embodiment of the Separation. In the eyes of the public, she was now as Byron had depicted her: a symbol of a purity of love that, thanks to the contamination of human weakness, ultimately escapes us all.

The final effect of the press coverage was upon Byron himself. In March the papers were reporting nothing more sinister than the effect of an earthquake in the Midlands on the fabric of Newstead Abbey, where the ceiling in one of the dining rooms had collapsed. By April all but a few were vilifying him daily, quaking far more violently than any earth tremor in their indignation at his behaviour.

Publicly, Byron responded to the assault with a devil-may-care attitude, continuing to lead his life just as he had led it before, going to dinners and the theatre, circulating satirical verses and airing his views. He also had another affair, this time with Mary Jane 'Claire' Clairmont, stepdaughter of the author of *Caleb Williams*, William Godwin. Claire had pursued him with a determination worthy of Lady Caroline Lamb. She was writing a novel about a woman brought up in the wild whose behaviour was governed by natural impulses – a theme that must have attracted Byron after his recent experiences with the less than impulsive Annabella. By the time his name was all over the papers, she was openly showing him off as her lover, bringing her stepsister Mary Shelley to the Green Room at Drury Lane to admire her trophy. She was also pregnant with his child, a half-sister for Augusta Ada that the little girl was never to meet.

Such distractions were, however, just a prelude for Byron's real response to the media frenzy: to make his dignified exit from the country that had now proved itself to be irredeemably corrupt and priggish. In a gesture of magnificent defiance, he proposed being carried off to the Continent in a replica of Napoleon's carriage, which he had commissioned at enormous cost from a London coachmaker. He made great play of his impending departure, selling all his possessions at public auction. And he made no attempt to reconcile old enmities, sarcastically delivering the final contract of Separation 'as Mrs Clermont's act and deed'.

Providing a glimpse of the feelings that lay behind this façade of facetiousness, he wrote a note to Augusta asking her to keep him informed about his daughter's welfare, and begging her never again to 'mention or allude to Lady Byron's name again in any shape – or on any occasion – except indispensable business'. He also wrote a final letter to Annabella, asking her to be kind to Augusta, and enclosing a ring for his daughter that would become a most treasured possession.

The day before he was due to depart for the Continent, the *Courier*, now relishing its role as Annabella's semi-official advocate, launched its most vitriolic attack on Byron. Acting as the self-appointed foreman of the jury of public opinion, it rejected 'his pathetic appeal to their passions – his elegant delineations of his own stormy feelings; they would demand the severity of plain sense, and not the brilliancy of a warm imagination'. 'The public have already decided the case, without waiting a reply to his tender reproofs, or his passionate invectives,' it continued. 'They have classed him in that rare order of minds, which possess the wildness of talent, without its utility – its sublime without its gentle.'

It followed these words with a verse lampooning 'Fare Thee Well' entitled 'Oh! Forget Me':

> If passion win thee to her gusts,
> Let not thy thoughts to home be turn'd –
> Bear not that doubting heart which bursts
> To think of peace despised and spurn'd –
> Oh! then forget me – and if time
> Pluck from thy breast this rankling smart,
> Uncheck'd by shame, unaw'd by crime,
> Cling to some warm and kindlier heart . . .

The only kind lines to be found among the papers in those, his last moments in his homeland, were in the pages of the radical *Examiner*. It, too, would publish a verse inspired by the incident, penned by the paper's editor, Leigh Hunt, who a few years earlier had been visited by Byron while he served a prison sentence for libelling the Prince Regent, and who a few years later would prove himself not quite so loyal to Byron's reputation.

> . . . For early storms, on Fortune's basking shore,
> That cut precious ripeness to the core; –
> For faults unhidden, other's virtues owned;
> Nay, unless Cant's to be at once enthroned,
> For virtues too, with whatsoever blended,
> And e'en were none possessed, for none pretended; –
> Lastly, for older friends, – fine hearts held fast
> Through every dash of chance, from first to last; –
> For taking spirit as it means to be, –
> For a stretched hand, ever the same to me, –
> And total, glorious want of hypocrisy.
> Adieu, adieu: – I say no more. – God speed you!
> Remember what we shall expect, who read you.

*

Byron set off early for Dover on 23 April 1816. Even as his huge new Napoleonic carriage trundled off down the road, the bailiffs moved in on 13 Piccadilly Terrace, the scene of a few happy days and so many miserable ones, and stripped it bare. His retinue, comprising his newly appointed personal physician, Dr John William Polidori, and his friends

Hobhouse and Scrope Davies, reached Dover in the evening, and the carriage was immediately lifted into the ship for fear that it would be seized by bailiffs while it remained on British soil. The following day the wind blew in the wrong direction, and the departure had to be postponed. Byron spent the time visiting the grave of Charles Churchill, who shared with him the distinction of being a famous satirical poet who had gone bankrupt and been separated from his wife. Byron lay down on his soulmate's grave, and donated a crown for it to be smartened up with a new blanket of turf.

According to rumours later reported by Annabella's lawyer, the inn where Byron and his friend spent their second night was infiltrated by a number of women dressed as chambermaids, hoping to catch a final glimpse of and bid their anonymous farewells to the author of 'Fare Thee Well'.

The following morning the wind had changed and Byron's ship prepared to sail. Byron walked arm in arm with Hobhouse to the quay. He boarded the ship, which had been waiting for him. As it weighed anchor and set sail, Hobhouse ran down the wooden pier as he had run alongside the bridal carriage at Seaham. Byron pulled off his cap to wave his final farewell. Hobhouse reached the end of the pier and gazed as his friend receded from shores to which he would never return alive.

*

Later that year, Byron would recall that departure in poetry. He too lamented the distance opening up between him and one he then held dear, but it was not Hobhouse:

> Is thy face like thy mother's, my fair child!
> ADA! sole daughter of my house and heart?
> When last I saw thy young blue eyes they smiled,
> And when we parted, – not as now we part,
> But with a hope. –
>   Awaking with a start,
> The waters heave around me; and on high
> The winds lift up their voices: I depart,
> Whither I know not; but the hour's gone by,
> When Albion's lessening shores could grieve or glad mine eye.

Man's Dangerous Asset

EVERYONE WANTED TO KNOW: what would the offspring of such a mutant match be like? What sort of creature would be assembled out of the remains of genius and depravity, poetry and geometry, loyalty and jealousy, liberty and licence, love and hate, virtues and crimes that rotted in the charnel house of her parents' marriage?

During a visit to Ely, where Annabella was being shown round the cathedral by the dean's wife, a crowd gathered at the inn where she and her baby daughter were staying, hoping to catch a glimpse of the infant Byron, wondering if they might spot the buds of angel's wings on her back, or of devil's horns on her brow. The same had happened at Peterborough. Annabella wrote home to Kirkby in her daughter's name, which was no longer to be Augusta but *Ada*, underlined to emphasize the fact: they were being treated as 'lionesses'.

In her earliest years it was, of course, impossible to tell what sort of person Ada would be. Her mother, for one, had very little idea, as she now embarked on a ceaseless circulation of the country, moving from one rented house to the next by way of her favourite spa towns and seaside resorts.

During these travels, Ada was usually left at Kirkby Mallory. There she was cared for by her grandparents with the help of a succession of nurses. By the age of two, she had developed into an exuberant little girl living happily in the protective custody of her doting grandmamma, skipping through Kirkby's echoing halls and elegant chambers quite unaware of the significance of the veiled portrait over the mantelpiece, the loaded pistols at the bedside, the burly servants watching the back door.

Annabella's dealings with her daughter were almost exclusively focused on her role in the still unfolding Separation drama. The little girl had a very big part to play in proving that Annabella had been on the side of virtue and justice. Social attitudes as well as religious law (marriage was still seen as more a Church than a civil matter) strongly favoured the father in any separation. Women were simply not expected to 'desert' their husbands, no matter how dreadful the domestic despotism they had to endure. The only possible mitigation was the welfare of a child. Thus Annabella had to prove that Ada's exposure to the values of a dissolute Regency libertine would lead to the little girl's ruination.

The inescapable consequence of this was that, publicly as well as personally, the child would become, as Byron put it, 'the inheritor of our bitterness'. Even the most intimate maternal gestures were deployed to strengthen Annabella's case, to promote her image as the doting, virtuous mother. For example, she sent touching letters back to Kirkby asking anxiously about Ada's welfare, together with instructions in a covering note telling Judith to keep the correspondence in case she needed to produce any evidence of her maternal concern.

It was a wise precaution, because the public image she needed to cultivate was so completely at odds with the private reality. Annabella experienced few maternal pangs during the first few years of her daughter's life. When she was still pregnant she made it a matter of policy not to develop any affection for the daughter she was about to have. She was, as she put it in the title of a poem written soon after Ada's birth, 'The Unnatural Mother'. 'I talk to it for your satisfaction, not my own, and shall be very glad when you have it under your eye,' she once wrote to Judith, 'it' being Ada.

For Annabella, motherhood existed only in the abstract, as a medium for expressing her virtues and justifying her actions. When she did consider it as something connected with emotion and succour, it was in the form of poetical symbolism and metaphor:

> *On Seeing A Flower Sheltered from the*
> *Storm by an Oak*
>
> (*the mother — to her child*)
>
> I, like this tree which spreads its shelter wide,
> 'The pelting of the pitiless storm' would bide

> To save the blossom of the flower below –
> And oh, my flower! if from the cup of woe
> Pour'd on my head, some lonely drop should fall,
> Escap'd the love that would have spar'd thee all –
> May it on thine descend, a softening dew,
> And add but tenderer grace – a lovelier hue!

Hardly surprising, then, that the little flower would, Annabella complained, emit a 'war whoop worse than the Americans' every time the nurse left the infant alone with her mother. In contrast, when the nurse left her alone with her beloved grandmamma she would coo with pleasure, and gripe when the nurse returned to take her away. When Annabella performed the then usual practice of lancing gums about to cut milk teeth, the pain Ada suffered from the operation caused the nurse to wince. When Judith did the same operation on some back teeth that broke through late, Ada thanked her.

Annabella made a few lacklustre attempts to engender some filial affection in the child, such as telling her parents that while she was away Ada should be encouraged to kiss a portrait of her absent mother. Her primary concerns, however, remained moral rather than maternal. The girl had to be made into a model of what Annabella represented, which meant everything had to be focused on the development of her intellectual and ethical powers, and the complete suppression of her imagination.

*

Novalis, the 'prophet of Romanticism', wrote that the imagination is '. . . most attracted by what is most immoral, most animal'. It is humanity's 'dangerous asset', as another critic put it. It gave ordinary humans the ability to make things up, to create, and put them on a plane with God, where they could challenge His divine law. To explore it was to explore the limits of human capability in an age when there seemed to be no limits – when humans could traverse the globe, fashion the landscape, plunder the earth.

Annabella's own view of the imagination was more akin to that of Francis Bacon, one of her favourite philosophers. In his great treatise on the scientific method, *The Advancement of Learning*, he wrote: 'The imagination, being not tied to the laws of matter, may at pleasure join that which nature hath severed and sever that which nature hath joined, and so make unlawful matches and divorces of things.' Bacon saw no great harm in this, but Annabella did. Let loose in Byron, it had made no end of unlawful matches and divorces.

Annabella could not risk the same freedom of thought being let loose in Ada's mind. The child had to be raised in an atmosphere of systematic and unremitting superintendence, designed to ensure the development of a self-controlled, dutiful character devoid of imagination. A succession of hapless nurses and governesses were charged with this weighty task, and none of them measured up to Annabella's exacting standards. The first, referred to only as 'Grimes', developed an unhealthy attachment for her charge, for example taking sole responsibility for nursing the child through a dose of chickenpox when Annabella, always sensitive about her own health, absented herself for fear of infection. This attachment compromised Annabella's own capacity to control 'the child's' temper, and this she could not tolerate. Grimes was dismissed. A Miss Lamont, appointed as a governess to begin the girl's formal lessons at the age of five, suffered in the opposite direction, failing to have sufficient influence. She was overpowered by Ada's 'masterly' mind, and so was unable to command attention or obedience. In Annabella's view, 'the only motive to be inculcated with a character like Ada's is a sense of duty, combined with the hope of approbation from those she loves'. This Miss Lamont was clearly incapable of achieving, so, by the following year, she too had gone, though accompanied by a warm reference as well as a warning never to talk to anyone about Ada.

Up until her departure, Miss Lamont had kept a journal for Annabella recording Ada's development. In its opening pages, it is full of admiration for the girl, reporting her proficiency at adding up columns of figures, her accuracy in drawing the parallel lines that her father had once pointed out were destined never to meet, her understanding of geometrical concepts such as perpendicularity, her fluency in reading books such as *Early Lessons* by Maria Edgeworth, one of Annabella's favourite authors.

The curriculum was set by Annabella, who was nurturing a growing interest in education as well as philanthropy and the Co-operative Movement – subjects that took her as far as it was feasible to go from the Byronic Babylon she had been forced to inhabit the year before. At the time, there was no educational system that would be recognizable today, but with the rapid spread of new agricultural and industrial practices there was a clear demand for one to be set up. Annabella decided to use some of her own money to found an experimental village school at Seaham, which would be run according to new principles being pioneered by educationalists such as Johann Heinrich Pestalozzi and Emmanuel de Fellenberg. Pestalozzi had set up a school in Berthoud, Switzerland where children were taught in accordance with the laws of nature, which meant that all that they learned was to be gained through observation. Annabella ordered the unfortunate Miss Lamont to try out some of his ideas on Ada, such as exploring the physical properties of wooden bricks by using them to construct boxes of various designs. Unfortunately for an increasingly exasperated Miss Lamont, Ada got quite carried away with the idea, and far from confining herself to observations about the bricks and boxes, began to imagine how she might turn them into cities and towers.

The little girl was also inclined to be wilful and distracted during other lessons, and in an effort to control her a system of rewards and punishments was instituted. If she did well at any of her tasks, she was awarded tickets, which she could later redeem for improving gifts such as books on botany. If she misbehaved, tickets were confiscated – two forfeited for carelessness in her sums, two for being inattentive during a music lesson. Sometimes even the threat of taking all her tickets failed to bring her into line, in which case corporal remedies were used, such as locking her in a room, forcing her to lie still on a wooden board, or tying bags around her fidgeting hands. These remedies, too, could fail, such as when carelessness during a French lesson caused her to be sent into the corner, which she protested against by taking a bite out of the dado rail.

As the pages of Miss Lamont's journal are turned, the story gets progressively worse: confiscations, corners, bags, all were tried with increasingly frequency until, finally, Miss Lamont took her leave. Her successors had to implement the same regimen, and faced the same challenges trying to enforce it on their spirited charge. The method of teaching each subject was minutely specified: history, for example, had to be strictly chronological, concentrating on the modern era because the value of individual actions is usually assessed according to Christian principles. On no account was Ada to be taught about anything that might arouse her 'fancy', such as ghosts. She was to imbibe only facts and moral principles.

Annabella, who confessed to being so self-controlled as to be a perfect candidate for the job of police officer or prison governor, became increasingly concerned with the failure of her employees to control Ada. She outwardly attributed this to the girl's innate intellectual superiority, but there were deeper anxieties about other possible causes. Was this really precociousness, or evidence of a Byronic temperament beginning to emerge? It was horrifying to see the child turning into her father when she had in her mother such a model of human virtue and purity to emulate. If this was allowed to happen, it would undermine the precarious edifice of self-justification Annabella had constructed to vindicate her actions against the still popular Byron: her decision to leave him, her legal efforts to deny him his child.

*

When mother and daughter were walking together in the garden one day, Ada piped up in the innocent way little children do, 'Mamma, how is it that other little girls have got Papas and I have none?' According to Ada's own account of the incident confided to a friend years later, Annabella responded in such a 'fearfully stern and threatening manner from ever speaking to her again upon that subject' that she 'acquired a feeling of dread towards her mother which continued till the day of her death'.

The same touchiness was evident in Annabella's attitude to Augusta, who was the only permitted intermediary between Byron and his child. In a candid letter to a mutual friend of hers and Augusta's, Theresa Villiers, she wrote that

> Ada's intellect is so far advanced beyond her age that she is already capable of receiving impressions that might influence her – to what

extent I cannot say – and without supposing that A[ugusta]. would mean to injure me, her delusion [that Byron was anything less than irredeemably evil] might tempt her to put ideas in Ada's mind that might take root there – something of 'poor Papa,' &c. My apprehensions may be exaggerated – but I have an insurmountable repugnance to Ada's being in her company.

In these years immediately following the Separation, Annabella's hostility towards her husband and desire to distance her world from his became more and more pronounced, as is revealed in the journals that she kept at the time. Never has a series of supposedly private papers been so full of such pure thoughts. They contain an exhaustive reading list of formidable moral texts: treatises on subjects such as the 'Duties of Children to their Parents' (which Ada was obliged to read as soon as she was old enough to do so); lives of radical clergymen like John Barclay, the Presbyterian minister who Annabella approvingly noted had walked through Aberdeen covered only in sackcloth and ashes; the collected sermons of divines like Joseph Butler, whose influential ethical theory of 'objective intuitionism' saw passion and reason, the two fundamental forces of human nature, as controlled by a third, conscience. Interspersed among all this are little attempts of her own at ethics and theology (for example, she details three reasons why the disciples knew that the resurrected Jesus was not an impostor), reports of discussions with her friends Francis and Selina Doyle on the meaning of memory and the imagination, passing references to Ada, and, just occasionally, glimpses of Byron, which appear encased in ice: a notice about a recent work spotted in the *Edinburgh Review*, a letter by him forwarded to her by her lawyers, a new poem.

In 1821 Ada began to accompany her restless mother as she journeyed around the country. Letters arrived at Kirkby written by Annabella but supposedly dictated by Ada, itemizing for the doting grandparents Annabella's characteristic itinerary of industrial sites, friends' homes, and the very occasional entertainment. There would be a trip to a glass factory in Birmingham one day, a visit to the Gosfords the next (whose kindnesses were flatteringly noted), and perhaps a donkey ride in between. Ada would sometimes offer striking descriptions of the scenery she beheld along the way – such as the white and yellow hue of the sea – and sign the letters herself in carefully drawn capital letters: ADA.

Then, on 28 January 1822, Judith died. The effect on Ada is not recorded, but it must have been devastating, as Judith was the only member of the family who had shown her any real devotion. It also marked the end of her days at Kirkby, as Annabella decided to move away. In her will, Judith left strict instructions to ensure that the little girl had no contact with anything relating to her father until she reached adulthood. The shrouded portrait of Byron in his Albanian costume that had hung over mantelpiece at Kirkby was duly boxed up and put into storage, to remain unseen by Ada until her twentieth year.

The first letters Ada writes in her own hand appeared later that year. The earliest to survive, in alarmingly constrained copperplate, contains a detailed timetable of her lessons together with two columns headed 'G' and 'B' for good and bad, which itemize for each subject the number of tickets awarded and forfeited. That day the 'B' column is blessedly empty, which is presumably why she spent so much effort compiling the table. In subsequent letters, she concentrates more on her failings than her attainments, minutely recording her lapses into disobedience and her determination to correct them.

Very soon, another theme emerges, that of loneliness. Without her beloved grandmother, without a father, without her nomadic mother most of the time, she was completely isolated. On the orders of her mother, she was prohibited from having any contact with the estate workers or local villagers. Augusta, the laughing loveable aunt who she occasionally glimpsed, was off limits too, as Augusta herself noted with a pang of regret. Her only society was her governess, some servants and Annabella's support group, most of them spinsters or widows, and all of them selected for their unstinting support for Annabella's cause and opinions.

To provide Ada with some company of her own age, a small circle of approved friends and relations was assembled. Among them was George Byron, son of George Anson Byron, the poet's cousin and eventual inheritor of his title. Though young George was two years her junior, she clung to him the moment he came within reach, appointing him an honorary brother, confiding in him her most private feelings and hopes. In one particularly heartfelt letter, written in an unstoppable stream of consciousness, she imagines them together

> like the Brothers and sisters of a family who were going abroad with their Mama and papa. They were grown up and left a brother and a married sister [who] did not much care, but afterwards he said he was very glad, because he was able to take care of his sister. Do [you] not think that it was and indeed must have been a very affectionate family?

Like Christian in *Pilgrim's Progress*, she seems overwhelmed by the burden of perpetual wrongdoing, and yearns for George to help lift the load from her shoulders. She has been untruthful, she has been disobedient, and only love can absolve her – her love for him, his love for her. And she wants to show how much love she has in the only way she knows, by offering up herself to his judgement, and by embracing the reproofs that will inevitably follow.

She was eight the year she wrote that letter, and it is apparently her last to him. From then on, references to her beloved George simply evaporate from the archive. It can only be guessed that the expression of such powerful sentiments to a member of the Byron family was regarded as dangerous. Ada was never to know anyone she could call brother again.

*

One of the events she wished she had been able to share with George, to witness with a consoling hand to hold in hers, was a visit to a ship called the *Florida* – 'Papa's ship,' Ada called it.

On 24 May, the *Florida* had sailed from the Ionian island of Zante (now Zákinthos) carrying the body of Byron. The poet, now hero, had died of marsh fever in Missolonghi, where he had been fighting the cause of Greek nationalism against the Turks. He was reported to have called out to his faithful valet William Fletcher on his deathbed:

> Oh my poor child. My dear Ada! My God! could I but have seen her! Give her my blessing, and my dear sister Augusta, and her children – and you will go to Lady Byron and say – tell her

everything – you are friends with her ... My wife! my child! my sister! – you know all – you must say all – you know my wishes!

These words were uttered on 17 April 1824. Twenty-four hours later, he said to Fletcher, 'I want to sleep now,' turned on his back, closed his eyes, and died.

That last reference to Ada seems, perhaps, fancifully sentimental, but Byron had, during his entire exile in Italy and Greece, continued to enquire after her via Augusta. For the sake of form, Annabella had obliged with occasional reports. Six months before his death he had written to Augusta:

> I wish you would obtain from Lady B some account of Ada's disposition – habit, studies, moral tendencies, and temper, as well as of her personal appearance. When I am advised on these points, I can form some notion of her character, and what way her dispositions or indispositions ought to be treated ... Is the girl imaginative? At *her* present age I have an idea that I had many feelings & notions which people would not believe if I stated them *now* ... Is she social or solitary, taciturn or talkative, fond of reading or otherwise, and what is her *tic*? I mean her foible. Is she passionate? I hope that the Gods have made her any thing save *poetical* – it is enough to have one such fool in a family.

A few days later, he also heard that she was ill, and sent an anxious letter asking for news of her health.

Annabella replied, via Augusta, with a detailed, carefully composed portrait, describing her as cheerful (which she seems to have been, but not generally in her mother's company), happy, lively, natural, clever, observant and with an imagination that was (mercifully, she implied) 'exercised in connection with her mechanical ingenuity'. She preferred prose to poetry (another mercy), enjoyed music and drawing and had learned a little French. That was on the plus side. On the minus side she lacked application and was prone to being impetuous, but these vices were under control. Physically, she was 'tall and robust' with irregular features, implying that she was not particularly beautiful. He had nothing to fear about her health, as she was under the medical care of Drs Warner and Mayo, two leading physicians who prescribed a regimen of mild medicines and leeching.

Annabella also enclosed a 'profile' or silhouette, which she assured Ada's father was a good likeness. Byron was delighted by it, and showed it proudly around the garrison at Missolonghi. He replied with an anxious note about a seizure he had recently suffered, alerting Augusta to look out for signs of congenital epilepsy in his daughter's manner. It seems more likely the seizure was a precursor of the disease that would, a few weeks later, kill him.

His death was, as he anticipated in his last lines, a good one, at least with regards to its reception at home. 'If regret'st thou Youth, *why live?*' he had asked rhetorically in a poem entitled 'On This Day I Complete My Thirty-Sixth Year', written just three months before he died. Now no one regretted his youth, and everyone his death.

He was mourned across the world. Harriet Beecher Stowe, the American novelist who much later published the allegations of Byron's incest to the world in her book *Lady Byron Vindicated*, was just a child at the time, but would nevertheless recall vividly her reaction to the news of Byron's death, how she stopped playing and took herself off to a lonely hillside, where she spent the afternoon thinking of the man she then regarded as a hero and would later attack as a demon. A teenage Tennyson mourned Byron by taking a solitary walk through the Lincolnshire countryside and inscribing on a rock the stark news 'Byron is dead!'. Goethe, perhaps the greatest European writer then alive, was asked to write about the loss, but was too distressed to do so.

What was Byron's legacy? Like that of all the great Romantics, the 'bold assertion that the play of emotions was quite acceptable, even necessary, to one's full humanity', as the critic Peter Gay put it. The emotions now played across not just England but all Europe and as far as North America, and everyone seemed to feel for a moment their full humanity.

Not Annabella. She just noted that his life in Greece had always been precarious, and it was perhaps just as well that he died as he did ... He duped the 'romantic' reader with the 'mysterious grandeur' of his poetry, and the pretend candour of his confessions. 'When

dismantled — what remains? — Selfishness — Cowardice — Cruelty —
Effeminacy —.'

Ada cried when she first heard the news of her father's death, but
Annabella assumed the tears were for her, as the girl knew nothing
about her father, so could not possibly have any feelings for him. She
was allowed no part in the public grief that greeted Byron's return
home. The visit to 'Papa's ship' three months after its return with
Papa's body, and by then empty of its cargo, was the only indulgence
of her feelings allowed.

The *Florida* had arrived in the Thames Estuary on 29 June 1824.
John Cam Hobhouse came three days later to escort his old friend's
remains up the river. The experience was a terrible one for such a
faithful friend, made all the worse by the sight of Byron's dogs playing
on the deck.

The body's arrival in London was, according to the *London Magazine*,
'like an earthquake'. It was taken to the home of Sir Edward Knatch-
bull in Great George Street, where it would lie in state for several days.
There Hobhouse finally summoned up the courage to look at the face
he had last seen disappearing towards the horizon from that pier at
Dover eight years before. Now it was barely recognizable.

Byron's body had been mutilated by an autopsy carried out by his
doctors in Greece, who were overcome by a curiosity to find out what
a man of such genius was made of. His heart, brain and intestines had
been removed and placed in separate containers. His face, disguised by
a moustache he had grown while in Greece, had become distorted
during its long journey in the cask. It came almost as a relief to
Hobhouse that the man he saw lying there was not the beloved friend
he remembered.

The corpse became an object of intense clinical interest. The *Morning
Chronicle*, which had played such a part in the Separation scandal in
1816, published a detailed account of the autopsy, noting the enormous
size of the lungs and the unusual thickness of the skull. The paper also
observed that Byron might have survived had he agreed in the earlier
stages of the illness to being bled regularly.

Meanwhile, a debate raged about where Byron's body should be
interred. Should his past be forgiven and forgotten? Did he now

deserve the full pageant of a state funeral? Did the man who in life had been driven out of the country as a traitor and a devil deserve to be welcomed back in death as a great patriot? *The Times* seemed to think so. 'There were individuals more to be approved for moral qualities than Lord Byron – to be more safely followed or more tenderly beloved; but there lives no man on earth whose sudden departure from it . . . appears to us more calculated to impress the mind with profound and unmingled mourning,' it stated in its obituary.

> Lord Byron was doomed to pay the price which Nature sometimes charges for stupendous intellect, in the gloom of his imagination, and the intractable energy of his passions. Amazing power, variously directed, was the mark by which he was distinguished far above all his contemporaries. His dominion was the sublime.

Only *John Bull* seemed unable to suppress a sourer tone. It allowed only that he had died 'in the most unfortunate manner . . . in voluntary exile, when his mind, debased by evil associations, and the malignant brooding over imaginary ills, has been devoted to the construction of elaborate lampoons.'

Privately Annabella might have been in sympathy with *John Bull*; publicly she felt compelled to match the public mood and support a memorial in Westminster Abbey. It was a mistake, as Dr Ireland, the imperious Dean of the Abbey, refused to countenance Byron's presence in the Abbey in name or body, which made her momentarily look as if she was no longer on the side of the angels.

In the end, to the great relief of a political and aristocratic establishment that was still uncomfortable with such a popular figure, it was agreed that he would be buried in the Byron family vault at Hucknall Torkard church, near Newstead Abbey. It had been Augusta's suggestion. Any hopes, however, that this meant he would go quietly were to be crushed under the stampede of mourners that gathered at the house where he lay in state on 9 and 10 July, the only two days when the public were allowed to see the body.

An equally substantial number also lined the streets on the sunny summer's day when the coffin and the urns containing his brain and heart were finally put on a black hearse to begin their journey out of

London. It was drawn past Westminster Abbey, which had conceded a
tolling bell to mark the occasion, and up Whitehall and Tottenham
Court Road. It was followed by a cortege of no fewer than forty-seven
carriages until it reached the tollgate leading out to the Great North
Road. There, the carriages stopped, and the hearse proceeded accompanied only by the crowds of less exalted people that had gathered
along the way.

Something extraordinary then happened. The carriage began to
make its slow progress along the Great North Road towards Nottingham, and as it went

> through all the villages crowds flocked round the hearse; and every
> demonstration of respect and feeling was paid by these [middle
> and lower] classes; the village bells tolled the 'passing knell'; and
> wherever the funeral stopped, the greatest anxiety was exhibited to
> manifest every attention and respect . . . The body lay in something
> like state each night when the funeral halted

a contemporary account recorded. In other words, the people took
over the event. Though Byron had been born into the elite, it had
spurned him, so the public adopted him as theirs, and, in a spontaneous
act of collective will, carried him to his grave.

Over 170 years later, a few days after the burial of Princess Diana,
the similarities between Byron's life and death and hers were noted by
at least one newspaper:

> If there is a parallel for that mixture of collective grief and guilt that
> seized London last week, it is not with any royal occasion or the
> great state funeral of a Churchill or Wellington, but in the final
> homecoming of another thirty-six-year-old aristocrat: Lord Byron.
> Byron was, in many ways, the first modern superstar, the original
> creator and victim of that meteoric and insatiable celebrity that in
> the end killed Diana, Princess of Wales.

Stopping at Welwyn the first night, Higham Ferrers the second and
Oakham the third, the hearse finally arrived in Nottingham at five
o'clock on Friday morning. Thousands were there to greet it as it
proceeded through the streets to the Blackamoor's Head Inn. There

the coffin and the urns were placed in a room lined in black and decorated with the Byron coat of arms and six large candles. The public were then admitted twenty at a time, with a large body of constables struggling to keep the queuing crowd outside under control. Many thousands were admitted in the course of just one morning.

The body was then placed back in a hearse, and began its final eight-mile journey to Hucknall Torkard. 'The great bulk of the population', attired in mourning, followed the hearse to the city's limits on foot, where many stopped and watched it proceed down the road. Some continued to walk behind, and the number grew again as the procession passed through villages along the way. When the hearse finally arrived at Hucknall, the church was already crowded with people anxious to glimpse the open vault before the body was interred there and it was sealed shut.

At just after 3.30 p.m., the funeral took place. The contemporary account continues:

> While this last act of devotion was performing, the mourners advanced to the head of the steps conducting to the vault. Mr Hobhouse appeared much affected, and intent upon seeing the coffin deposited with the utmost care next to that of the last Lady Byron [Byron's mother]. All the mourners were deeply affected, and the domestics of Lord Byron, particularly Fletcher, who had been about twenty years in his service, were overpowered with grief.

Memories flashed through Hobhouse's mind – of Newstead, of Albany, of foreign escapades, of the great dinners with Byron surrounded by his friends, Shelley, Polidori, Scrope Davies, all now dead or exiled. He was the only one left. The conclusion was inescapable: a wonderful Byronic era had drawn to a close, and a new, very different one was about to begin.

Byron composed a fitting epitaph for this finale in *Childe Harold*:

> My task is done – my song hath ceased – my theme
> Has died into an echo; it is fit
> The spell should break of this protracted dream.
> The torch shall be extinguish'd which hath lit

My midnight lamp – and what is writ, is writ, –
Would it were worthier! but I am not now
That which I have been – and my visions flit
Less palpably before me – and the glow
Which in my spirit dwelt is fluttering, faint, and low.

*

Ada, isolated from all this, with only Annabella's approved guardians
for company, now began to sing her own song. She embarked upon an
extraordinary set of projects and ideas which she was to pursue in the
coming years.

Her first, mentioned in several letters written to her mother in 1826,
concerned the founding of a 'colony'. The inspiration for this idea
remains unknown, but there are several possible sources. The idea of
colonization was firmly embedded in the nineteenth-century consciousness as a means of escape, of overcoming the problems of the old world
by creating a new one. In 1795 the poets Samuel Taylor Coleridge and
Robert Southey developed the idea of setting up an ideal community
or 'pantisocracy' on the banks of the Susquehanna River in America,
which would combine the 'innocence of the patriarchal age with the
knowledge and refinements of European culture' and where Coleridge
imagined living out 'the sober evening of my life' gazing at 'the cottages
of independence in the undivided dale of industry'. Nothing came of
the idea, but as a result of publicizing it, Coleridge, whose career as a
dramatist Byron would help develop at Drury Lane, reinforced the link
between the ideas of Romanticism and liberty. That is no doubt why
this would not have seemed a suitable model for Ada to copy in
Annabella's eyes. Instead, she would have encouraged the girl to seek
inspiration from a very different source, the educational theorists who
were preoccupying her own thoughts at the time.

Pestalozzi, for example, had tried to create a colony among the
Swiss mountains, as Annabella related in an essay she wrote on the
'History of Industrial Schools'. She tells how, having abandoned a
career in the Church to follow a path of more 'practical' Christianity,
Pestalozzi gathered around him a group of 'helpless little ones . . .
converted his little property into money, tied up his bundle, and set off'
to found a new home for them all in the ruins of a village in the canton

of Unterwald called Stanz. Despite dreadful weather, he managed to raise a shelter for his 'new family', and to feed and clothe those pitiful specimens who daily flocked there to join it. The only help he had was from the children themselves, so he selected those with the right leadership qualities to act as his assistants, with responsibilities for teaching lessons, cooking, housework, maintenance and growing produce. 'The little colony then assumed the character of an orderly community,' Annabella observed.

It is easy to imagine how the idea of creating a 'new family' for little orphan children would attract a girl who was so lonely. In her letters to the absent Annabella, Ada's excitement about the colony idea is entangled with entreaties to be able to spend more time with her mother. She asks permission to wake her up earlier in the morning, she looks forward to her early return from the latest trip to town, or friends, or a health resort, any contact at all so she can share her latest thoughts on the project. She even promises to improve her understanding and character to make herself fit for the 'happy day' five years hence when the colony will be opened, the greatest incentive she can think of to get her mother's notice and support.

The last letter to appear on the subject is a strange one, expressing her concern that the colony might turn into Vanity Fair, the mythical bazaar in Bunyan's *Pilgrim's Progress* set up by Beelzebub, Apolyon and Legion in the town of Vanity, on the outskirts of the Eternal City, where all manner of temptations and allurements are offered: palaces, honours, kingdoms . . . She abruptly breaks off the speculation, and returns to more familiar matters, the welfare of her cat Puff, who she reports was the previous day given a dinner of vole and oyster specially prepared. Ada asks Annabella to bring back the finest turbot for her beloved pet.

Annabella, who unlike Byron never much liked animals, had more pressing issues to consider than catfood. In 1818, while living in Venice, Byron had started writing his memoirs, which according to a letter to his publisher John Murray contained his own account of the Separation. He had sent Annabella a copy the following year, asking her to 'mark any part or parts that do not appear to coincide with the truth'. 'The truth I have always stated,' he added in his covering letter, but there are two ways of looking at it – and your way may not be mine. I have never revised the papers since they were written. You may read them – and mark what you please ... You will find nothing to flatter you – nothing to lead to the most remote supposition that we could ever have been – or be happy together. But I do not choose to give to another generation statements which we cannot arise from the dust to prove or disprove – without letting you see fairly and fully what I look upon you to have been – and what I depict you as being.

Annabella did not care to find out how Byron had looked upon her, and refused to read the manuscript on the grounds that this would amount to endorsing it, which might encourage its circulation – again, a very calculated response, and also confounding to Byron. He, like so many, was left bewildered by Annabella's lifelong (and very effective) policy of returning any provocation to act or speak with imperturbable silence.

Byron also handed a copy of the manuscript to Thomas Moore, the Irish poet whose literary fame and financial acumen rivalled his own. Moore was, Harriet Beecher Stowe would later euphemistically remark, bewitched by Byron 'as ever man has been by woman', and from the perspective of the Annabella camp, a man to be no more trusted than her husband. He had not only sold the memoirs to Murray for £2,000 with the aim of posthumous publication, but allowed numerous copies to be made, which were now thought to be circulating around London. A group of Annabella's friends were immediately mobilized to buy back the manuscript from Murray, a move which proved unnecessary, as Murray himself was coming round to the view that it should never be published. Annabella kept her distance from the events that followed, allowing the ever-loyal, no-nonsense, I-speak-as-I-find Colonel Francis Doyle to protect her interests. He attended a key meeting held at Murray's offices in Albemarle Street on 17 May 1824. Five others were there, including Hobhouse and Moore, who bickered over the pile of papers that had been placed before them. Doyle's mission was simple: to witness the memoirs' immediate destruction. His only ally was Murray, who in an act of inexplicable sacrifice was now intent on the destruction of a work that assured sales as sensational as its subject matter. The others searched desperately for a compromise that would save the manuscript without publicizing it. Perhaps excerpts could be preserved. It could be deposited in a bank vault, and then used only to check against any spurious copies that came into circulation – a fine idea, several agreed, if it received Augusta's agreement . . .

Doyle had no time for this. He snatched the papers and, with the help of Wilmot Horton, a relation of Byron's who had acted as an intermediary of Annabella's during the Separation negotiations, he tore them up and threw them onto a fire burning in the grate. The two men apparently performed the auto-da-fé without the help of the others. Horton offered Hobhouse a few sheaves to consign to the flames, but Hobhouse demurred.

That, of course, was not an end to the matter, merely a beginning. The papers were incredulous when news of the destruction leaked out. 'It *must* be taken for granted that the Memoirs were utterly unfit for publication in any shape,' one editorial observed. There was also a huge vacuum of information now left to fill, and, to Annabella's growing alarm, it sucked in a succession of revelations that for sensationalism would surely have left even Byron's offering in the dust. As early as June 1824, before his body was buried, the newspapers were already carrying advertisements for a memoir containing the 'PRIVATE CORRESPONDENCE OF LORD BYRON'. The author was Robert Dallas, a distant relative of Byron's who had arranged the publication of *Childe Harold*. Hobhouse, Byron's executor and the self-appointed keeper of the Byronic shrine, was appalled that Dallas intended to exploit his connection with the poet in this way. His efforts to dissuade Dallas from going ahead with the project were repulsed, so he went to the courts and secured an injunction preventing publication. But Dallas and his son, the Reverend Alexander, were not to be put off, and fought for the injunction to be lifted, helping to exacerbate a row over the will by securing the backing of George Anson, now the seventh Lord Byron. After much legal letter-writing, and a drawn-out hearing at the Court of Chancery which led to an important copyright precedent being set by the Lord Chancellor concerning the ownership of letters, the Dallases eventually decided to escape English jurisdiction and publish the book in France.

In November, Robert Southey finally managed to get in his retaliation for Byron's lampooning reference to him in the opening lines of his great epic *Don Juan*. In the pages of the *Courier*, Southey, the Poet Laureate, accused Byron, the very opposite, of committing 'a high crime and misdemeanour against society, by sending forth a work in which mockery was mingled with horrors, filth with impiety, profligacy with sedition and slander'. At the same time, Hobhouse was also fighting off another attempt at publishing a biography, this time by Thomas Medwin, who announced a 'Journal of the Conversations of Lord Byron, Noted During a residence with His Lordship at Pisa in the Years 1821 and 1822'. It was published by Longman's in October 1824, complete with a disclaimer by the publisher, endless errors, several lies, one or two juicy revelations (such as the first substantial account in print of Byron's relationship with Lady Caroline Lamb) and what was convincingly claimed to be Byron's own assessment of Annabella. 'She married me from vanity and the hope of reforming and fixing me,' Medwin reported Byron as saying in one conversation.

> She was a spoiled child, and naturally of a jealous disposition; and this was increased by the infernal machinations of those in her confidence . . . She thought her knowledge of mankind infallible . . . She had a habit of drawing people's characters after she had seen them once or twice . . . She was governed by what she called fixed rules and principles squared mathematically.

Hobhouse's fight to keep Byron's life a closed book was now defeated, and the steam presses that were then being installed in the printing rooms of London's newspapers and book publishers were called upon to feed a public yearning for more. Annabella, meanwhile, retained her position of aloof detachment. When Augusta attempted to recruit her for a joint attack on Medwin's 'Vile Book', Annabella wrote back suggesting that they should not appear to have even read the volume. She had, as she later wrote, withdrawn from the conflict, 'convinced that my duty imperatively enjoined that step'. She did not, however, find it an easy step to take. She felt as though she was in the middle of a desert, a 'burning world within me which made the external one cold . . . I returned kindness heartlessly & mechanically.'

This feeling of being marooned in a salt lake of sanctity helped reinforce the image already established that she was haughty and detached. Her station, sex and class meant that no one would come straight out and attack her character, but there were plenty of hints in the compliments paid to her that amounted to the same. It must have been particularly punishing for her to read the discussions about her relationship with Byron that began to appear in the 'Noctes Ambrosianiae'.

The Noctes was a notorious column published in *Blackwood's Magazine*, a very successful Tory rival to the Whig *Edinburgh Review*, then the leading literary journal. In many respects, the column was the model for future satirical parodies, the spiritual ancestor of *Private Eye* and its like. Mostly written by the brilliant Scottish lawyer and writer John Wilson ('Christopher North'), it took the form of a series of imaginary conversations between a variety of boisterous Regency revellers enjoying a convivial drink at Ambrose's Tavern, a real bar situated in the backstreets of Edinburgh. Inevitably, the thoughts of these genial men would stray upon the subject of Byron, such as in the November 1825 issue, published at a time when the Byron story was still dominating the headlines:

> *James Hogg.* – 'Reach me the black bottle. I say Christopher, what, after all, is your opinion o' Lord and Leddy Byron's quarrel? Do you yoursel' take part with him, or with her? I wad like to hear your real opinion.'
>
> *Christopher North.* – '. . . At present we have nothing but loose talk of society to go upon; and certainly, if the things that are said be true, there must be thorough explanation from some quarter, or the tide will continue, as it has assuredly begun, to flow in a direction very opposite to what we were for years accustomed. Sir, they must explain this business of the letter [the one Annabella sent to her 'Dearest Duck' just after her departure from Piccadilly]. You have, of course, heard about the invitation it contained, the warm, affectionate invitation, to Kirkby Mallory' –
>
> *Hogg.* – 'I dinna like to be interruptin' ye, Mr. North; but I must inquire, Is the jug to stand still while ye're going on at that rate?'
>
> *North.* – 'There, Porker! These things are part and parcel of the chatter of every bookseller's shop; *à fortiori* of every drawingroom in May Fair. *Can* the matter stop here? Can a great man's

memory be permitted to incur damnation while these saving clauses are afloat anywhere uncontradicted?'

The saving clauses were not to remain afloat for much longer. They were soon swamped by contradictions, validations, revelations and fabrications. By 1826, just a year and a half after Byron's death, there had already appeared countless articles, a spurious chapter of the destroyed memoirs supposedly describing Byron's wedding night, several biographies, now including one by Leigh Hunt, who had so touchingly waved a poetic farewell to Byron as the latter left the country following the Separation – and nothing that had the imprimatur of Hobhouse, Augusta Leigh or Annabella to give it any sort of authority. The strategy of silence adopted by Byron's friends and family had, in other words, hardly produced a corresponding reticence among his critics and biphers.

Eventually, Hobhouse relented. He grudgingly agreed to Thomas Moore writing a Life, and Murray agreed to publish it, if only to thwart Leigh Hunt's book, which he detested. It appeared in January 1830, and despite all the obstacles put in his path, Moore managed to write what was and is generally agreed to be a brilliant biography. Hobhouse carped at some of the details, but even he had to admit that its portrayal of Byron was fair.

But hovering overhead was a vengeful angel who now, for the first time, swooped into the fray. 'I have disregarded various publications in which facts within my knowledge have been grossly misrepresented,' wrote Annabella in a pamphlet she had begun to circulate among her friends,

but I am called upon to notice some of the erroneous statements proceeding from one who claims to be Lord Byron's confidential and authorised friend. Domestic details ought not to be intruded on the public attention: if, however, they are so intruded, the persons affected by them have a right to refute injurious charges. Mr. Moore has promulgated his own impressions of private events in which I was most nearly concerned as if he possessed a competent knowledge of the subject. Having survived Lord Byron, I feel increased reluctance to advert to any circumstances connected with the period of my marriage; nor is it now my intention to disclose them further than may be indispensably requisite for the end I have in view.

The silence was not just broken, it was shattered.

'Self-vindication is not the motive which actuates me to make this appeal,' Annabella continued, 'and the spirit of accusation is unmingled with it; but when the conduct of my parents is brought forward in a disgraceful light by the passages selected from Lord Byron's letters, and by the remarks of his biographer, I feel bound to justify their characters from imputations which I know to be false.' This, then, was the motive for going public: defending her parents against the charge of being in some way responsible for the Separation.

Whether Annabella was genuinely outraged or merely using a vague allegation as an excuse to get her side of the story into print, the slight against her mother and father shook her into producing what Byron-watchers of the time must have considered a windfall of confirmations and revelations. She gave details about her consultation with Dr Baillie concerning Byron's supposed insanity and his advice that she get out of the house. She revealed that the affectionate tone of the 'Dearest Duck' letter she had written after her departure was one she had affected. She wrote about how Byron was found to be sane, how this made her feel she could no longer live with him, as he was deprived of the only mitigation for his cruel behaviour.

And then she produced what was the most suggestive revelation of all, a letter from her solicitor Dr Lushington, in which he recalled the events leading up to the Separation being made final. At first, he wrote, he had assumed a reconciliation was possible, and Annabella's mother had said nothing to change this view. Then he had been paid a visit by Annabella who informed him of 'facts utterly unknown' to her parents. 'On receiving this additional information, my opinion was entirely changed: I considered a reconciliation impossible,' he concluded. So, whatever it was that finally put an end to Annabella's relationship with Byron, it had nothing to do with Sir Ralph or Judith. They were innocent of the charge of interference made by Thomas Moore. But never mind that, there was something else, some '*additional information*' which was responsible, and by admitting this, Annabella let loose a scintillating new line of enquiry. Those in the know would have immediately guessed what this 'additional information' might be – Byron's rumoured incest and homosexuality. But the little phrase created such a void, anything, no matter how hugely improbable or scandalous, could be deployed to fill it.

Colonel Doyle advised Annabella against publishing her pamphlet on Moore's memoirs, as it was eventually bound to be read by Ada. So, ever observant of the good Colonel's advice, she sent out copies privately – to friends, writers, librarians, peers, King George IV . . . anyone she could think of. She dispatched packets to a John Hill of Leicester, who promised to distribute it among members of his library, and to Fanny Bowdler, a relative of Dr Bowdler of bowdlerizing fame, who pledged to keep the original in her hands, and to ensure the widest possible circulation of its bowdlerized contents.

One name on the circulation list was Thomas Campbell, a poet and journalist who had once been a Byron supporter, but who was instantly transformed into one of Annabella's most enthusiastic champions. He visited her and offered her his services, which she seemed to accept, following up the visit with a friendly letter to him in which she made it clear that the 'additional information' explaining Byron's behaviour towards her had nothing to do with his debts, one of the many theories then being circulated.

The next thing she knew, this letter was published in Campbell's paper the *New Monthly* as part of a detailed (and inaccurate) account of the Separation. In one of its many attacks on Moore's biography and eulogies about the character of Annabella, it picked up on the idea that Lady Byron may have been an unsuitable match for her Lord. 'A woman to suit Lord Byron,' fulminated Campbell, his points now punctuated with multiple exclamations, 'Poo! Poo! I could paint you the woman that could have *matched* him . . .'

The damage this article caused was manifold. To begin with, it revealed Annabella's furtive role in the orchestration of the anti-Byron campaign. And there was that '*matched*'. Everyone knew what those suggestive italics not-so-subtly suggested. *John Bull* certainly picked up on the reference, writing knowingly of how the 'dreadful allusions' the article contained would probably have Byron's relatives 'writhing'. One well-matched relative in particular: Augusta.

During the period this whole affair raged, relations between Augusta and Annabella had been deteriorating over the choice of trustees for Augusta's portion of the Byron estate. Annabella had been canvassing her own lawyer, Dr Lushington, a name tartly rejected by Augusta. Annabella was not used to such treatment from her 'sister', and immediately climbed up onto her highest horse to issue a warning that she would break off relations with Augusta if she continued in this uncooperative vein. To add a little carrot to the stick, she did, however agree to provide Augusta with reports of Ada's health, to which, since Byron's death, she no longer had any entitlement. Augusta had clearly developed strong feelings for the little girl she was so rarely allowed to see, and was likely to cling to Annabella's offer.

Augusta would not adopt a suitably compliant attitude, partly because of her growing annoyance at what she saw as the 'despicable tirade' contained in Annabella's pamphlet on the Moore biography. Nevertheless, when it came to Ada, she tried to maintain a more convivial atmosphere.

On 10 December 1830, for Ada's fifteenth birthday, Augusta carefully packaged up a beautiful little prayer book which she had specially bound and engraved with the word 'Ada' in Old English characters. She addressed the package 'To the Hon. Miss Byron, with every kind and affectionate wish', adding the note 'With Lady Byron's permission'. She sent the package off by specially booked coach . . . 'and never heard one word since'. She was not to communicate with Annabella for twenty years, not even with the girl who had been baptized her godchild and namesake. Ada would later know her estranged godmother only as the 'wicked' woman who came between her parents.

*

For Ada, the furore surrounding Byron's burial and biography would have been detected as no more than a rumble from a distant battlefield. Nevertheless, it was to change her life in several practical ways. The first was a foreign tour. Annabella had wanted to take Ada abroad for some time, but Byron had refused to give his permission for such a hazardous expedition. Now he was no longer around to stop her Annabella felt it was time for the two of them to go, which they did in 1826.

Ada's letters back to her mother's friends in England are her earliest writings of substance to survive, densely packed and written in a copperplate hand now beginning to break free of its restraining neatness. The correspondence reveals a girl who is fascinated by the world unfolding before her, a world which she is so impatient to describe she cannot pause for punctuation. Each report begins with a medical bulletin about her mother, requests for news about her cat Puff, and a few pleasantries about the latest group of friends they have met along the way. Then she commences a haphazardly ordered list of the latest places, sights, people, animals, houses, trees, lakes, thoughts that have passed before her.

The first place she visited to leave a deep impression was Lake Geneva, written about by Byron under its French name, Lake Leman. Little did she know how closely her footprints along its shores came to crossing those left by her father. She almost certainly did not visit Diodati, the villa along the lake's banks where Byron had challenged the Shelleys and Dr Polidori to compose a ghost story, a challenge that resulted in Mary Shelley writing *Frankenstein*. But she would at least have seen, looming on the shore, the castle of Chillon, underneath which lay the dungeon that inspired Byron's great poetic study of imprisonment and despair, *The Prisoner of Chillon*.

> Lake Leman lies by Chillon's walls:
> A thousand feet in depth below
> Its massy waters meet and flow . . .

Ada spent hours staring out over those same massy waters, trying to understand the changing colours that played across the surface. With the help of Miss Stamp, the governess accompanying her on the trip, and in accordance with Pestalozzi's educational principles, she struggled to relate what she saw to the physical phenomena that gave rise to it. To achieve this, she drew a simple picture of the scene from the shore illuminated by a full moon, showing the pools of 'dead' and 'living silver' light that glimmered across it. The dead silver, she explained, was where the water was motionless, the living silver, the glitter of ripples where the water moved. And then, interrupting her analysis, her attention is caught by a boat sailing across the lake, its hull, sail, crew, and oars forming a black silhouette against the beautiful curtain of light.

During October she sent some letters to Joanna Baillie, the writer who Annabella had so heartily recommended to Byron as an example of the sort of woman he should get to know. Ada reports her tour through Switzerland and Germany, 'tracing the deep valleys between the mountains', and visited a series of cities, including Heidelberg, which she particularly admired.

At the beginning of 1827 she stayed in Genoa, at first at the Hôtel d'Amérique, which overlooked the sea, from where she witnessed beautiful sunsets and a flotilla of ships bringing the king to the city. She imagined what it would have been like to live there in the time of Christopher Columbus, an age when the city must have been magnificent, a northern Naples. As spring arrived, she started to hunt for insects. She found a scorpion in the house; locusts, lizards and beautiful snakes outside. The garden was covered in large anthills, and crawling with fat woodlice. At night, the scene was filled with the speckled light of innumerable fireflies, and the croaking choruses of black and green frogs, which annoyed her. She was also taking music lessons, and enjoyed singing more than anything.

In the summer they started touring again, visiting Turin before crossing the Alps back to Geneva, and then on up Mont Blanc, where she was excited to catch her first glimpse of glaciers. She stayed at Ouche for a while, and then visited some more German cities, Stuttgart now being her favourite, before returning to Turin to prepare for the journey back to England at the end of October 1827.

Ada returned home to Bifrons, a house near Canterbury rented by Annabella just before their departure. While they had been away, Annabella had loaned it to Georgiana, one of Augusta's daughters, who had just married Henry Trevanion, a handsome poet manqué who boasted a distant kinship to Byron. He was adored by Augusta and loathed by her increasingly tyrannical and unstable husband Colonel Leigh.

These temporary guests, a residue of Annabella's connection with the Byron family, were cleared away by the time Ada was back in residence, and she once again found herself confined to her own little chamber of isolation. At times, it must have seemed as imprisoning as the dungeon beneath Chillon. However, with the help of her benign gaoler Miss Stamp, for whom she had developed a strong affection, she now appeared willing to reconcile herself to her highly regulated existence, like the Prisoner her father had imagined manacled 'to a column stone':

> My very chains and I grew friends,
> So much a long communion tends
> To make us what we are . . .

In this new mood of acceptance, she began to build castles in the air, and then imagined the apparatus she would need to fly among them.

*

The first mention of a flying machine appears in a letter to her mother dated 8 January 1828, written a few weeks after her thirteenth birthday. By that time, she seems to have been well advanced with her plans, promising her mother a demonstration when she returned home, hoping to be rewarded with a 'crown of laurels' if Annabella were pleased with the work. The exact mechanism she intended to build to achieve this feat is unknown, but some suggestions remain in subsequent letters.

As a sort of feasibility study, she began to design a set of paper wings, to work out the proportions they would need to support a human. She also announced that she planned to write a book about a new science of artificial flight to be called 'flyology'. She set up a lab at Bifrons she called the 'flying room', which was kitted out with ropes, pulleys and a 'triangle', presumably a mounting for the wings she was trying to design.

As her designs progressed, she started to think about applications of the 'art of flying', and was particularly intrigued by the possibilities for the post. It would be feasible, she pointed out, for her to become like a 'carrier pigeon', delivering and collecting her mother's mail far faster than '*terrestrial* contrivances'. All she would need was a letter bag, a small compass and a map, enabling her to take a direct route across the country irrespective of physical obstacles such as hills and rivers. She was so taken by this idea that for a while she signed all her letters to Annabella 'Carrier Pigeon'.

The scheme developed over the coming months, with Ada embarking on a detailed study of the anatomy and in particular the wings of birds, focusing on those of a dead crow she found in the fields. She then planned to create a pair of wings for herself with exactly the same proportions as the bird's, which she would make out of paper stiffened with wire and attach to her shoulders.

She also consulted Dr Mayo on the feasibility of her plans. Mayo was a member of Annabella's growing team of high-powered medical advisers. He had acquired a name for research on disorders of the nervous system. He would also have been Ada's only contact with the scientific world at that time, and she was keen to canvass his expert view.

Whatever it may have been, her train of thought was now unstoppable. As soon as she had perfected flying with her artificial wings, she wanted to explore the possibilities of powered flight – something more wonderful than the steam packets and steam carriages that at the time represented the latest in transportation technology. She imagined a body in the shape of a horse with a steam engine inside which would be connected to an immense pair of wings fixed to each side. The passenger would then sit astride the horse and be lifted aloft.

This extraordinary idea entangled her yearning to fly with a fascination and fear of horses, which she would not be able to conquer fully until she reached adulthood. She had asked her mother if she could have a small pony, hoping that such an innocuous creature might help her overcome her phobia. But in her schemes and ideas, she had no need to be so tentative. She could fearlessly mount a steam-powered steed that would soar through the skies, overcoming her fears and her confinement in a single leap of the imagination.

At first her mother indulged these grand plans, while not encouraging them. Ada was continuing with her studies to the general satisfaction of Miss Stamp (though she was occasionally forced to report lapses in her behaviour in letters to her mother), reading with satisfactory thoroughness various approved texts such as *Useful Knowledge*, a 'Familiar and explanatory account of the various productions of nature, mineral, vegetable, and animal, etc.' by the Reverend William Bingley, the *Encyclopaedia Edinensis, or, Dictionary of Arts, Sciences and Literature* by J. Millar and a book of geometry which she particularly enjoyed written by one Mr Paisley. She was also putting a lot of effort into building a

'Planetarium', a map of the stars, which she had nearly completed by the beginning of 1829.

She embarked on several literary projects, too. Some of these formed part of her education, such as an essay on Sir Walter Scott's *Heart of Midlothian*, which combined a fresh and intelligent critical style with a dutiful acknowledgement of the story's moral, viz. that the main character's 'filial piety cannot be too much imitated by all who have parents'. Other works were apparently extra-curricular, though still aimed at pleasing their audience, such as a study entitled 'On Human Nature', which identified the four principles of human nature to be an inclination to gratify the senses, goodwill and affection towards others, a disposition to consider our own good (also known as 'self-regard'), and an ability to reflect. It is the latter, she wrote reflectively, that regulates the others.

There are also two fragments of creative writing to be found among her juvenile papers. One is a series of three tales about a heroine. The first tale concerns a Princess Isabella, daughter of the King of Denmark (perhaps an allusion to her being a female Hamlet). The princess arrives at a hotel in the mountains at the same time as two ladies, both daughters of an English baroness transparently identified as 'Lady NB' (as in Noel Byron, the name Annabella had adopted following the death of her mother). The ladies have been made very ill by their journey, and are desperate for rest. Unfortunately, the hotel has only one room with two 'dirty little beds' vacant, and the two invalid ladies insist upon the princess having it. She accepts their offer, but only so that she can gain entry to the room and make up the beds for her two fellow guests. Her task completed, she leaves to sleep in the hayloft. The second tale once again concerns three ladies, who are now fleeing before an enemy. They find two horses, which one lady insists the others take to make their escape. She stays behind and is captured by the enemy, who imprison her for the rest of her life. The final tale concerns a large party dining upon a green bank in the mountains. A maid is taken very sick, and only a particular herb will save her. A lady among the party immediately leaves the dinner, mounts her horse and gallops off into the mountains, finds the plant and saves the maid's life.

The other surviving fragment is entitled the 'Story of Sophia'. At first it appears to be yet another homily to filial duty, concerning as it does a girl who takes great delight in pleasing her mother. However, reading between the almost scrawled lines, it can be seen as a terrifying vision of what she fears she is becoming: a thirteen-year-old girl who is neither pretty nor clever, who takes such care to do exactly as her mother and governess tell her so that there is never any need to reprove her for anything.

These fantasies came to an abrupt end in 1829. A number of events seem to have been responsible. First her governess Miss Stamp left Ada to get married. Then Ada was struck down by a mysterious illness.

What apparently began as a dose of measles developed into a disease that attacked both her sight and her limbs. She soon became paralysed and almost blind. For weeks she could not walk or even stand, according to her mother.

There is no possibility of reaching an accurate diagnosis as to what afflicted Ada so badly, but her symptoms seem very symbolic for a girl who so recently had been obsessed with flying. It is hard to ignore the possibility that it was a sort of mental as well as physical breakdown, a psychosomatic acknowledgement of defeat, of being unable to break free of the restrictions she had been forced to endure since her birth.

It is also possible, even likely, that the illness would have then been seen as having hysterical origins. Annabella had close connections with a number of physicians who were studying the nervous system, a new area of medical research. As already noted, one of them was Dr Herbert Mayo. Mayo had a knack of reflecting exactly Annabella's opinions in his own medical theories. For example, he confirmed her (and, indeed, the prevailing) view that in the order of nutrition, 'the flesh of the adult, warm-blooded quadrupeds' occupies the summit, followed in descending order by game, poultry, 'farinaceous' food (cereals), arrowroot, bread, fish and, last of all, fruits and vegetables. He also believed that a 'reduced diet' helped produce hysteria, and it is possible that Annabella, who had a peculiar mania for mutton, would have seized upon a diagnosis along these lines to explain her daughter's extraordinary symptoms. After some initial anxiety, she certainly did not appear to take the condition all that seriously, as most of the letters

she wrote at the time show far more concern for her own ever-
vulnerable state of health than her daughter's.

*

In the summer of 1830 Ada staged a fight back, against her illness and
then, more formidably, her mother.

She was by now living in Mortlake, having been moved there from
another of Annabella's rented houses at Hanger Hill near Ealing.
Bifrons was once again occupied by Augusta's daughter and son-in-
law, Georgiana and Henry Trevanion. Georgiana was pregnant, and
had been joined for her confinement by her beautiful sister, Elizabeth
Medora Leigh, who as a little girl had been observed by Annabella
arousing mysteriously paternal feelings in her uncle Byron.

The sexual intrigues and violent jealousies that overtook these three
in the months they were there would have dreadful repercussions not
just for Annabella but Ada too. These are for a later episode in this
story. For now, Ada was to remain ignorant of the activities of her
relatives, her mind being occupied by a renewed campaign of education
and improvement launched by her mother.

A whole army of educational advisers was now deployed to help
her, all gathered from the same species of Nonconformist intellectual
nurtured by the industrial revolution now in full tilt. There was William
Frend, the reformer and mathematician who had been forced to give
up his living as a vicar after converting to Socinianism, the austere
religious doctrine also favoured by Annabella. At the time he became
interested in Ada's education, Frend was also a member of that
peculiarly nineteenth-century fraternity of actuaries, mathematicians
employed by insurance companies (in his case the Rock Life Assurance
Company) to assess risks. William Frend's advice about Ada's education
was supplemented by contributions from his very censorious daughter,
Sophia, who overcame a deep-seated dislike for Ada in order to indulge
Annabella's desire that she befriend her daughter.

Dr William King, the man Annabella would later recruit to watch
over Ada when she first saw the portrait of her father, was also enlisted
around this time. He was a devout Unitarian who came to be called
'Co-operation's Prophet', having brought the Co-operative Movement
to Brighton. He and his wife Mary were to continue keeping an eye on

Ada, twitching their curtain whenever she strayed from the path of righteousness.

And there was Arabella Lawrence, another Co-operatively minded Unitarian educationalist, who ran a school in Liverpool. Ada wrote several letters from Mortlake to Arabella in 1830, while she was still recuperating from her illness. Some of them are in pencil, as Ada had to lie flat on her back for most of the day and could not use a pen. By the summer, she was allowed out on short excursions in a wheelchair, and occasionally to rest on the bank of the River Thames, which passed close to the house. On one such occasion she saw a couple wandering along the towpath. They stopped to look at the house, and Ada overheard the man mention that it was Lady Byron's. The woman said how much she would like to meet her, and continued to talk about her, but out of earshot, much to Ada's frustration.

Indeed, frustration was to be Ada's constant companion as the illness dragged on. Hints of rebellion began to appear – refusing to sleep in her bed but instead wrapping herself in a blanket and sleeping on the floor or a sofa; refusing to kiss anyone but her mother, who would then only receive a grudging peck; developing fickle tastes when it came to food, which reminded Annabella of the girl's father. She became argumentative, turning each trifling disagreement into a 'French Revolution'. She admitted as much to Arabella, and, acknowledging Arabella's philanthropic concern for the happiness of people much worse off, wearily accepted the need for a sense of proportion. Annabella also complained to Arabella about Ada's habit of 'conversational litigation', which she found disrespectful and demanding immediate correction. In the same letter she protested about the contents of one of Ada's, revealing that, as even Ada probably suspected, her communications were being vetted.

The sense of oppression was exacerbated by the constant presence of Annabella's friends, a circle of cronies handpicked for their respectability, soberness, earnestness, restraint and piety. They were, in other words, models of the middle-class values that would come to dominate the Victorian age, and of what Annabella required her daughter to become.

Following the Separation, Annabella wanted nothing more to do with the aristocratic world. Though she kept her title of Lady Byron, and retained links with various aristocratic families such as the Gos-fords, she increasingly portrayed herself as bourgeois, and wanted Ada brought up in that milieu. She would jokingly (by her standards) refer to her peers as belonging to the 'inferior classes', and made it her business to visit reform schools, penitentiaries, hospitals and madhouses to expose herself to the opposite, more deserving end of the social scale.

This put her a world away from Byron. As the essayist William Hazlitt once explained, the imagination is an 'aristocractical faculty', 'exaggerating and exclusive ... it takes from one thing to add to another: it accumulates circumstances to give the greatest possible effect to a favourite object'. Byronic, in other words. The understanding, in contrast, was more bourgeois, indeed republican, as it was about 'measuring and dividing', 'it judges of things not according to their immediate impression on the mind, but according to their relations to one another.'

Measured, divided, judged not according to immediate impressions but according to her relations to others – this had been Ada's life up until now and what, as she slowly recovered from her illness, she became less and less able to tolerate. There is a distinct change in the tone of her letters, the admissions of weakness and pleas for approval giving way to a certain petulance, a flash here and there of wilfulness and wit. She seems to be only half joking when she refers to Arabella Lawrence's 'habitual maliciousness', which she writes will be frustrated by the discovery that she is actually enjoying a particularly difficult text. And her feelings for her mother seem to have frozen at this time into a state 'more akin to awe and admiration than love and affection', as she later confided to a friend. This is reflected in the few surviving letters that they exchanged. The nearest to the relaxed, confiding communication that one might expect between a mother and her only daughter is a report about the 'unspeakable agitation' aroused by a 'very handsome' man riding up to the door on a black horse – and here she was teasing, as the man was the Reverend Francis Trench, another of Annabella's Nonconformist associates. She evidently made an impression on him, as he would later comment on Ada's 'fine form of countenance, large expressive eyes, and dark curling hair' and the similarities of her features to her father's.

By 1832 a sixteen-year-old Ada, though still walking on crutches, was deemed sufficiently mobile to be moved yet again, to the fourth house that she was to try to make her home in as many years. Fordhook was an elegant villa on the edge of Ealing Common. Even though it had once belonged to Henry Fielding, the writer of such deplorable manifestos of eighteenth-century manners as *Tom Jones*, Annabella appears to have become quite attached to the place.

Like so many of the houses associated with Annabella and Ada (a melancholy list that includes Kirkby Mallory, Halnaby, Ockham Park and Ashley Combe), Fordhook was demolished at the beginning of the twentieth century. The neighbouring estate of Hanger Hill has become the Hanger Lane Gyratory System (part of London's gridlocked inner ringroad system), and the nearby village the conurbation of East Acton. But in the 1830s it was a perfect suburban retreat from the city of London, which seethed like a termite's nest six miles down the road.

Ada, however, did not settle into her new home. Though the occasional trip to the metropolis was now countenanced, her sense of isolation and oppression escalated. It was like being in Jeremy Bentham's panopticon, the new prison design admired by Annabella, in which every cell is completely visible, in which there is no privacy or individuality.

Ada had a team of tutors attending to her education, and was kept under the constant supervision of a coven of Annabella's closest and most trusted spinster friends: Sophia Frend (later to marry Augustus De Morgan), Selina Doyle, Mary Montgomery and Frances Carr. The latter three in particular took it upon themselves to keep an eye on Ada, which they passed among themselves as they stirred a cauldron of gossip and recrimination.

Ada found their surveillance intolerable. She dubbed the watchful witches the Furies, and would harbour for the rest of her days a special hatred for them. In the latter years of her life she told a friend, in a tone of bitterness otherwise remarkably absent from her correspondence, that these women would 'exaggerate or invent' stories about her behaviour, and hated her like 'poison'. They were, reported Woronzow Greig, who later became Ada's lawyer and occasional confidant, 'constantly with Lady Byron, who was entirely led by them, and as the daughter informed me, they took the most unwarrantable liberties with

Lady Byron and interfered in the most unjustifiable manner between mother and daughter.'

Some time in 1832 or early 1833 Annabella engaged a new tutor for Ada, a member of a 'humble family' living in the neighbourhood. His identity remains a mystery, but he may have been William Turner, brother of Edward Turner, the first Professor of Chemistry at University College, London. He had been taken on to teach Ada shorthand around this time. Ada developed an attachment for this young man, which was evidently mutual.

At first the relationship went unheeded by the Furies. Perhaps sensing an opportunity to disturb the dreadful calm of her life, this seems to have provoked Ada to be increasingly open about her interest in him. Eventually, one of the Furies (it is unclear which; in Ada's eyes they were obviously interchangeable) realized Ada was taking far more notice of the young man than of her studies. She reprimanded the girl and told her to get on with her work. Ada ignored the instruction, sending the Fury into a fury which resulted in her expulsion from the room. She left in a state of 'high indignation'. She returned later, perhaps putting on an air of contrition, which would certainly have been expected, saying that she wished to take some books to read. This she was allowed to do, which gave her the opportunity to slip a piece of paper into the tutor's hand proposing that they meet in secret in one of the outhouses at midnight.

At the appointed hour, as the eye of the Furies was temporarily closed in sleep, Ada stole out of the house and met him. There followed a passionate encounter during which the two young lovers 'went as far as they possibly could without complete penetration', as Woronzow put it in his account of the incident, based on Ada's own recollections. He later lightly crossed out the phrase 'complete penetration' and replaced it with the word 'connexion'.

Up to this point, Ada was probably acting out of a mood of defiance. Now it must have been different. The experience reawakened her alienated body, revived her repressed senses. As she clung to that young man, she must have thought, as her father put it in a letter celebrating the lustiness of *Don Juan*, 'is it not *life*, is it not *the thing*?'

As the days went by, further assignations were apparently arranged, and they were inevitably caught. The man was immediately sacked and Ada confined to the house. But Ada would not now be constrained. She fled from Fordhook and made her way to her lover's house, presumably on foot, despite the residue of paralysis that still weakened her legs. From there the couple planned to elope. Ada was finally to escape the Furies, the strictures, the sermons, the tutors, the tickets, the tours, the improving texts, the whole panoptical apparatus that Annabella had assembled to keep her daughter under control.

But there was no escape. The tutor's relations, who knew Annabella, recognized Ada, and immediately alerted the girl's mother. Soon, perhaps within hours, Ada was back at Fordhook.

Since it would have undoubtedly featured prominently in the newspapers if it had got out, a concerted and successful campaign was launched to hush the whole episode up. Ada was back at home, her studies were resumed, and no one – other than Lady Byron at a later and altogether crucial moment – referred to the matter again except in the most oblique terms. Over forty years later, Sophia Frend (then De Morgan), mentioned Ada's 'heedlessness & imprudence'. 'I do not think this matter need be further entered into. There was, I hope, no real misconduct at that time and an open scandal was prevented,' she delicately observed.

In many respects, Annabella and her Furies could congratulate themselves on their handling of the whole episode. True, there had been a lapse in security, but the fugitive had been caught and returned. Girls of Ada's class and age were prone to such things, and just as long as it did not compromise their eligibility and reputation no real harm was done. Sophia Frend herself suggested the incident need only be treated as an 'imprudence'.

Ada, however, was different. Her actions had revealed something that Annabella had spent the girl's upbringing trying to annihilate: a lack of control, a taste for independence, a dislike of regulation, *passion*. Sophia Frend helpfully spelled out the implication on everyone's lips: 'It was very evident that the daughter who inherited many of her father's peculiarities also inherited his tendencies'. Despite all that Annabella had done to make her daughter a Milbanke, Ada had turned out to be a Byron.

CHAPTER FOUR

The Devil's Drawing Room

The sun went down, the smoke rose up, as from
    A half-unquenched volcano, o'er a space
Which well beseemed the 'devil's drawing room',
    As some have qualified that wondrous place . . .

A mighty mass of brick and smoke and shipping,
    Dirty and dusky, but as wide as eye
Could reach, with here and there a sail just skipping
    In sight, then lost amidst the forestry
Of masts, a wilderness of steeples peeping
    On tiptoe through their sea coal canopy,
A huge, dun cupola, like a foolscap crown
On a fool's head – and there is London town!

IT WAS THE SCENE CONFRONTING DON JUAN, the fictional hero of
Byron's last and greatest work, when he approached London from
Shooter's Hill. A visitor approaching from the same direction in the
1830s would have beheld a similar vision – the forest of masts along
the busy docks, the wilderness of steeples in the East End and the City,
the foolscap crown of St Paul's dome and the canopy of coal dust and
smoke – but now supplemented by a bristling stubble of chimneys and
a suffocating pall of smog.

Juan was standing on a spot that in the opening years of the
nineteenth century was isolated from the city. Footpads and highway-
men still skulked in the woods, the rotting corpse of a criminal still
swung from the gibbet on the summit. Our 1830s visitor, in contrast,
would see industrial and suburban development seeping round the side
of Greenwich Park, and the glinting steel tracks of the London and
Greenwich railway line cutting like a razor through Bermondsey. He
or she would behold the glittering, gross Regency city described in

Byron's other great epic, *Childe Harold*, transforming itself almost before their very eyes into the world's first modern metropolis.

'It is a monstrous Wen!' exclaimed the writer Thomas Carlyle to his brother during a visit from his native Scotland in the 1820s, a mad mêlée of

> coaches and wains and sheep and oxen and wild people rushing on with bellowings and shrieks and thundering din ... The thick smoke of it beclouds a space of thirty square miles; and a million vehicles, from the do-or-cuddy-barrow to the giant wagon, grind through its streets forever.

It was like a hothouse, a totally artificial environment, in which strange new species sprouted:

> the carman with his huge slouch-hat hanging half-way down his back, consumes his breakfast of bread and tallow or hog's lard, sometimes as he swags along the streets, always in a hurried and precarious fashion, and supplies the deficit by continual pipes, and pots of beer. The fashionable lady rises at three in the afternoon, and begins to live towards midnight. Between these two extremes, the same false and tumultuous manner of existence ...

It terrified him, but he could not stay away. 'There is an excitement in all this,' he confessed, and it was to draw him back.

Among those do-or-cuddy-barrows and giant wagons came a cargo of exotic new ideas, shipped in by Nonconformist, free-thinking and predominantly bourgeois thinkers. This was the era of reform, of pocket boroughs and Poor Laws, terms which might now be only distant echoes from half-forgotten history lessons but then reverberated through the streets of cities being shaken apart by industrial and commercial turbulence. In 1831 and 1832 the first Reform Bill dragged through parliament, taking government after government with it as supporters battled to bring the vote to the middle classes and abolish ancient anomalies and traditions. In 1834, the Poor Law Amendment Act was passed, which brought new scientific thinking into the heart of public policy by enshrining in law Thomas Malthus's theory of population growth: the need for moral restraint to check the 'contagion of numbers', as one commentator characterized the population explosion then beginning to blow apart Britain's cities.

In Bloomsbury, Jeremy Bentham's University College, known as 'that godless institution in Gower Street', had just opened. With its secular admissions policy, it provided a refuge for the non-Anglicans excluded from Oxford and Cambridge, people like William Frend, Annabella's friend and Ada's adviser, who had been thrown out of his college at Cambridge for promoting Unitarian views. At University College, Frend and his like suffered no such restrictions, which no doubt encouraged the development of the sorts of scientific ideas that would later so dangerously threaten the established religious order: the idea that the universe ran according to the laws of physics rather than God, the idea that the world and its inhabitants had evolved over millions of years rather than in seven days.

Such notions were not just for academic consumption. The London Zoological Society and the Surrey Literary, Scientific and Zoological Institution set up their animal collections during this time, the former in Regent's Park, the latter in Vauxhall, provoking an explosion of interest in the natural world fast becoming a bricked and cobbled one.

With tickets for sale at a shilling each, the Surrey Zoological Gardens, set up as a purely commercial venture, managed to attract huge crowds (Ada happily losing herself among them), and quickly established itself as one of London's most popular attractions. It even rivalled the neighbouring Vauxhall Gardens, which just a few years before had been staging some of London's most extravagant spectacles, such as a re-enactment of the Battle of Waterloo featuring a thousand soldiers.

London Zoo, too, proved to be an instant success. Thirty thousand visitors came in the first seven months, and the crowds continued to pour through its gates as its collection was augmented by the transfer of the Royal menageries at Windsor and the Tower of London. By 1834, it boasted beasts gathered from the four corners of the empire: bears, emus, kangaroos, llamas, turtles and zebras. On seeing its famous monkeys, Theodore Hook, the editor of *John Bull* who had published the spurious wedding night chapter from Byron's memoirs, was moved to compose some lines of doggerel:

> The folks in town are nearly wild
> To go and see the monkey-child
> In gardens of Zoology
> Whose proper name is Chimpanzee.
> To keep this baby free from hurt
> He's dressed in a cap and Guernsey Shirt;
> They've got him a nurse and he sits on her knee
> And she calls him her Tommy Chimpanzee.

As the crowds marvelled at Tommy, a ship called the *Beagle* was setting sail to map the distant continents, carrying among its crew one Charles Darwin. During his travels he would see a forest of petrified trees seven thousand feet up the Andes, wild Aboriginals in Tierra del Fuego, and giant turtles on the Galapagos Islands, where the volcanic cones reminded him of the blast furnaces at Wolverhampton. And when he returned to London in 1836 he would fashion the theory that addressed the troubling questions then being comically posed by Tommy's mimicry of man.

Zoology was not the only subject that so successfully combined science with spectacle. Panoramas, scenic paintings stuck to the interior of immense cylinders, opened up new, uncannily realistic vistas on a world becoming re-enchanted by science and exploration. One of the most spectacular was sited in the Colosseum, a miniature Pantheon in Regent's Park which opened in the late 1820s and which an awe-struck Ada would visit in the following decade. The scenes were the height of a house and witnessed from a tall viewing platform ascended by steam lift.

Exeter Hall, a non-sectarian meeting place for religious, philanthropic and scientific gatherings, opened up a few years later on the Strand, and would later feature such engineering marvels as the electric telegraph. Ada went to see that, too, but found that it was she who had become the exhibit as a presumptuous visitor pushed to get a glimpse of Byron's daughter, and followed her back to her carriage when she tried to escape him.

Further down the Strand was the new and bustling glass-domed Lowther shopping arcade, which featured the Adelaide Gallery. More formally known as the National Gallery of Practical Science, Blending

Instruction with Amusement, the Adelaide was famous for showing the very latest in technology, such as a pocket thermometer, a gas mask, a steam gun and, attracting the particular attention of that same young celebrity daughter during an incognito visit, a Jacquard loom, a revolutionary device that managed to produce patterned materials from punched cards.

The genteel setting of a Georgian house in Dorset Street, Marylebone provided the venue for demonstrating a yet more extraordinary contraption. Its inventor was the engineer Charles Babbage, and he extravagantly claimed to the crowds of intellectuals, politicians and businessmen who came to see it – including, of course, our insatiably curious heroine – that it would propel the industrial revolution into a new technological realm.

*

The Dorset Street house was Babbage's home and workshop, and he held regular 'soirées' there to show off the fruits of his work. These became, as Ada's friend Woronzow Greig put it, the 'fashion' in the London social calendar during the 1830s. Everyone came, from earls to engineers. Charles Darwin, just returned from his world tour and still recovering from his sea legs, was told by the geologist Charles Lyell that Dorset Street was *the* place for someone who had been out of circulation to meet 'the best in the way of literary people in London', as well as plenty of 'pretty women'. (Lyell appears to have been an incorrigible philanderer; he tried to persuade Babbage to invite Colonel Codrington to one of the parties because the colonel had such a pretty wife.)

There were two attractions this august crowd flocked to see. One was on display in his drawing room, a beautiful silver lady Babbage had himself restored, a mechanical *danseuse* who would perform little pirouettes and arabesques for the benefit of her audience.

The other attraction was of a more serious nature, and the one Babbage really wanted his audience to see. It was on display in a specially modified dust-free room. A 'thinking machine', they called it, because it was rumoured to be able to perform mental arithmetic without any human intervention. Babbage, who had been working on it for more than a decade, had given it a less sensational but equally enigmatic name: the Difference Engine.

It was an astonishing piece of engineering, a demonstration version of a machine that once completed would be among the most complex in the world. Nobody had ever seen anything like it before: two thousand interconnected components of glimmering brass and steel. His audience may have seen complex clocks, perhaps the most complex of all, the great chronometers built by John Harrison to solve the problem of finding the longitude at sea. Indeed, Babbage had proposed his Difference Engine as a sort of successor to Harrison's clocks. As he pointed out in an open letter to Sir Humphry Davy, President of the Royal Society, announcing the invention of the Difference Engine, it could be used to produce nautical tables to unprecedented levels of accuracy. These could be published by the Board of Longitude, the body set up under the Longitude Act of 1714 to search for a solution to the longitude problem, which Harrison's clocks ultimately provided. Babbage even tried to become a member of the Board so he could promote his idea before it was disbanded in 1828.

However, though both Babbage's engine and Harrison's clocks brought a new level of mechanical sophistication to solving the same problem, they were fundamentally different both in terms of their design and in the principles they embodied.

Harrison's earliest clocks, developed in the mid-eighteenth century, were like miniature models of the clockwork universe they were designed to measure, meditations on balance, regulation and rotation captured in metal. His last two timepieces hid their wonderful mechanisms away altogether. The inscrutability of their beautiful clockfaces demonstrated that their maker was so confident of their astronomic precision he no longer needed to show how it was achieved.

Babbage's demonstration Difference Engine appeared sixty years after H-5, Harrison's last great clock, but was a world and an age away in terms of design. It was an immense block of metal about the size of an upturned luggage trunk, completely devoid of ornamentation. Set between two thick plates held in position by strong brass columns was an array of interlocking gears and levers, the only decipherable parts of which were wheels bearing numerals neatly engraved into their rims.

These displayed the terms and results of the machine's calculations. There was no centre, no face, no one point of focus, other than a crank handle incongruously placed at the top with its handle facing downwards. Where Harrison's clocks embodied symmetry, reciprocation and equipoise, Babbage's Difference Engine was the product of mechanical repetition, iterations of the same identical assemblies of cogs and wheels running up vertical axes embedded in the mechanism, each axis repeated three times across its width. A completed version of the machine was not actually built until 1991, when the Science Museum of London created it from Babbage's original plans to commemorate the bicentenary of his birth. Only then was the grinding regularity of its design fully revealed.

Such functional uniformity is a familiar sight to twentieth-century eyes accustomed to council blocks, production lines and supermarket shelves. To these pre-Victorians, living in a world where nearly every object was the product of nature or a craftsman's art, it must have come as a shock.

However, to one seventeen-year-old woman who first saw this miraculous machine on the evening of Monday, 17 June 1833, it came not as a shock but as a revelation. 'While other visitors gazed at the working of this beautiful instrument with the sort of expression, and I dare say the sort of feeling, that some savages are said to have shown on first seeing a looking-glass,' wrote one observer, this woman, 'young as she was, understood its working, and saw the great beauty of the invention.'

The woman might have identified with as well as understood that 'beautiful instrument', for she too would have been an object of intense curiosity, regarded as something novel and different, as her appearance in Dorset Street that evening provided the public with one of its first chances to behold in adulthood the daughter of Byron.

*

The Ada who came to see Babbage's marvellous invention was a changed woman – or so she and her mother hoped. She had recently been dragged away from the arms of her lover and tutor and back to Fordhook. There she had been left to contemplate her faults, while the

Furies were no doubt whipping up a storm of dismay and recrimination around her.

After many inquisitions, accusations, charges and sentences, even she began to suspect there was something wrong with her, some innate perversity. She now became determined to do something about it. In a letter to her mother postmarked 8 March 1833 she announced that she had changed. She understood that she could only be fulfilled if she developed 'deep religious feeling' and allowed such feeling to guide her through life.

She repeated the pledge to Elizabeth Briggs, her old governess. Miss Briggs was unconvinced, and much finger-wagging ensued. Merely saying you have changed is not enough, she warned Ada. She would have to tread a long, difficult path to redemption, and her only chance of reaching her destination would be to follow her mother who, in Miss Briggs's opinion, provided 'all that can be offered as an example short of the character of Our Saviour'.

Co-operation's prophet and the man who would later oversee Ada's first glimpse of her father's portrait, Dr William King, was also called upon to aid the girl's correction. He and Ada went on long walks to dissect what she had done, and they were obviously gruelling experiences. He painted a vivid picture to Ada of the perils associated with that 'dangerous asset', the imagination, how it was imperative that it was controlled, how thoughts could not be allowed to roam free in the mind, but had to be regulated.

For Annabella, achieving this regulation was now a matter of great urgency. The approaching month of May marked the beginning of the 1833 London Season, and, ready or not, the seventeen-year-old Ada had to be launched into society, just as Annabella had been twenty years before.

On 10 May Ada underwent the initiation rite every young aristocratic girl had to endure in order to pass into womanhood. She was to be presented to King William IV and Queen Adelaide at the Royal court. It was a tense occasion for Annabella. If Ada decided to misbehave, it would ruin not only her own reputation and prospects, but those of the Christlike mother, who was even now promoting her discreet campaign to vindicate her role in the Byron affair.

As the crucial moment approached, Ada had an attack of nerves. Someone, perhaps another debutante piqued at being eclipsed by the famous daughter of England's most celebrated poet, unsettled her by dwelling on the 'difficulties and dangers of the moment'. Annabella must have held her breath, wondering what her unpredictable daughter would do next. She was still physically as well as morally vulnerable; she might collapse, run from the room, conduct herself improperly towards one of the courtiers, anything.

In the event, Ada behaved. Annabella reported to Dr King's wife how her 'young Lioness' managed to acquit herself 'tolerably well' (tolerably high praise from Annabella). Ada curtsied to the King and Queen, and met the Duke of Wellington as well as an august gathering of visiting foreign dignitaries and dukes, including the Duke of Orléans, the Duke of Brunswick and the veteran French statesman Talleyrand, whom Ada disrespectfully thought a bit of an 'old monkey'.

The following week she attended the Court Ball, the highlight of the Season. She had been looking forward to the event, her excitement tinged with trepidation at becoming the centre of attention in a scene 'where there are so many others to attract observation', as Annabella put it. Her mother had selected her ballgown out of concern over her becoming 'occupied with frivolities'.

There is no record of how the Court Ball went, but it no doubt inaugurated a period when Ada was to become besieged by suitors, attracted not only by her celebrity, but also her position as heiress to the enormous Wentworth estates inherited by her mother, said to be worth £8,000 a year. Inevitably, she became the focus of attention of a number of 'notorious fortune hunters', as Woronzow Greig called them. One was Charles Knight Murray. He was a budding railway investor and would soon be appointed the first secretary to the Ecclesiastical Commissioners, a body representing the crown's interest in the Church of England.

Murray must at first have seemed a promising prospect, compensating for his lack of titles with powerful connections and considerable charm. However, it soon turned out he was not the gentleman he pretended to be. Just a few days after meeting her, he made a 'daring attempt' at Ada of some kind. Annabella no doubt sensed that

something was going on, and intervened. This provoked a direct
confrontation with Ada, who in response penned one of the most
caustic letters she would ever dare write to her mother, in which she
argued that her duty of obedience was diminishing as she grew older,
and would disappear altogether once she reached maturity. 'Till
twenty-one, the law gives you a power of enforcing obedience on all
points; but at that time I consider your power and your claim to cease
on all such points as concern me alone,' she wrote.

This cut no ice with Annabella, who castigated the girl for the
'baseness' of her no doubt flirtatious behaviour towards Murray, and
insisted that Ada saw no more of him.

The intervention proved to be justified. Murray was a crook on
the make. In the 1840s he tried to combine his secretarial post at the
Ecclesiastical Commission with a directorship of the South Eastern
Railway Company, and ended up losing both jobs after defrauding the
Commission of at least £6,000 (one estimate put the amount at
£25,000 – a vast sum in those days). To have yet again come so close
to disaster shocked Ada, and the pendulum duly swung abruptly from
what Annabella would call her sense of infallibility to a belief that she
was unable to trust herself.

By November 1833 her attitude to her mother had been transformed. They attended another Court engagement together, this time
held at the Royal Pavilion in Brighton, and Ada was full of pride for
her 'illustrious parent' who, dressed in a 'beautiful' satin gown, sat until
midnight with Queen's private circle.

This change of mood persisted into the following spring. That was
when the passage of her own personal Reform Bill through the rowdy
parliament of her passions was approaching its conclusion. As her own
letters confirm, the following provisions were duly enacted, with the
aim of improving the conduct of her self-government:

— all forms of excitement to be excluded from her life except
'intellectual improvement' because, as Dr King had advised,
only by applying herself to scientific ideas could she keep her
imagination and passions from running wild;
— mathematics to be concentrated upon because, as the good

doctor also diagnosed, 'her greatest defect is want of order, which mathematics will remedy', and because the subject has no connection with feelings and therefore could not possibly excite 'objectionable thoughts';
— her relationship with her mother to be improved;
— her ability to deceive herself to be recognized;
— greater circumspection to be shown when she found she had a liking for something, bearing in mind that her partiality probably meant it was bad for her.

Annabella herself gave the bill her support, but a letter to Dr King reveals she was sceptical about its chances of enforcement. She began the letter with a preamble on current interests: phrenology, asylums and prisons. She also commented on the irredeemable nature of the inmates of the new Millbank Penitentiary, which had been built according to the principles of Bentham's panopticon. From this observation, it was, as she put it, an 'easy transition' to the 'object of our anxiety': Ada. Since the elopement she had lost her sense of infallibility – that was the good news. However, she had not yet substituted it with what Annabella called the 'Infallible Guide' (by which she meant God, to whom she often referred in the abstract terms preferred by many Unitarians). And there was another, perhaps more alarming, cause for concern: Ada had shown a lack of any sort of moral sense – in those days a condition regarded as a form of insanity. When the girl tried to observe the principles of virtue and truth, she did so simply because it was expected of her, not because she had an innate understanding of their goodness.

In an attempt to correct this problem, all Annabella could recommend was that Ada list Dr King's moral maxims in a book, and Dr King obligingly provided a plentiful supply in his letters to the young woman. He also encouraged her studies of Euclid's geometry, which she pursued by working through a six-volume primer written by the famous scientific popularizer Dionysius Lardner.

How keen she was on copying morals remains unrecorded, but when it came to the rest of her studies Ada needed little encouragement. She threw herself into her new interests with a passion, and they were bountifully supplied. In fact, there was so much new scientific and mathematical material for her to draw upon, it must be providential, she proclaimed. The elopement, the arguments, the illness – it was all meant to be!

This characteristically excited reaction no doubt unsettled Annabella as well as the Kings, as it seemed horribly like wishful thinking, another species of self-deception to which Ada was heir. But there was now nothing they could do but buckle down (or, more precisely, buckle her down) and hope for the best.

*

Providence really did seem to be smiling on Ada at that moment – not just because she was in London at a time of a great intellectual revolution, but because her mathematical studies led to her being introduced to one of its most interesting if restrained revolutionaries: Mary Somerville.

By the time Ada got to know her in 1834, Mary Somerville had become established as an important scientist and mathematician. When she died in 1872, she was hailed 'The Queen of Nineteenth-Century Science', according to the *London Post*. She had won medals for tackling various 'Diophantine' equations. She had studied the 'magnetising power of the more refrangible solar rays' using only some paper, a prism and a pin. She had performed experiments into the 'transmission of chemical rays of the solar spectrum across different media'. And she had written about the action of those same rays on vegetable juices.

The summit of her achievement was the translation of Laplace's *Mécanique Céleste*, which was published to great acclaim by John Murray, Byron's publisher, under the title *The Mechanism of the Heavens*. Laplace's work was a monumental and famously complex five-volume survey of celestial mechanics covering everything from equilibrium to fluid dynamics. Even Laplace admitted that it was, like all his books, unreadable.

His great contribution to astronomy concerned the interaction of the planets in the solar system. Newton believed that the gravitational pull of one planet passing another in its orbit round the Sun would ultimately destabilize the solar system to such a degree as to bring about the end of the world – at least it would without the steadying influence of divine intervention. Laplace showed that this was not so, that the world could survive quite happily without the hand of God to guide it through the planets. Indeed, in the Laplaceian scheme of things, there was no need for God at all, something that Napoleon was said to have complained about.

It was this Godless universe that Mary opened up to the English-speaking world, making it all the more accessible with her substantial 'Preliminary Dissertation', which included an introduction to the mathematics the reader needed to know in order to understand what followed.

Despite all this work, and its enthusiastic reception, Mary Somerville remained self-effacing, prone to talk down her abilities, as well as those of her gender, much to the gratification of her male colleagues. She emphasized the dedication and persistence demanded by science, and criticized the notion that brilliant ideas leap out of great minds just as Archimedes leapt out of the bath.

John Adams, an English astronomer, confirmed that science was, indeed, done in the manner she suggested. Her second substantial work, *On the Connexion of the Physical Sciences*, was a manifesto for the idea that all the limbs of science then growing so vigorously sprang from the same body of knowledge. To illustrate this, she included in each new edition of the work discoveries and data that had arisen since the last, including an observation that the tables she had originally supplied for the motion of the planet Uranus had been shown to contain anomalies. This correction, Adams later told her, inspired him to see if the orbit of another planet beyond Uranus might explain the anomalies – whereupon Neptune was discovered. Mary's reaction to Adams's revelation was gratitude tinged with regret. 'If I had possessed originality or genius I might have [discovered Neptune],' she noted, adding parenthetically 'a proof that originality in discovery is not given to women???'

She developed this theme of gender bias in the draft of her autobiography. 'I was conscious that I had never made a discovery myself, that I had no originality. I have perseverance and intelligence but no genius,' she wrote, concluding, in a tone of defeat, 'that spark from heaven is not granted to [my] sex, we are of the earth, earthy, whether higher powers may be allotted to us in another existence God knows, original genius in science at least is hopeless in this'.

*

The young woman who bounded into her life in 1834 knew nothing of such limitations. Wherever Ada's spark came from, be it heaven or earth, she definitely had one, and its crackling energy would have momentarily illuminated, and occasionally irritated, Mary's quiet, studious existence.

At the time, Mary lived in the magnificent setting of the Royal Hospital, Chelsea, the army pensioners' establishment founded by Charles II and designed by Christopher Wren. Her husband, William Somerville, was a physician there. He was also, by the standards of the day, an exceptionally sympathetic partner. He encouraged his wife's interest in science and mathematics by editing her manuscripts and used his Fellowship of the Royal Society, which then excluded women, to get her books by and introductions to the most interesting scientists.

Woronzow Greig also lived at the Royal Hospital. He was Mary's son by her first marriage, to Samuel Greig, who died in 1807. Samuel was a sailor (son of the famous Scottish-born Russian admiral, also called Samuel), and had named his son after an acquaintance in the Russian navy. Woronzow was ten years older than Ada, yet her appearance at the Somerville home left a deep impression upon him. He recalled seven years later in a flattering, almost drooling letter to her how he 'in common with the rest of your countrymen felt a peculiar interest for one who was cradled in celebrity, and I often used to speculate upon the character tastes habits – pursuits and talents of Ada Byron'. What he beheld, however, confounded all his expectations, a woman whose extraordinary intellect and character differed 'not only from persons of your own rank and condition but from the generality of mankind.'

He first saw her some time in 1833, and then she was still suffering from the after-effects of her prolonged illness, being overweight and clumsy. Now, having got to know her, he found her 'moral courage and determination' remarkable. She was also amiable and unaffected, as well as a little proud, characteristics that seemed to both fascinate and frighten him.

Ada quickly developed a strong affection and respect for Mary, who must have seemed as stable as Laplace's universe – the perfect influence to steady Ada's eccentric orbit. Annabella was certainly pleased as well as relieved by the connection. Years later, Mary would receive a letter from her son about a visit he had made to Annabella: 'She spoke of you with deep affection, and with tears in her eyes she mentioned the heavy debt of gratitude she owed you for your kindness to her daughter at a most precarious period of her life and for the beneficial effects which your precepts and example had produced upon her daughter's character.' Annabella is very rarely mentioned as having tears in her eyes, and the fact that she did on this occasion gives an indication of how desperate she had been to see Ada's programme of reform succeed.

Mrs Somerville was born Mary Fairfax in Scotland in 1780, and, like Ada, suffered a severely restricted upbringing. She described how at Miss Primrose's boarding school, where she spent her one and only year of formal education, her posture was corrected using a corset featuring rigid stays fixed to a steel mount, a metal bracket to support her chin in a suitably elevated position, and braces which were used to strap back her shoulders so that the blades met.

She learned the rudiments of mathematics not at this torture chamber but by studying puzzles in women's journals, and later moved on to a more advanced level by borrowing books on geometry and algebra from her brother. Her parents tried to discourage her by removing the candles from her room, so she could no longer read at night. She responded by memorizing the texts during the day, and working out the problems in her head during the night.

Ada had obviously enjoyed a much easier route to mathematical enlightenment, and Mary may have sometimes thought that an only child taught by the best teachers and governors aristocratic wealth could buy was somewhat intellectually indulged. Nevertheless Mary, too, was prepared to indulge, just as long as Ada worked at her sewing as well as her sums.

Ada obliged, after a fashion. She made a cap for Mary, which was received with delight because 'it shows me a mathematician can do other things besides studying x's and y's'. Indeed she could, though most of those other things were far less domestic than Mary would have liked. She copied papers about steam engines, visited the Hydro-oxygen Mirrorscope in Bond Street, got one of Mary's science books autographed by the heir presumptive Princess Victoria, studied new mathematical theories, criticized the bust of Mary Somerville by Francis Chantrey just placed in the Great Hall at the Royal Society (an unexplained prejudice which might have some connection to Chantrey's association with her father, whose lower face, the sculptor famously observed, had a soft voluptuous character). She did endless other things besides studying x's and y's: she studied books, diagrams, models, minerals . . .

At the time, this thirst for intellectual stimulation was taken to be proof that she was truly committed to her reform programme. Little did anyone realize the true reason for her interest: exposure to the new, scintillating world that science was then beginning to uncover. Enquiring minds were dangerous things and, as Lucifer said in Byron's drama *Cain*, 'Nothing can / Quench the mind, if the mind will be itself'.

Her interest in minerals, for example, was no doubt aroused by the large collection the Somervilles had on display in a cabinet at their home. But it was not the glitter of crystals or the sheen of semi-precious stones that would have caught her eye. It was the movement of a new idea just beginning break out of them, like some mutant creature from its shell.

*

Between 1830 and 1833, a book had appeared which literally changed the course of earth's history. It was called *Principles of Geology*, and seemed to contain little more than a series of observations about mineral samples and rock formations to be found in various inaccessible parts of the Continent. In excavating through this mountain of facts, however, the book's readers uncovered an astonishing, even terrifying idea, one that did for the earth what Darwin would do to its inhabitants: showed that the world was not created in six days but over millions of years.

The author was Charles Lyell, the man who attended Charles Babbage's soirées to admire the women as well as the machines. Until he wrote *Principles*, the prevailing view was that the earth had stayed more or less the same since the Creation. Various theologians had even estimated when that might be. James Ussher, the Archbishop of Armagh, writing in the mid-seventeenth century, had put the date at 23 October, 4004 BC. There were certain anomalies that seemed to undermine this view, such as seashells found embedded in mountain tops, but they could be explained away by reference to such biblical catastrophes as the Flood.

Lyell's work suggested a very different world that had developed according to a very different timescale. Everything he saw during his field trips, even strata found high up in cliffs that contained the remains of marine creatures from the bottom of the sea, could be explained by natural processes. Mountain ranges could be worn down by the slow, relentless flow of water, and pushed up by the fierce pressure of subterranean fire – given time, far more time than the five thousand or so years imagined by Archbishop Ussher. To give an indication of the sort of time it takes, Lyell described how a particular type of 'thinly foliated' rock called a marl, which could be found in layers more than seven hundred feet thick, was made up entirely of scales annually moulted by barnacles, a crustation of crustaceans accumulated over aeons.

Such speculations had a profound effect on the way Victorians thought about things. They were already used to seeing buildings spring up all around them. Now they learned that the landscape itself was in a state of constant change, as Tennyson noted:

> The hills are shadows, and they flow
> From form to form, and nothing stands;
> They melt like mist, the solid lands,
> Like clouds they shape themselves and go.

Nothing was permanent, not the mountains, not the valleys, not the fields, not the villages. This was the first generation to feel the full impact of rapid changes in both ideas and technology, to feel the power of unstoppable geological and industrial forces tearing the world apart.

Solomon Gills in *Dombey and Son*, Ada's favourite book, which Dickens himself would on one very significant day read to her, was overwhelmed by it all: 'competition, competition, competition – new invention, new invention – alteration, alteration – the world's gone past me.' Many were as terrified as old Solomon. 'If only the Geologists would let me alone,' cried John Ruskin, 'those dreadful Hammers! I hear the clink of them at the end of every cadence of the Bible verses'. All that was left to hold on to was the least substantial thing of all, faith. This alone drowned out the tinnitus of tapping hammers, strengthened with such stirring words as those written by H. F. Lyte for the Victorian hymnal:

> Change and decay in all around I see,
> O Thou who changest not, abide with me . . .

Ada sought no such reassurances – not from God, at least, though she was supposed to be cultivating her religious feelings as part of her rehabilitation. She finally got her minerals, together with a rather more symbolic gift, a puzzle made up in the shape of a wooden cross, which she wrote she had not yet had the courage to dismantle, fearing she would not have the ingenuity to put together what had been taken apart.

The man who had given her these gifts was Charles Babbage. Mary Somerville was a regular guest at his soirées, and Ada now exploited the connection to see as much of him as she could. Almost every visit to Chelsea seemed to be followed by a carriage ride with Mary to Babbage's Dorset Street home, where Ada could once again catch a glimpse of his Difference Engine.

Her fascination for the machine deepened. She badgered Babbage for copies of his plans and diagrams, which he duly supplied. She studied them closely, trying to understand the mechanism and the ways that it could be used.

Babbage had boasted in his letter announcing its invention to the Royal Society that it would abolish the 'inattention of, the idleness or the dishonesty of human agents'. It was, in other words, a self-regulating mechanism. It was devoid of the human weaknesses which had brought her to the brink of the moral void that had consumed her father. No wonder she marvelled.

*

Charles Babbage was bewitched by technology, believing that there was no aspect of life that could not benefit from its empowering potential. He was a passionate advocate of scientific education, and was a leading supporter of the British Association for the Advancement of Science, that quintessentially Victorian institution lampooned by *Punch* and Charles Dickens, the latter renaming it the 'Mudfog Association for the Advancement of Everything' which he imagined holding meetings on matters such as 'Umbugology and Ditchwateristics'.

Dickens exaggerated – only a little. There was no subject too obscure or unappetizing to pique Babbage's curiosity. The disposal of horse carcasses, for example: a report about a visit to a knacker's yard near Paris is cited by Martin Campbell-Kelly, editor of Babbage's collected works, as a perfect example of Babbagisticology. He was fascinated by the various means used to extract value out of the bodies of old nags already stripped of their hair, hide, bones and hoofs:

> Even the maggots, which are produced in great numbers in the refuse, are not lost. Small pieces of the horse flesh are piled up, about half a foot high; and being covered slightly with straw to protect them from the sun, soon allure the flies, which deposit their eggs in them. In a few days the putrid flesh is converted into a living mass of maggots. These are sold by measure; some are used for bait in fishing, but the greater part as food for fowls; and especially for pheasants. One horse yields maggots which sell for about 1s. 5d.
>
> The rats which frequent these establishments are innumerable, and they have been turned to profit by the proprietors. The fresh carcass of a horse is placed at night in a room, which has a number of openings near the floor. The rats are attracted into it, and the openings then closed. 16,000 rats were killed in one room in four weeks, without any perceptible diminution of their number. The furriers purchase the rat skins at about 3s. the hundred.

What makes this passage so characteristic is that it reveals Babbage's love of abstraction, turning even this most carnal of phenomena into clean numerics. In Babbage's world, every man, maggot and rat has its price, every thing its number, from the quantity of material required to smelt a ton of pig iron to the statistical probability of death being followed by resurrection. Even jokes, he speculated, could be scientifically analysed, and with that in mind he had collected a large number of 'jest-books' as source material for an examination into the 'causes of wit'.

He was probably only half joking when he wrote to Tennyson in 1851 about the great poet's verses on *The Vision of Sin*, which feature the famous line 'Every moment dies a man, Every moment one is born'. He pointed out that this was impossible, as it would mean the population of the world would be static, which it so obviously was not. He therefore proposed that Tennyson change the line to read: 'Every moment dies a man, Every moment $1^{1}/_{16}$ is born', adding that $1^{1}/_{16}$ was an approximation, but should be sufficiently accurate for the purposes of poetry.

This sort of thing earned Babbage a bad name in literary circles. Thomas Carlyle, though himself once a maths teacher and the translator of a book on geometry, disliked him on sight, describing him as 'a mixture of craven terror and venomous-looking vehemence; with no chin too: "cross between a frog and a viper", as somebody called him; forever loud on the "wrongs of literary men" though *he* has his £20,000 snug!' These feelings were aroused when Carlyle approached Babbage about the formation of what was to become one of the former's greatest legacies, the London Library. The Library, later located on St James's Square (coincidentally next to what would be Ada's London home), aptly came to embody the culture clash that inspired Carlyle's dislike of Babbage. Its books relating to the humanities – literature, history, art, biography, religion – are arranged across eight floors. The rest, though still a superb collection, are ignominiously lumped under the heading 'science and miscellaneous' in a single room and an antechamber.

The antipathy between Carlyle and Babbage was mutual. Darwin recalled a 'funny dinner at my brother's' attended by Babbage and

Carlyle. Babbage, Darwin observed, liked to talk but was silenced by Carlyle 'haranguing during the whole dinner on the advantages of silence. After dinner, Babbage, in his grimmest manner, thanked Carlyle for his very interesting Lecture on Silence'.

In the animosity between these two can be seen early signs of the growing antagonism between art and science. This was the moment when each started to acquire its individual identity, when art became, thanks to the Romantics, more clearly identified with exploring the self, and science with exploring the universe.

*

Charles Babbage was born on 26 December 1791 in Walworth, Surrey, the son of a well-off banker. He had a conventional upbringing in relative comfort. He was raised as a Protestant because his parents had been 'born at a certain period of history, and in a certain latitude and longitude', as he provocatively put it, no doubt to annoy Christians of all denominations.

Walworth was then a country village near the Elephant and Castle, a well-known junction and tavern that had recently become much busier with the building of Blackfriars Bridge over the Thames. In his anecdotal memoirs *Passages from the Life of a Philosopher*, Babbage recalled an incident that captured how much the world around that small village was to change during his lifetime.

One day, when he was about five, he went for a walk with his nurse which took him across London Bridge. At the time, the bridge was still the medieval one so prominently featured in early paintings of the city, though now shorn of the houses that once perched on its piers (removed in the late 1750s and early 1760s to prevent it from collapsing), and deprived of the tarred heads of decapitated traitors that had once been stuck on pikes over its gatehouse.

As he was crossing the bridge, the little boy's attention was diverted by the sight of the ships swarming in the docks downriver, which he stopped to watch for a while. When he turned to speak to his nurse, he suddenly realized she was no longer at his side. Any other boy might have then panicked, run screaming tearfully in all directions to find the secure refuge of his nurse's skirts. Not young Babbage. In his recollection of the incident at least, he stayed calm, determining that the best course of action was to turn back towards home.

When he reached Tooley Street, which ran along the south bank of the river, he stopped and waited to be retrieved, amusing himself by watching the passing traffic.

His nurse, meanwhile, had been far from calm. She roused the city crier, and got him to ring his bell and call out the news of her loss to passers-by, offering a five shilling reward to anyone who found the boy.

The boy was eventually spotted by a linen draper whose shop was at the corner of Tooley Street. Discovering the lad to be the son of one of his clients, the linen draper sent off a messenger 'who announced to my mother the finding of young Pickle before she was aware of his loss'.

He revisited the sight of the incident half a century later and noted the changes since he had last been there. The houses on Tooley Street were now being demolished to make way for London Bridge Station, one of the capital's major new railway termini. He sought out the very spot where he had sat so many years before, and found it to be unrecognizable. The linen draper's shop had long gone, so had the crier and his bell, even the little that remained of old London Bridge, which had now been replaced by an elegant five-arch stone construction further upstream.

Babbage was quite unsentimental about such change. Indeed, he made a principle of fanatical rationality, responding to every event as though it was a crank of a calculating engine, trying to figure out the cause and compute the consequences. In August 1814 he wrote to his friend John Herschel, the astronomer to be, announcing that he had got married, and that the marriage had resulted in him falling out with his father. He supplemented this news, reported in a mere paragraph, with thoughts about his future career (he favoured a job connected with the mines, where he could exploit his knowledge of chemistry) and a selection of mathematical theorems he thought Herschel might find interesting. An astonished Herschel replied a week later: ' "I am now married and have quarrelled with my father" – Good God Babbage – how is it possible for a man calmly to sit down and pen those two sentences . . . and pass off to functional equations?'

Some found this aspect of his character amusing and unaffected. There seemed to be an innocence about him, an incorrigible curiosity that extended to every subject of study and aspect of life, from telegraphy to tick-tack-toe. He was a shameless self-publicist, a name-dropper, a flirt and, as we shall see, sometimes underhand; he could also be irritable, impatient and capricious. But all of these shortcomings seemed to spring from the same virtue, a yearning to grasp the world and all that was in it.

Others found him annoying, Annabella Byron among them. She did not trust him. He was 'finical', pushy. He would do just about anything – such as, perhaps, flatter a potential patron like herself – to achieve his ends.

*

A man with such a head for numbers and flair for flattery was bound to end up in life insurance. The profession first beckoned him in 1824, when he was invited to participate in setting up a new assurance company to be called The Protector. He was much taken with calculating mortality rates and tables of 'the value of lives'. Even though the venture fell through, it inspired him to pen a small monograph entitled *A Comparative View of the various Institutions for the Assurance of Lives*, which he proudly claimed in his memoirs led to the foundation of the Great Life Assurance Society of Gotha in Germany.

Such an actuarial approach to life stood in stark contrast to the Regency passion for gambling, the urge to increase risk rather than reduce it. In this, as in so much, Babbage anticipated the mood of the Victorian age. During the two-year period in which he dabbled in lives, the number of assurance companies doubled, creating an industry that left London dotted with such great Gothic cathedrals of commerce as the Prudential Building in Holborn.

During this period, Babbage also made a number of visits to the Continent. These were not Grand Tours to see the glories of Greece or Rome and explore the roots of civilization; more like the roadshows of a travelling huckster. He raced from country to country in a carriage he had specially designed, complete with bed, storage compartments, pockets and a lamp that doubled as a stove. Clattering in the cupboards and dangling from straps was the strangest assortment of gifts and mementoes. There was a set of Barton's then renowned gold buttons, which, Babbage explained to anyone interested enough to listen, acquired their iridescent hue from lines drawn four to ten thousandths of an inch apart. These baubles he kept to delight the natives, to be given away as a reward to anyone who saved his life. He also stashed several copies of a paper about the Thames tunnel, the world's first to pass under a river, which was finally completed in 1843 after nearly twenty years of work. The paper describing this extraordinary and very risky venture (there were repeated inundations during its construction, and many lives were lost) had been translated into French and German, and if he had twice the number 'I should have found that I might have distributed them with advantage as acknowledgements of the many attentions I received'. He had specimens of the eyeless cave-dwelling creature *Proteus anguineus* swimming around in bottles filled with river water, these bottles 'placed in large leather bags lashed to the barouche seat' of his horse-drawn Dormobile. There was a kit comprising a diamond, a small disc of window-glass and some Canada balsam glue, which he used to demonstrate the manufacture of glass that was transparent in only one direction. And he had a stomach pump. This recently invented device was still a novelty on the Continent, and it so impressed Dr Weisbrod, the King of Bavaria's physician, that the chief surgical instrument-maker was commanded to produce a replica.

Of all the places Babbage visited, he seemed to warm particularly to Paris, not as a city of romance but, quite the contrary, as a post-Revolutionary centre of science. Here was point zero, the prime meridian of the metric system, the axis of a new standard matrix for measuring time and space that could extend through the universe. It was this system of measurement that interested Babbage, and which was to inspire the invention that made him famous.

In 1793, the French Revolutionary Convention abolished the Gregorian calendar introduced in 1582 by Pope Gregory XIII and substituted it with a secular one which had ten days to the week and replaced the saints' days with ones named after plants and tools. This new calendar only lasted until 1806, when Napoleon re-established its predecessor. But the other great revolutionary innovation, metric measures of distance and weight, survived and provided scientists with the sort of standardized system they so badly needed to compare the results of their experiments – a virtual impossibility when basic units were derived from the length of poppyseeds or the number of oxen needed to plough a field.

The metrication initiative had come from Talleyrand, the 'old monkey' Ada had met during her presentation to the Royal Court. In 1790 he started a debate in the National Assembly that led to the French Academy of Sciences being directed to produce a report on the subject. It recommended that a new unit be created, to be called 'metre' after the ancient Greek for 'measure', reflecting the taste of scientific institutions to draw inspiration from the pre-Christian classical age. One metre was to be one ten-millionth of the distance from the North Pole to the Equator, as measured along a meridian that passed, naturally, through Paris. All other measures were to be derived from this, so the unit of weight would be determined by the weight of one cubic metre of water, and so on. All larger and smaller units were, furthermore, to be decimal divisibles and multiples identified by prefixes that the Academy once again drew from ancient Greek: kilo- for a thousand, milli- for a thousandth, etc. Thus, from direct measurement of the earth and water, both of them global constants, all weights and measures could be objectively determined.

The report was ratified, and engineers were dispatched, including the astronomer and member of the French Bureau of Longitudes Jean-Baptiste-Joseph Delambre, to measure the all-important meridian. After years of arduous surveying which took them through war zones and across mountain ranges, they presented their results and the metre was formally established in June 1799 'for all people, for all time'.

The new system, however, would be of no use without new tables, as tables were then relied upon to make the sorts of increasingly complex scientific calculations in which decimal measures would be used. The problem was: how do you achieve x, the drawing up of the tables, when y, the number of calculations required to draw up such tables, is greater than z, the total number of calculations that could be done by all the mathematicians in France? It simply did not add up.

The French mathematician Gaspard Riche de Prony volunteered to find a solution to this conundrum. He was admired on both sides of the Channel as an example of the sort of clear-thinking, rational man of numbers the industrial world needed. He was a friend of Maria Edgeworth, the writer and educationalist whose novels Annabella had so eagerly read before her marriage to Byron. He was also a fan of Adam Smith's great work of free-market economics, *A Treatise on the Wealth of Nations*. He was particularly struck by the chapter on the division of labour. There he read Smith's famous example concerning the manufacture of pins, a quite complex procedure that industrialization had rendered into a sequence of mindless steps: 'One man draws out the wire, another straightens it, a third cuts it, a fourth points it, a fifth grinds it at the top for receiving the head . . .' Smith approvingly observed that a 'manufactory' could produce twenty pounds of pins a day with just ten 'very poor' (i.e., uneducated and therefore cheap) men.

De Prony was very struck by this extraordinary level of productivity, and the humble nature of the labourers needed to achieve it. 'I conceived all of a sudden,' he wrote in 1824, 'the idea of applying the same method to the immense work with which I had been burdened, and to manufacture logarithms as one manufactures pins.'

It was ambitious idea, and took the support of Delambre, the scientist who had helped measure the metre, to pull it off. The technique he developed was to create a sort of mathematical manufactory divided up into a series of sections. The first section had the fewest employees with the scarcest (and therefore most expensive) skills: professional mathematicians. They were responsible for deciding on the formulae to be used, and reducing them to the simplest possible forms of expression so that they could be used by non-mathematicians. A second, slightly larger section of 'calculators' worked out the range of values to be calculated and the layout of the tables. The third section, by far the most numerous, comprised sixty to eighty human 'computers', the cheap, unskilled mathematical line workers, who had the job of working out the results using the given values and formulae.

When de Prony set up his factory, the status of each section was quickly established. The mathematicians in the first section were drawn from the committees that had helped set up the metric system, men favoured by the republican establishment for their dislike of tradition and devotion to the new order. Members of the third section, the computers, were in contrast from rather different stock, the outcasts of the post-Revolutionary era with minimal arithmetical skills and economic power: hairdressers. They had found themselves unemployed after so many pompadours had become so spectacularly detached from their aristocratic bodies by that instrument of scientific decapitation, the guillotine. Now there was no longer any call for their skills, since the preferred hairstyle was, a contemporary observer noted, one reduced 'as the geometers say, to its simplest expression'.

Babbage first became aware of de Prony's work during his visits to Paris in the early 1820s, and was very excited by it. 'I wish to God these calculations had been accomplished by steam,' he noted.

Then it dawned on him they could be. The army of 'computers' de Prony employed to produce these wonderful figures were no mathematicians. They could only add, subtract and cut hair. A machine could do that, certainly the adding and subtracting. Thomas de Colmar had demonstrated as much with his 'arithmometer', a mechanical calculator which first appeared in the 1820s.

The key, Babbage realized, was the use of the method of 'finite differences'. This technique can create any number of numerical tables using simple addition and subtraction. Suppose, for example, you were wanting to create a table of the squares of a series of numbers. Each row of the table would contain two columns, one showing the number to be squared, and the other its square. The first row, for the square of 1, would be 1, the second, for the square of 2, would be 4 ($2 \times 2 = 4$), for 3 it would be 9 and for 4 it would be 16. Drawing up such a table for a few numbers would be relatively easy for most of us – until we reached the higher values, beyond the square of 12, for example. Then we would start to struggle, and errors would creep in.

However, using the method of finite differences, you, or rather the army of ex-hairdressers you employed, could continue indefinitely using the simplest arithmetic. You would start with the sequence of numbers representing those first four squares – 1, 4, 9, 16.

First, an ex-hairdresser would be instructed to subtract each number in this sequence from the next – to take 1 from 4, 4 from 9 and so on

– to produce another sequence (a new column in the table): 3, 5, 7. These are the 'first differences'. Three is the difference between the square of 1 (1) and the square of 2 (4), five the difference between the square of 2 and the square of 3 (9) and so on.

The next ex-hairdresser along the line could work out the next set of differences by performing the same job of subtracting each entry in the new column from the next, producing yet another, but much more significant sequence: 2, 2, 2. If the same operation is performed once more on the second differences, the underemployed third ex-hairdresser would immediately discover he or she was repeating the same sum, $2 - 2$, and producing the same result for each entry: 0, suggesting that the job was done.

This end, however, is also a beginning, the basis of a process of reverse engineering that will lead to the next number in the original sequence. The next hairdresser, aided by fingers made nimble by years of scissor-work, takes the last entry in the list of first and second differences, 7 and 2, and this time adds them together, which gives the result 9. This number can now be handed along the line to be added to the last number in the original sequence of squares, 16, to give us the next number in the sequence, 25, which, the hairdresser may not realize, is the square of 5 (5 $\times$ 5), the desired result. The same operation can then be used to produce subsequent squares: add 2 to 9, add the result, 11, to 25, to produce 36, the square of 6, and so on, as the table opposite illustrates. Using a machine that can perform the steps that make up the method of finite differences – a 'Difference Engine' – you could continue this operation for ever, and with complete reliability. 'One of the great advantages which we may derive from machinery is from the check which it affords against the inattention of, the idleness or the dishonesty of human agents,' as Babbage put it misanthropically in his influential 1832 survey on British industry, *On the Economy of Machinery and Manufacturers*. Furthermore, with the engine printing as well as calculating the results, even typesetting errors could be excluded. Ever larger numbers of newspapers were now churned out by steam presses to feed the public appetite for scandal and sensation, why not tables, publications of unsurpassed truth and utility, devoid altogether of human contamination?

Number to be squared	Square	First Difference	Second Difference
1	1		
2	4	3	
3	9	5	2
4	16	7	2
5	25	9	2
6	36	11	2
7	49	13	2
8	64	15	2
9	81	17	2
10	100	19	2
11	121	21	2
12	144	23	2

Back home in London, it took just a few hours of sketching for Babbage to think of a way of turning de Prony's *coiffeurs* into cogs. Over subsequent months he worked out the details, considering various methods of storing values, and of carrying over numbers from units to tens, tens to hundreds and so on. His friend William Hyde Wollaston, the distinguished scientist who discovered the chemical element palladium, encouraged him to throw all his energies into this calculating engine rather than any of the other fabulous and batty schemes ticking away in his busy brain, and he had soon built a small working model.

In 1822, he announced his invention in an open letter to the President of the Royal Society, Sir Humphry Davy, claiming with typical bravura that his machine could perform of all sorts of wonderful mathematical tasks, including the greatest of all, creating a table of all the prime numbers between 0 and 10,000,000.

A copy of the letter was sent to Robert Peel, who was then Home Secretary. Like his Tory successors, he was at the time busy implementing a policy of getting tough on criminals, the result of which was the foundation of the Metropolitan Police a few years later. He therefore had little time for Babbage's proposals, which he regarded with a mixture of suspicion and condescension. 'It is an engine designed against our walls or some other mischief hides in it,' he announced, quoting from the *Aeneid*. A Trojan horse, in other words.

Despite this reaction, Babbage secured the backing of the Royal Society, which in turn aroused the interest of the Treasury. Babbage had a meeting with the Chancellor of the Exchequer, who was so impressed that he offered £1,000 of government backing there and then. It was not just a generous offer but an unusual one, demonstrating that the government was beginning to recognize a need to take a role in industrial development. Though state support had been provided for technological projects before, it was usually justified in defence terms. The Difference Engine had some attractions for the navy, which needed the accurate navigational tables that Babbage promised, but its real potential was commercial. Steam-driven machines had automated physical effort, producing increases in productivity never before contemplated. This machine would automate mental effort, taking the benefits of industrialization into a new realm.

Babbage was ecstatic, as he made obvious in a letter to John Herschel far less restrained than the one reporting his marriage. Imagine it, being able to produce 'stereotyped logarithmic tables as cheap as potatoes'! In just a few years! A 'frolic' was called for to celebrate the success, which a friend had proposed should take the form of a visit to see a telescope.

With the money secured, Babbage immediately set about putting together the resources he needed for such an ambitious project. He built a second workshop at his home and converted another room into a forge. He also hired Joseph Clement, a brilliant precision engineer recommended by Marc Brunel, designer of the Thames tunnel that had so impressed Babbage and his Continental hosts.

In the following years, Babbage was to discover that working with the government would not be easy when it came to supporting such a speculative venture. On a number of occasions he would have to draw on his own resources to keep the project from collapsing, as he would not get a proper flow of funding until the early 1830s, when the Duke of Wellington, now Prime Minister, gave it his backing. On, off, fast, slow, hot, cold, Babbage found himself working with a government machine apparently gone haywire. Charles Dickens satirized the process in *Little Dorrit*, with the How Not To Do It Office, which specialized in making a meal of 'mechanicians' and 'natural philosophers' like Babbage.

Such political problems were exacerbated by personal ones, principally a deteriorating relationship with Clement, whom Babbage suspected of overcharging, particularly in respect of the costs he demanded for moving his workshop to one Babbage had specially built at great government expense in the stables behind his house.

Nevertheless, by the early 1830s he had succeeded in getting Clement to build a demonstration model of his machine. The problem now was persuading not just the government but the public that it was worthy of their interest.

*

Babbage may have been mechanically and mathematically minded, but he was by no means as dull as his Ditchwateristics. He, like the learned professors in the London Zoological Society, knew that even the most compelling scientific ideas needed a bit of showmanship to attract public attention. This is what brought him to the closing down auction of Weeks's Mechanical Museum.

Weeks's was just round the corner from the Adelaide Gallery of 'Practical Science, Blending Instruction with Amusement'. It was the last establishment in London to serve the once hugely popular fashion for automata, intricate clockwork creatures and dolls that were celebrated for reproducing lifelike actions. The museum had been founded by Thomas Weeks in 1803 in Tichborne Street, just off London's Haymarket, famous as the capital's entertainment centre, as well as notorious for being 'the great parade ground of abandoned women'. Weeks had set up the museum as the successor to John Cox's Museum at Spring Gardens, the other side of Pall Mall, which before its closure in 1775 was renowned for its screeching mechanical peacock and gliding silver swan. Weeks had also acquired the collection created by John Cox's assistant, the suggestively named engineering wizard John Joseph Merlin. Merlin had set up his own Mechanical Museum in Hanover Square, which had closed following his death in 1803.

The main gallery in Weeks's Museum was a room a hundred feet long lined with blue satin and featuring a variety of figures 'inert, active, separate, combined, emblematic and allegoric, on the principles of mechanism, being the most exact imitation of nature'. A particular draw was the tarantula spider made of steel, which, propelled only by the mechanism contained within its body, could crawl out of its box and scurry around apparently at will, terrifying the audience.

However, in the Utilitarian 1830s, even a steel tarantula was not enough to reverse the public's diminishing interest in such essentially useless toys. In 1834, it was Weeks's turn to close, and an auction was held to sell off the collection.

Babbage had come along to the auction in search of a long-lost love. He showed no interest in the jumble of self-opening umbrellas, musical clocks, adjustable bedsteads and automated arachnids that formed the bulk of the lots. Rather, he had his eye on another object, a particular 'silver lady' who had captured his attention during his youth.

He had first seen her when his mother had taken him on a visit to Merlin's Mechanical Museum. Merlin had noticed the child's precocious interest in mechanics, and invited him to come up to his attic workshop where he was building his latest clockwork marvels. There, the wide-eyed schoolboy saw Merlin's greatest creation, the work of years of labour: two 'uncovered' female figures made of silver, each about twelve inches high.

Babbage was mesmerized by what he beheld, certainly by the intricacy of the mechanism, surely by the anatomical parts the mechanism moved – arms, legs, head, neck, fingers, eyelids, breasts. One of the ladies glided around, bowing and occasionally raising a pair of glasses to her eye, as if trying to identify someone she had just seen. The other, the one that captured his heart, was a dancer, with a bird poised on her forefinger that wagged its tail, flapped its wings and opened its beak. He was excited by the pose of the lady's body, and her eyes – those eyes, he later recalled, were 'full of imagination, and irresistible'.

It was that lady he found for sale at Weeks's Museum. Merlin had died before he could bring his miniature Galatea to life, and she was inherited by Weeks as part of a job lot. She had since been left in

pieces in an attic, 'utterly neglected'. A bid of £35, and the lady
together with a box of her beautiful parts was in the hands of her new
Pygmalion. Babbage took her home to Dorset Street and lovingly
brought her to life. He put her on display in his drawing room and
dressed her up, wrapping strips of pink and light green Chinese crêpe
around her torso, which provoked Lady Morgan, who had called to
see her one evening, to observe that she was 'rather slightly clad' and
in need of a petticoat. 'My dear Lady Morgan, I am much indebted
for your very considerate offer, but I fear that you have not got *one* to
spare,' Babbage replied, saucily. (He helpfully points out in his memoirs
that this rejoinder was an example 'double-entendre' – which shows
how much progress he had made with his study into the 'causes of
wit'.)

Having enticed audiences sometimes running into the hundreds
with his Silver Lady, Babbage often found it a struggle to transfer their
interest to the Difference Engine next door. On one occasion he
ruefully noted that only foreigners seemed to be interested in the
Engine, all his English visitors being bewitched by his mechanical
marionette.

The solution, one he had developed before he had the lure of the
Silver Lady to attract audiences, was to assemble his guests before the
Engine and announce that he was about to make it perform a miracle.
This is what he did the evening Annabella brought Ada along to see
the Engine for the first time.

Before beginning his demonstration, he would explain that he had
set up his machine to perform a simple mathematical operation, say
one that incremented a number by two. Thus, as he cranked away at
his machine, the machine would display on its figure wheels a predictable sequence that the audience could read off for themselves: 0 . . . 2
. . . 4 . . . 6 . . . 8 . . . 10 . . . So it would continue, perhaps for fifty
iterations, apparently labouring the point that it could faithfully and
endlessly perform the mathematical calculation it had been set up to
compute.

Just as his guests were beginning to wonder what the point of all
this might be, and start to drift back to take another look at the Silver
Lady, Babbage would crank the handle once more and, without him

intervening in any way, the number would suddenly leap to a new value, say 117. Then it would carry on from there: 119 ... 121 ... 123 ... 125 ...

To the gazing eyes of those pre-Victorians, Ada's included, this apparently erratic behaviour from a machine would have seemed strange. Machines did not behave in this way, not unless they had gone wrong, which, Babbage reassured them, his had not, a fact he would prove by making the machine behave in exactly the same way again.

The hard-headed Annabella, who admitted to having 'but faint glimpses' as to how the machine actually worked, regarded such anomalous behaviour as positively supernatural – the machine was operating according to an 'occult principle', she wrote. This was just the sort of reaction Babbage was after, and not just because it piqued his audience's interest. He wanted to show that the Engine was about more than nautical tables and metric measurements. It demonstrated not just a new type of mechanism; not just a new sort of mental manufacture, but a new model of the cosmos.

*

In every age, machines, those most physical of artefacts, have a metaphysical significance. The monumental medieval cathedral clocks, for example, were much less about telling the time than chiming the divine music of the spheres, showing through their elaborate mechanical tableaux how God had ordered His universe. Babbage's hope was that his Difference Engine would perform a similar philosophical function, providing a model of how the post-medieval, scientific cosmos worked.

There was certainly a need for such a model. In the first half of the nineteenth century God was under considerable pressure. As the historian David Newsome noted, 'For the first time in English history the phenomenon of "Unbelief" emerged as so evident an intellectual and emotional problem that the whole question of what lay beyond one's life on earth, hardly ever before considered to be a matter for speculation or debate, became the subject of fierce and bitter dispute.' Science was making it more and more difficult to make sense of the divine order of things, since science seemed to be making perfectly good sense of it without the need for any divinity at all. In his masterpiece *Sartor Resartus*, which was being published in *Fraser's Magazine* at the time Babbage was performing his demonstrations, Thomas Carlyle satirically pointed out that the Creation itself had been rendered no more mysterious than the cooking of a dumpling. 'Our Theory of Gravitation is as good as perfect . . . Man's whole life and environment have been laid open and elucidated; scarcely a fragment or fibre of his Soul, Body, and Possessions, but has been probed, dissected, distilled, desiccated, and scientifically decomposed . . .'

As Byron put it in *Don Juan*, ever since Newton had discovered gravity,

> . . . man hath glowed
> With all kinds of mechanics, and full soon
> Steam-engines will conduct him to the moon.

The only respite God now had from the remorseless march of the machine was the existence of miracles. These events were, by definition, beyond the reach of science and mathematics. While people continued to believe in them, there was hope, hope for religion, hope for morality, hope for the powerless atom caught in a clockwork universe with which Ada's father had so despairingly identified.

The philosopher David Hume had tried to deny the world its miracles in his essay on the subject published in 1748. It was a work well known to the early Victorians, since it effectively argued that all miracles were lies or delusions. Such extreme rationality was easy to dismiss. There are more things in heaven and earth, Hume, than are dreamt of in your philosophy, the pious could protest. Even scientists refused to countenance Hume's unmiraculous world. Science can take you only so far, argued Babbage's more emotional friend John Herschel in his 1830 *Preliminary Discourse on the Study of Natural Philosophy*, to the edge of knowledge where you can but glimpse the 'boundless realms beyond'. Only after death can the scientists really get to the truth, 'drink deep at that fountain of beneficent wisdom for which the slight taste obtained on earth has given him so keen a relish'.

French scientists and mathematicians, with their taste for a more analytical, materialistic philosophy, were less prone to such sentiments.

In 1814, Laplace, whose godless *Mécanique Céleste* had just been translated by Mary Somerville, wrote a paper on probability. He argued that 'the most important questions of life . . . are indeed, for the most part, only problems in probability'. Furthermore, this probability could be mathematically measured. He gave the example of an urn filled with black and white balls. If the urn contained sixty balls in total, of which twenty were black, then someone about to draw a ball from the urn would have a 20 in 60, or 1 in 3, chance of picking a black ball.

In Laplace's view, all decisions are a matter of plucking balls from the urn of life, and can therefore be rationally determined using probability theory. All you need is 'great precision of mind, a nice judgement, and wide experience in worldly affairs. It is necessary to know how to guard oneself against prejudice, against illusions of fear and hope, and against those treacherous notions of success and happiness with which most men lull their [vanity].'

Babbage, as one might expect, was much taken by this view. He also saw it as providing a means of both supporting Hume's argument, which he thought to be self-evidently true, and the existence of miracles – a miraculous compromise. His machine could demonstrate how it could be achieved.

As he told his audience during his demonstration of the Difference Engine, miracles are really about probabilities. Hume had argued that it is more probable that a witness is lying or mistaken when he or she reports a miracle than that the miracle actually happened. So, for a miracle to have really happened, you have to prove that the miracle is more likely to have occurred than that those who witnessed it are lying or mistaken.

By means of some basic mathematics, which included such assumptions that a witness of 'tolerably good character and understanding' would speak the truth ninety-nine times out of a hundred, Babbage showed that if just six such people witnessed a miraculous event, the chance that they were all deceived or lying was 1 in 1,000,000,000,000 – a much more remote possibility, Babbage, calculated, than the most improbable miracle.

Furthermore, he could show that even a mechanistic universe of the sort described by Laplace did not preclude the possibility of such improbabilities. To prove this, he turned to his Difference Engine, now acting as his model of a mechanical universe, and proceeded to crank the handle. Imagine the Engine to be the universe, he told his audience, as the number displayed on the results wheels once again counted through the sequence, 0, 2, 4, 6, 8 ... Imagine, further, that mathematical functions it performs are equivalent of the laws of physics. Now, assuming it did not break down, you would expect the machine to carry on in the same way for ever – the equivalent of a universe devoid of miracles. At this point the machine would suddenly produce its 'miracle', a number completely out of sequence. Then, equally suddenly, the machine would revert to its old sequence, counting through the numbers two by two.

This was, of course, no miracle. In contemporary terms, Babbage had programmed the machine to perform one routine (*take the number currently displayed, add 2 to it, then display the result*) for so many turns of the crank, whereupon it would jump to another, a 'subroutine' (such as *add 17 to the current number, then display the result*), for so many turns before reverting to the original routine.

The trick may have been simple, but the point was meant to be profound. A machine such as the one he was demonstrating could in theory perform such improbabilities no matter how long it had been running in accordance to one law – it could happen when it reached 100, or not until it reached 1,000,000,000,000. It might only deviate once even if it continued performing more operations in the same manner 'than the world has known days & nights', as Annabella put it.

Thus, if the universe were like one of Babbage's calculating engines, it could both be mechanistic and miraculous. Even better, God still had his role to play, albeit a diminished and rather passive one, as the 'superintendent' of his divine machine – a cosmic Babbage, who sets up his celestial mechanics and leaves them to run into eternity.

*

A machine that was rigidly regulated by mathematical laws as precise and unyielding as machined steel, yet which was capable of behaving as though with a will of its own – the idea turned in Ada's mind like one of the whirring cogwheels in Babbage's engine. It obsessed her.

She pestered Mary Somerville to chaperone her on yet more visits to Dorset Street to see it, read every article she could find that discussed it.

In the autumn of 1834 her mother took her up to the Midlands to see friends and take in the sights. She enjoyed the chance to see 'many new & beautiful specimens of nature (both human & inanimate)', but regarded the real highlight to be an industrial tour, because there the new mass-producing machines she saw glittering with grease in the print rooms and ribbon factories of Coventry reminded her, she wrote to Dr King, 'of Babbage and his gem of all mechanism'.

Dr King was beginning to get concerned at her enthusiasm for such radical and novel developments. It indicated that her course of intellectual study was not having the anaesthetizing effect on her it was supposed to have. 'Loose reading does no good,' he warned. Perhaps in a year's time she could be allowed a glimpse of a wider intellectual terrain. He might even favour her with an account of his own system of Logic & Morals. Until then she must stick to her Euclid.

Dr King's anxieties were apparently well founded. In early 1835, soon after her nineteenth birthday, Ada suffered a nervous attack while she was visiting the Somervilles at Chelsea. It frightened her, and the Somervilles too, who saw her suddenly overcome by an 'agitated look & manner'. Ada was sent home, together with an anxious communication from Mrs Somerville to Annabella suggesting that life in London might be too much for her.

This terrified Ada. Though barely strong enough to write, she managed to get a letter to Mary Somerville. Adopting a tone that was at times pathetically apologetic, she promised to be a 'very good little girl' and pleaded not to be kept in the 'desperate tight order' she had so recently escaped. She was, she explained, susceptible to such turns. They frightened her, which is why she behaved in the manner that the Somervilles had found so alarming. But she was in control of them, she could cope, she promised . . .

Mary Somerville obviously accepted Ada's explanation, as there seems to have been no curtailment of her London activities in the following months. Nevertheless, the event reinforced the suspicion that there were faultlines in Ada's character that could crack at any moment, producing the sort of fatal eruption that her father had himself only managed to prevent through his poetry. Since she had shown no interest in becoming a poet, she now needed more desperately than ever to concentrate on developing alternative means of expressing her individuality, her unique 'genius'.

CHAPTER FIVE

A Deep Romantic Chasm

IN THE SPRING OF 1835, following her nervous attack at Mary Somerville's, Ada was packed off to Brighton, the Sussex seaside resort. At the time it was still a two-day carriage trip from the capital, one Ada had to endure in the company of the 'Fury' Frances Carr, who was no doubt sent to monitor Ada's recovery.

Ada spent her time horse riding, the 'finest of all medicines', as she put it in a letter to Mary, designed to reassure her mentor that she was fully recovered and ready for her return to London. 'If I am to believe your daughters' own account of their feelings on this tender subject,' she added teasingly,

> I am afraid I shall excite in them hatred, & malice, & envy, & all manner of bad passions, when I say that I generally ride in the riding school everyday, and – best of all – leap to my heart's content. I assure you I think there is no pleasure in way of exercise equal to that of feeling one's horse flying under one. It is even better than waltzing. I recommend it too as a nervous medicine for weak patients.

In her youth, Ada had been frightened of horses, but here she overcame her phobia and developed what was to be a lifelong love of riding. Tam O'Shanter was to become her favourite horse. He was a stallion, sometimes difficult to handle, but she claimed to like him best when he was unruly, galloping full tilt across the Surrey downland.

Most men and many women of that period would agree with Ada's view that horse riding was an excellent treatment for female nervous tension. But it provided only temporary relief. A more permanent solution was marriage and pregnancy, and Woronzow Greig, Mary Somerville's son, began to develop a scheme that might provide it.

The year 1835 was full of Williams and Kings for Ada – Dr William

King, her moral mentor; King William IV, to whom she had made her curtsies at court. Now another came into view, William King, eighth Baron of Ockham. William had been a friend of Woronzow's at Trinity College, Cambridge. Soon after going down from Cambridge, he had left the country to pursue a career as a diplomat, ending up as secretary to the Governor of the Ionian Isles, then a British protectorate. There he must have been aware of Byron's legacy, as he was to be based at Cephalonia just a decade after Byron had arrived there in his ship *Hercules* on the first leg of his final, fateful trip to help liberate Greece from Ottoman domination.

The Byron connection is eerily captured in a portrait of William painted in the early 1830s. In an almost exact echo of the famous picture of the poet in Albanian dress (the one Ada was soon to see for the first time), William poses in traditional Greek attire. The similarities, at least in dress and attitude, are striking – the magnificent headdress, the ornate jacket, the draping silks, the ceremonial sword held across the midriff, the same gaze into the middle distance. There are differences too: where Ada's father's face is handsome and heroic, William's appears more boyish, almost androgynous. It was a look that he would soon abandon, preferring in subsequent portraits to adopt the grander pose more fitting to a peer of the realm, Justice of the Peace, Lord Lieutenant, Fellow of the Royal Society and Governor of the Royal Agricultural Society.

In 1833 William's father died suddenly, cutting short a promising career as a Whig politician. William was forced to return to England immediately to assume his title as the eighth Lord King. He re-established his friendship with Woronzow, who soon saw the opportunity for a match with the extraordinary young woman who had been in and out of his mother's house for the past two years.

In Woronzow's eyes, William was perfect for Ada. He had the title, the career prospects, the connections and the personal accomplishments, which included fluency in Greek, French, Italian and Spanish. He also knew something of science and philosophy, his interest perhaps encouraged by a the fact that he was a descendant of the great philosopher John Locke. There was another connection with philosophy through the family estate, Ockham, which had been the home of the fourteenth-century thinker William of Ockham (or Occam), famous for his 'razor', the principle that philosophical arguments should be kept as simple as possible – a notion that the straight-thinking William would have wholeheartedly endorsed.

William's intellectual interests were of a more practical nature than Ada's – his imagination was captivated by such matters as crop rotation and animal husbandry. He was also rather taciturn. However, such differences in character were an advantage, as William's down-to-earth nature would provide a good solid grounding for a woman Woronzow evidently regarded as a bit of a live wire.

Woronzow broached the subject with his friend in the spring of 1835. Even though, as Woronzow put it, he and his friend were 'on the most intimate terms', William did not at first respond to the suggestion. 'Even to me in whom he confides more than in any other person living, he is not very communicative,' Woronzow observed. He heard nothing more about the matter until 12 June 1835, when William wrote to his friend suggesting dinner at his London house in St James's Square. There he sprang the news that he and Ada were now engaged to be married. He had apparently contrived a meeting with her at the Warwickshire house of Sir George Philips, a mutual friend of his and Lady Byron's.

The courtship was conducted with the full knowledge of Annabella, who would have been delighted that a man with such a distinguished title, unblemished record and straightforward manner should show an interest in her unruly daughter.

Ada welcomed the interest as well, telling her mother that if William proposed to her, she would accept. Annabella may have taken this as decisive evidence that her once unruly daughter had finally decided to submit to her mother's wishes. However, it seems Ada's motives were in quite the opposite direction. Marriage now provided her with her only means of escaping her mother and the meddling Furies, and she could only hope to marry a man who had her mother's approval. William would have seemed harmless enough to her.

Perhaps she looked upon the reassuring, undemanding character of a man nearly eleven years her senior as a sturdy vehicle to convey her into a new era of adult independence. Aristocratic marriage was, after all, then regarded as a commercial and genetic transaction rather than a romantic one, a means of maintaining estates and bloodlines, which was of particular importance in an age when the primacy and power of ancient noble lineages were being challenged by new bourgeois ones. Ada was not particularly concerned with such issues, but they did allow, even encourage, a sort of acceptance in her conduct during the courtship.

The match's attractions for William were more obvious. Ada was to many men extremely alluring. She was a Byron – a very outgoing, often familiar, sometimes downright flirtatious one. She was also pretty, with cascades of dark hair and large expressive eyes (though, like her father, her weight seemed to fluctuate from one month to the next, and the last remnants of her juvenile paralysis sometimes made her physically awkward). Of more material significance to an ambitious young aristocrat, she came not just with the certain prospect of inheriting the Wentworth estate, but with connections to some of the most important people in the country, such as William Lamb, the widower of Lady Caroline (she died in 1828) and, as the second Viscount Melbourne, Prime Minister of Great Britain.

When William did propose in June 1835, within days of their first encounter, he was instantly accepted and, with a haste that might have aroused suspicions in a less enthusiastic suitor, the wedding day was set for the following month.

Despite the lack of time, there was to be no whirlwind romance, more a light summer breeze, one too soft to raise so much as a flutter in Ada. The first personal letters that their state of betrothal allowed them to exchange were, on William's part, full of pent-up emotion and expectation, and on Ada's almost languid casualness mixed with a little indulgent tenderness. 'I look upon such happiness as too excessive to be enjoyed otherwise than in a dream, as too splendid & too overcoming for a reality,' William panted. Ada replied by saying she felt 'calm and peaceful'. The strongest emotion she managed to arouse was gratitude that he had accepted her – a more truthful sentiment than he probably realized, because had he not asked her to marry him it might have sentenced her to years more under her mother's thumb, even to the prospect of spinsterhood.

Nevertheless, the engagement had its moments, such as a ride they took together in Chelsea a few days after their engagement had been announced, which probably provided them with their first opportunity to be alone together.

When he came to pick her up from Mary Somerville's house in the magnificent setting of Wren's Royal Hospital, they were still virtual strangers. William would have been at his most reserved and silent, though as her letter mentioning the event makes clear she was very much at her ease. Ada had an unaffected, confiding manner in the company of men she liked, and it is easy to understand how it captivated a man of William's reticence. Each time she lightly touched his arm to emphasize a point, or leaned over to exchange an indiscreet observation, she would have melted his reserve just a little more. Perhaps he suddenly found himself talking and joking too, for the first time enjoying his own company as much as his wife-to-be seemed to.

As they trotted along Royal Hospital Road, Ada no doubt talked excitedly about her ideas and plans, her love of mathematics, machines and music, while William listened with earnest attentiveness. As they passed the tall cedars of Lebanon that stood at the entrance to the Chelsea Physic Garden, the first to be planted in England when the Apothecaries Company founded the garden in 1676, perhaps he began to loosen up a little and offer a few observations of his own. By the time they were down by the Thames, passing along the elegant terraced Queen Anne houses lining Cheyne Walk, he might have started to talk about his love of agriculture and his plans for his estates in Surrey and Somerset. On such a warm summer's day, in such a setting of urban gentility, with the hood of the barouche drawn back to reveal the leafy canopy of London planes overhead, they might even have exchanged a kiss – a discreet reference to the ride in one of Ada's subsequent letters suggested something of the sort happened. Whatever occurred, when they returned to the Royal Hospital, William knew that he had found the woman of his dreams, and Ada the man for her needs.

A few days later William departed for Ashley Combe, his Somerset estate perched on the steep hills leading down from Exmoor to the sea. It was to be the venue for their honeymoon, and he was determined to make it ready for his bride. Despite suffocating heat, he laboured hard, supervising the restoration of the interior of the house, and himself helping to hack down the huge branches of overgrown arbutus, bay and myrtle trees to open up vistas over the Bristol Channel and the distant shore of Wales. He found moments amidst the work to pen some letters to Ada, each one suffused with love and anticipation. He envied her her calmness; his had deserted him. He was in a fever of excitement, positively absent-minded (his mind being in the company of Ada, in the garden at Fordhook), a condition which delighted him nonetheless.

And so an accommodation was established with which everyone could be satisfied: William was to have a wonderful, funny if rather eccentric celebrity wife to dangle from a branch of his family tree, Ada would have a sturdy, undemanding husband to take her away from Fordhook, and Annabella was to acquire a son who was prepared to assume responsibility for her wayward daughter.

*

Without warning, Annabella intervened in the last few days of her daughter's engagement with an act that can be interpreted in a number of ways – as sensibly pre-emptive, as vindictive and demeaning towards her daughter, as reflecting what Woronzow called her 'stern sense of justice and right'. In a carefully worded letter to Woronzow, she instructed him to tell William about Ada's attempted elopement with her tutor. Like any good lawyer, Woronzow suppressed any personal scruples he may have felt about revealing a friend's confidences to her prospective husband, and broke the news to William.

It was a nasty moment, revealing to William perhaps for the first time a waywardness in Ada's character that he might have anticipated one day putting the marriage at risk – which indeed it would. But he did not flinch. 'He knows *all*,' Annabella wrote with relief to Sophia Frend, who knew *all* as well, 'and is most anxious for the marriage.'

News now leaked out about the marriage, and letters began to arrive at Fordhook offering congratulations, among them one from Lord Holland, the great Whig peer who had known William's father, and a message communicating the pleasure of the King, who had 'never heard anything but good of Lord King'. Ada herself received a rather different letter from the wife of the other William King in her life, the good Doctor.

Mrs King wrote to announce that she was surrendering her role as guardian of Ada's morals, but did not intend to dismiss herself without a parting volley of admonitions and advice. The first item in the endless list that then unfolded down to the floor was Ada's duty to reciprocate Lord King's generous gift of his affection, a sentiment which Mrs King felt Ada lacked (because she did not show any towards Mrs King, presumably). She must nurture this feeling, and express it through the 'thousand delicate but invaluable attentions' with which a good wife will always endear herself to so estimable a husband. She must hide nothing from him, as she did so destructively from her saintly mother; he must be confident in his belief that he knew everything he needed to know about her.

The next piece of advice was that Ada should display female charms – charms which a woman like Mrs King would have believed she most definitely lacked, being outgoing and opinionated. There was a veiled injunction here that Ada should not yield to her habit of flirting, particularly not now, except towards, of course, her husband, who was to be the continuing beneficiary of her sexual attentions.

She must draw up a timetable of her day and stick to it, as well as ensure that she keep control of the domestic accounts. And she must say her prayers, and pay heed to the scriptures.

Finally, there was the inevitable reminder of her sinful behaviour towards her tutor. 'Gracious God who has so mercifully given you an opportunity of turning aside from the dangerous paths you were treading has given you a friend and guardian in whom you may thoroughly confide to assist you in the new one His providential care has opened before you,' wrote Mrs King, many of whose sentences read like clauses in a legal contract. 'Bid adieu to your old companion Ada Byron with all her peculiarities, caprices and self-seeking,' she added, 'determined that as Ada King you will live for others'.

Ada, showing just how in command she now was of her feelings, managed to pen a thankful reply to this insufferable missive. She had spent her life being told to deny her father. Now that her name was finally to be severed from his, it must have come as a relief – at least, she may have hoped, it would free her from that legacy and let her get on with the life she wanted to lead. So she may have even agreed with Mrs King's parting sentiments. Goodbye to Ada Byron, and good riddance.

*

Having cleared away the obstacle of Ada's history, Annabella was now unstoppable, arranging immediately for the marriage settlement to be drawn up. This was a complex document, and in Ada's case involved what were then very substantial sums of money. Sixteen thousand pounds of Annabella's £20,000 dowry on her own marriage to Byron had been settled on the issue of the marriage, which meant that in effect it was to act as Ada's dowry. Annabella added a further £14,000 to this sum, which meant that William was to get a hefty £30,000 for the marriage, along with the almost certain prospect of inheriting the Wentworth estate.

William's lawyers chose not to quibble over such munificence, but it was to come at a price; Annabella was generous with her money, but she always held the belief that accepting a gift from her implied an obligation of loyalty and even obedience. This was something that William was to discover, though he would, at least initially, happily accept it from a woman he regarded with undisguised awe.

All the money and property that were settled on Ada would actually belong to her husband. She would have no access or right to them. Instead, she was to receive an income, so-called 'pin money', which was supposed to pay for clothes and other feminine necessities. The amount was negotiated as part of the marriage settlement, and Ada would have had little or no say in determining how much it would be. The sum was eventually fixed at £300 per year, regarded as more than adequate by William. Annabella thought it sufficient too, as she had received the same amount during her marriage to Byron. For Ada herself, it was inadequate – not just the amount, but the whole idea of pin money, of having to go to her husband cap in hand if she wanted any more, which, after the cost of books and ballgowns, she knew she inevitably would. It came to symbolize her powerlessness. Of course, she could have accepted payment for her work. This was strictly disallowed by the conventions of her class, but Ada was quite happy to challenge convention. However, on this matter she chose to conform, perhaps revealing a hint of her father's haughty disdain for earned income.

For the time being, however, Ada had more pressing anxieties to deal with. On the Sunday before the wedding she went to church with one of her favourite men, the Reverend Samuel Gamlen, a Yorkshire minister with a less than reverent sense of humour. As she sat in her pew she began to think that in just a week's time she would be going to the church that nestled next to William's house in Ockham as his wife. The thought seemed to sow seeds of self-doubt in her mind. Perhaps the voice of Mrs King insinuated itself into the prayers being recited . . . was she worthy of William . . . could she be trusted . . . did she possess the qualities of perseverance and self-denial that, as Annabella's friends had never tired of reminding her, a wife needs to fulfil her duties?

Such prenuptial anxieties were soon forgotten in the whirl of more pressing concerns that surrounded the Big Day, such as keeping control of the publicity it was beginning to attract. On 29 June, with just a week to go, she spotted stories in both the *Morning Post* and the *Morning Herald*. 'Only a slight notice,' she wrote to William with relief later that morning, 'nothing that need annoy us.'

It was no doubt to avoid publicity that the decision was taken to hold the wedding at Fordhook. If Ada and William had agreed to a more conventional church ceremony, it would have attracted large crowds.

The decision turned out to be a fortunate one, as it was not just the newspapers who were kept away. Teresa Guiccioli happened to be in London at the same time. Nearly fifteen years before, Byron had met the petite, voluptuous Italian countess in Venice when she was just nineteen and married to a dull and sinister count in his fifties. Teresa was, Byron had sighed to his friend Douglas Kinnaird, 'as fair as sunrise, as warm as noon', and a passionate love affair ensued which inevitably became public. She had now come to England to make her first pilgrimage to Byron's homeland, a well-publicized trip which would include a visit to Newstead Abbey and the Byrons' grave at Hucknall Torkard.

In London an eager queue of men, including many of Byron's old cronies and some of his daughter's new admirers, lined up to meet her. The unavoidable Babbage was among them, together with Dionysius Lardner, whose lectures Ada had attended the year before, and Edward Bulwer (later Bulwer-Lytton), a popular novelist who would strike up a friendship with Ada in years to come. Babbage obligingly promised to point out Lady Byron to Teresa at the theatre.

The Contessa learned, possibly from Babbage, of Ada's impending marriage, and decided that she would attend the ceremony. She seemed unaware that her presence at such a moment would result in a fracas.

In the event, she got the wrong venue. She had assumed they would be married before a large gathering at St George's Church in Hanover Square.

So, as six miles away one of her father's lovers stood disconsolately in the empty nave of St George's, wondering where everyone had got to, Ada quietly, privately, inconspicuously cast herself adrift from Byron and became a King. Like Annabella's own wedding at Seaham, it was strictly a family affair, the only friends being those within Annabella's close circle, Ada selecting Olivia Acheson, Lady Gosford's daughter, as her maid of honour.

The event was duly reported in the newspapers, with Ada's paternity and fortune providing the focus of interest. Throughout her life, and following her death, just about every story concerning her would begin with the famous words introducing the third Canto of *Childe Harold*, 'Ada! sole daughter of my house and heart . . .'. So began the report in *World of Fashion*, which, having drawn its readers' attention to the £30,000 dowry (apparently handed over by Annabella 'in cash'), was moved to imagine Byron's ghostly presence at the giving away of his daughter:

> The prayer breathed by the departed poet over his child . . . seems
> to be realised; she has escaped the dangers and woes which enclosed
> her footsteps, the clouds that gathered round the morn of her life

... Is it not to be regretted that the father lived not to see his wishes realised, and write his daughter's bridal song?

Thus the love/hate relationship the papers and public had with her father continued with her – not helped by her behaviour being measured against that of her mother and the 'strict religious principles' by which Annabella, as the *World of Fashion* reminded everyone, was known to live. Ada had to be a Byron without being Byronic. What made this worse for her was that, at the time, she did not want to be a Byron. She felt no pride at the connection, which only served to act as an instrument of control over her. Ironically, it was her mother who kept the link alive. Annabella never relinquished the title 'Lady Byron', and decided that Ada should 'reserve' her right to it as well.

*

After the wedding, Ada departed with William for his Surrey estate, Ockham Park. She was to spend a week there, beginning to familiarize herself with her new household, before the couple departed for their honeymoon at Ashley Combe.

At Ockham she quickly discovered that marriage, far from providing her with a means of escaping her mother, bound them together more tightly than ever. This was not altogether unwelcome, as Annabella underwent a Damascene conversion concerning her daughter. A year earlier, Annabella had advised Dr King not to form too positive a view 'lest you should be disappointed'. Now she was applauded for her 'moral courage'. 'Ada Byron that was, or is . . . has applied the singular combination of powers with which she was endowed to the best and most Christian purposes,' Annabella wrote in a sparkling new 'character' she now drew up of her daughter. These unprecedented sentiments were supplemented by endorsements that Annabella had presumably gathered from the friends who, in other circumstances, were so quick to condemn the girl: 'Ada teaches so that one cannot help learning,' said one. 'She repays one for the drier part of the tasks she enjoins by singing like the bonnie bird on the banks of the Doone,' chirruped another.

Annabella was overwhelmed by maternal feelings for her new

paragon, and showered her with advice about running a house and keeping a husband happy (despite a certain lack of expertise in both departments). Ada was told to be careful what she said in front of her lord, to show gratitude and humility towards him and to get to know his relatives (even though he did not get on with them). There was more practical advice, too, about keeping the cook happy, how to welcome guests, how to conduct dinner parties. 'I have made a discovery,' Annabella wrote to Ada while this new mood flushed, 'that in consequence of your marriage I am become amiable, affable, and all sorts of pretty epithets.' She even resorted to giving herself a pet name, something she had not done since she signed the letters she wrote to Byron after she left him 'Pippin'. Now she was the 'Hen', and the other members of her brood became birds, too. Ada, who adopted the scheme enthusiastically, was her 'dear little Canary bird' who Annabella, becoming momentarily carried away by the metaphor, imagined twittering away in the new 'cage' of matrimony (that is exactly how Ada would come to see it). Later, Ada would sign herself variously 'Thrush', 'Bird' or 'Avis', depending on her mood, while William, being prone to black moods which neither woman took very seriously, became teasingly known as 'Rook' and 'Crow'.

As the ultimate gesture of the trust her daughter now deserved, Annabella started to pass on a few of Byron's possessions which she had acquired from Augusta. Within a few months of marrying, the girl who had before been met with icy silence or erupting rage at the very mention of his name now received an inkstand that he had owned, and, of course, the portrait of him dressed in Albanian costume as a Christmas present.

As Ada and William finally set off for their honeymoon at Ashley Combe these gifts were yet to come. However, the house in which she was about to stay had its own echoes of her father, or rather the poetic world in which he was to have such an influential role.

The journey to Ashley was then a long and arduous one, though soon it would be transformed by the opening of the Great Western Railway and a branch line to nearby Bridgwater. It would have required several days in a carriage, with stopoffs at inns along the way. As they got nearer to their destination, Ada would have also become aware that they were approaching a place barely on the map, a little outpost of civilization on the edge of the vast, empty wilderness of Exmoor.

The last village they reached before making their final approach to Ashley was Porlock, squeezed into one of the crevices that broke the indomitable row of hills lining the shore between Minehead and the Valley of the Rocks. 'This place is called in the neighbourhood the End of the World,' the poet Robert Southey had written in his diary thirty-five years earlier. 'All beyond is inaccessible to carriage or even cart.'

Since Southey's time, a track had been laid leading out of Porlock and up the west side of the valley, a precipitous climb that would become famous as one of the steepest roads in England. At the top lay Porlock Common, a windswept plateau of grass and gorse, with views of the wide Bristol Channel and the shores of Wales off to the north, and to the south the endless undulations of Exmoor.

This was the King estate, Ada's new rural domain. From here Ashley Combe could be reached via a path that disappeared into a thick wood. At first the path sloped gently, before a series of sharp turns announced a steep plunge to the sea, following the course of a stream. Crashing down the track like the water that cascaded through the cracked and tumbled rocks, the path made its final descent, its fall broken by a turning up to a gate.

Passing through the gate, along a gentle track that William had probably repaired during his previous visit to the house, the honeymoon couple could recover from the drama of their journey in the tranquillity of a dense wood, nothing to be heard but the rustling of the trees, the noise of the now hidden brook . . .

> . . . In the leafy month of June,
> That to the sleeping woods all night
> Singeth a quiet tune . . .

in the words of Coleridge, a description in his case of the wind in the sails of a becalmed ship in 'The Rime of the Ancient Mariner', words which, as we shall see, may well have been inspired by this same setting.

The track turned a corner, and the house of Ashley Combe was revealed, sitting on a shelf cut into the hillside, surrounded by terraces crowded with cedar, oak, bay and cork, the trees leaning over the house and waving in its visitors with their branches.

*

William had remodelled the house in the fashionable Italianate style. With its flat walls punctured by elaborate arches and finished with patterned brickwork, it could have been standing on the shores of Lake Maggiore. He was even in the process of building a clock tower and a water garden irrigated by a lake cut into the hill further up, which fed a fountain that he would dedicate to his new wife and a bathing pool he would build for her into the cliffs down by the shore.

In the coming weeks of their honeymoon, Ada developed a passion for the place, one which would stay with her throughout her life. After her death, William's most touching memory of her was here, riding off into the hills and walking along the shore. At any opportunity, she would take off on an expedition, on foot or the back of a wild 'Forester mare', 'over hills, valleys, moors, downs, every variety of wild or beautiful country', along the way forming 'castles in the air about Switzerland, the Pyrenees, or the Danube'.

Little remains to give an idea of that first honeymoon visit, but it can be easily imagined. She may have set off as soon as she got there, excitedly exploring the house and its surroundings, while William dealt with estate business and the continuing (indeed, never-ending) plans for improving the house. Inside, she would have found a large library stocked with the works of the great philosophers and writers. There was the *Essay Concerning Human Understanding* by William's ancestor, John Locke, stirring historical novels by Sir Walter Scott, the extraordinary, fantastical poetry of Samuel Taylor Coleridge. And among the volumes, a collection of works by Byron, including *Mazeppa* and *Childe Harold*.

This was almost certainly the first time Ada had unsupervised access to her father's works. There, on the opening page of *Childe Harold*'s third Canto, which was published in a separate edition to the previous two, she would have seen her own name . . .

ADA! sole daughter of my house and heart?

Flicking over to the final stanzas, she would have read of herself again . . .

> My daughter! with thy name this song begun –
> My daughter! with thy name thus much shall end –
> I see thee not, – I hear thee not, – but none
> Can be so wrapt in thee; thou art the friend
> To whom the shadows of far years extend:
> Albeit my brow thou never should'st behold,
> My voice shall with thy future visions blend,
> And reach into thy heart, – when mine is cold, –
> A token and a tone, even from thy father's mould.

'. . . my brow thou never should'st behold . . .' It reads almost like an instruction – or a warning, one that would not be heeded even by her mother, who a few months later would allow Ada to behold that brow so nobly painted in his portrait. And, just as he had anticipated, his voice would with her future visions blend – ultimately she could never escape it. Here her father's voice talked directly to her, his 'child of love', 'born in bitterness'.

Ada's response to these words remains undocumented, as, so frustratingly, do all her encounters with her father's work. But at Ashley, she found herself not only among the echoes of his voice, but those of two other great Romantic poets, Wordsworth and Coleridge.

When she first heard their words can only be guessed at, but had she been there a few decades earlier she might have literally heard them, as in 1797 the two men walked along the track leading up from Porlock Weir, past the house and on to the next combe.

Ada often followed in their footsteps, along the path that traced the precipitous coast from Porlock to Culbone. She, like them, penetrated the thick copses of oak, ash, birch and hazel, pushed through tangled bushes of bramble and whortleberry, past rocks covered with velvet pelts of lichen and the vivid but rare yellow reindeer moss, amidst woodland alive with the sounds of wild deer, foxes, badgers, martens, squirrels and pheasants scurrying through the undergrowth.

That hot summer of her first visit the thick canopy of trees would have provided welcome shade as she penetrated into the heart of her new husband's estate. After a mile or so the path turned inland, following the contours of another combe, and Culbone itself could at last be glimpsed, a glimmering pool of light amidst the darkness of the wood.

The path tantalizingly encircled it, before suddenly dropping down and breaking cover into a deep valley with a tiny, solitary building nestling in its nook: the parish church for William's estate. Culbone Church was little more than a stone hut, which at the time Ada first saw it did not even have a spire (one was later added, little taller than a doorway, perched on its roof like a crumpled fool's cap).

A geographer sent to map Somerset's wilder terrain had stood on the same spot in the 1780s, there to record the topography and population but moved by what he saw to lyricism:

> This spot is as truly romantic as any perhaps which the kingdom can exhibit. The magnitude, height & grandeur of the hills, rocks & woods at the back, & on each side of the cove, the solemnity of the surrounding scene – the sound of the rivulet roaring down the craggy channel; the steep impassable descent from the Church down to the beach; the dashing of the waves on a rough and stony shire at an awful distance below; the extent of the channel, and finely varied Coast of Wales beyond form in the whole a scene peculiarly adapted to strike the mind with pleasure & astonishment!

Ada would have had similar feelings during those first days of her honeymoon. The geographer was right: the spot upon which she stood was as romantic as any in the kingdom. Indeed, it had a special claim to be, as this was one of the birthplaces of that intoxicating mix of the sentimental, the supernatural, the passionate, the nostalgic, the wild and the wayward that became English Romanticism.

*

In the autumn of 1797, a twenty-five-year-old man was standing where Ada now stood, recovering from the long climb up from Porlock, feeling ill and in search of somewhere to rest. His name was Samuel Taylor Coleridge, and he had walked about twenty miles from Nether

Stowey, where he was living in a cottage near his friend William Wordsworth, who had taken up residence with his sister Dorothy in a smart Queen Anne mansion called Alfoxden.

Coleridge had been struggling to finish a play. It was a form of literature that he sometimes found difficult, as Byron would later discover when he commissioned a tragedy from him for Drury Lane. Feeling creatively blocked and suffering from chronic sickness, he decided to try and purge his coagulated mind and exhausted body by spending a few days exploring the local countryside.

He walked along the Great Track that crossed the Quantock Hills and arrived at Porlock Weir. From there he wandered up to Ashley Combe, and along the same zigzag path that Ada would take so many years later. He reached Culbone and decided to rest at Withycombe Farm, one of those dotting the headland of the King estate. A track then led to it from Culbone church, and it stood next to one of the fast-running brooks through which Exmoor empties its waters into the sea.

Settled into his retreat, overlooking Exmoor to one side and the thickly wooded combes sloping away on the other, Coleridge found his poetic block immediately cleared, and wrote some lines for his verse-drama *Osorio* (later and better known as *Remorse*) clearly inspired by the scenery he had just beheld:

> The hanging Woods, that touch'd by Autumn seem'd
> As they were blossoming hues of fire & gold,
> The hanging Woods, most lovely in decay,
> The many clouds, the Sea, the Rock, the Sands,
> Lay in the silent moonshine – and the Owl,
> (Strange, very strange!) the Scritch-owl only wak'd,
> Sole Voice, sole Eye of all that world of Beauty!

He had also brought some reading material with him, including a volume published in 1614 entitled *Purchas his Pilgrimage*, which he began to read. Still feeling ill, he decided to take a few drops of opium to settle his stomach. He soon found himself falling into a restless slumber, just as he reached a passage in the book about 'Xanada' where 'the

Khan Kubla commanded a palace to be built, and a stately garden thereunto'.

'The author continued for about three hours in a profound sleep (at least of the external senses),' Coleridge later wrote, referring to himself in the third person, 'during which time he has the most vivid confidence that he could not have composed less than from two to three hundred lines – if that indeed can be called composition in which all the images rose up before him as *things*, with a parallel production of the correspondent expressions, without any sensation or consciousness of effort.'

He awoke, and, finding he could recollect the 'expressions' he had just dreamed, immediately set about writing them down. 'At this moment,' Coleridge continued,

> he was unfortunately called out by a person on business from Porlock and detained by him above an hour, and on his return to his room, found to his no small surprise and mortification that though he still retained some vague and dim recollection of the general purpose of the vision, yet, with the exception of some eight or ten scattered lines and images, all the rest had passed away like the images on the surface of a stream into which a stone has been cast.

The resulting fragment is the poem now known as 'Kubla Khan', a work of wild, supernatural power which unforgettably begins:

> In Xanadu did Kubla Khan
> A stately pleasure dome decree . . .

It continues with lines that again powerfully evoke the surrounding countryside native to that person from Porlock, rather than the fantastical land of Xanadu . . .

> But oh! that deep romantic chasm which slanted
> Down the green hill athwart a cedarn cover!
> A savage place! as holy and enchanted
> As o'er beneath a waning moon was haunted
> By woman wailing for her demon-lover!
> And from this chasm, with ceaseless turmoil seething,
> As if this earth in fast thick pants were breathing,

> A mighty fountain momently was forced:
> Amid whose swift half-intermitted burst
> Huge fragments vaulted like rebounding hail,
> Or chaffy grain beneath the thresher's flail:
> And 'mid these dancing rocks at once and ever
> It flung up momently the sacred river.

No one knows who from Porlock interrupted this extraordinary poetical outpouring, and the anonymous stranger has now passed on into literary mythology as the man who may have deprived the world of a longer poem but probably gave it a more potent one. Prefaced by the account about its genesis quoted above, 'Kubla Khan' did not appear until 1816, when Coleridge decided to publish it 'at the request of a poet of great and deserved celebrity' – Lord Byron.

The circumstances of its authorship were to provide a model of the creative act that romantics, from poets to rock stars, have emulated ever since – the combination of drugs, dreams and poetry, the rude intrusions of the real world, these are the elements which enable art to touch the transcendent truths that are beyond science's reach.

Indeed, some scientists were even beginning to wonder whether there might not be a special, hallucinogenic state of consciousness that artists experienced which gave them this special form of perception. This was an idea explored by the physician John Elliotson in 1837, who cited 'Kubla Khan' and the story of the person from Porlock in a lecture about what he called 'diseased sleep', or somnambulism, the state that the mesmerists had recently discovered they could invoke. Ada was a friend of Elliotson's, and, as we shall see, became involved in his own mesmerical experiments. Perhaps part of her interest was a desire to see whether she, too, could explore the elusive transcendent state that Coleridge had discovered just a few hundred yards from her honeymoon home.

The links between Ada's Somerset surroundings and Romanticism were not confined to 'Kubla Khan'. Coleridge was so excited by the terrain that he had discovered he returned later the same year with his friend William Wordsworth, and the two of them spent days exploring the surrounding landscape. In the midst of the ancient hills and wild, windswept heaths, they found themselves tapping into the sort of rich stratum of fossilized folklore that was to become such an important source of inspiration for romantic fiction and poetry. During one excursion, for example, they encountered a famous gibbet from which had once hung the body of John Walford.

Walford was a humble charcoal burner who lived in the woods that frothed up on the lower slopes of the Quantocks. He fell in love with a local girl called Anne Rice, but was forced to marry a half-mad woman who got pregnant after she visited him in his isolated hut. One night, after a month of wedlock, Walford and his wife were walking to an inn when, overcome by a fit of despair, he murdered her. He tried to dispose of the body down a copper mine, but was too weak to lift the body into the mouth of the shaft, and ended up abandoning it in a ditch.

He was quickly accused and found guilty of the crime, and sentenced to be hanged in chains. Just as the execution was about to be carried out, he saw Anne Rice among the spectators and asked that she be brought forward. She came to the front of the crowd, and they talked for ten minutes before she was dragged away by guards. Walford, according to tradition, snatched her hand and kissed it, tears for the first time running down his cheeks. He then admitted his crime to the crowd and asked for forgiveness, before he was finally dispatched and his corpse hung from the gibbet that Wordsworth and Coleridge would later behold.

The mixture of beauty and tragedy provided by such stories enacted in such surroundings helped give birth to the *Lyrical Ballads*, published in 1798, which some herald as the work that founded English Romanticism. In his *Biographia Literaria* Coleridge recalled how he and Wordsworth divided the labour between them (very much along the industrial lines that would have been approved by Babbage). Coleridge dealt with 'persons and characters supernatural or at least romantic' while Wordsworth concentrated on giving 'charm and novelty to things of every day'.

Nothing could be more supernatural or romantic than Coleridge's 'Rime of the Ancient Mariner', set as it is in the 'silent sea' sailed by skeletal ships. But even in this there are echoes of Somerset, arising out of a walk Coleridge made with Wordsworth along the coast. Culbone is to be found in the hermit's woodland home that 'slopes down to the sea':

> He kneels at morn, and noon and eve,
> He hath a cushion plump;
> It is the moss that wholly hides
> The rotted old oak-stump.

And the primal power of the Valley of the Rocks, a great windswept stone fissure further down the coast, where Coleridge enjoyed watching 'the commotion of the elements' as a tempest broke across the sea, is echoed in the storm that threw the Mariner's ship on a course for the frozen South Pole . . .

> And now the storm-blast came, and he
> Was tyrannous and strong:
> He struck with his o'ertaking wings,
> And chased us south along.
>
> With sloping masts and dipping prow,
> As who pursued with yell and blow
> Still treads the shadow of his foe
> And forward bends his head,
> The ship drove fast, loud roared the blast,
> And southward aye we fled . . .

It was these early flirtations with the natural and the elemental found in places like the Somerset coast that helped establish the Romantics' love/hate relationship with science, their fascination in scientific discoveries, their terror that such discoveries might be the only valid ones, the only possible route to eternal truths. This ambiguity was later captured in Turner's roiling 1842 painting *Snow Storm*, of a steamboat trapped in shallow water signalling to be saved, as in the picture it is shown helplessly lost amid the waves and spray. The picture was inspired by what Turner had learned about the lines of magnetic force which it had been recently discovered girded the earth. Mary Somerville had explained the idea to him when he visited her in Chelsea around the same time as Ada. In her book *The Connexion of the*

Physical Sciences, Mary had speculated that a ship passing across the sea, like Coleridge's ghostly hulk, 'ought to have electric currents running directly across the path of her motion'.

*

Ada would soon have her own chance to explore all these ideas – indeed become as enraptured by them as the Romantic artists and poets. But for now she had only one task, which was to fulfil her duty as a wife and start producing children. She returned to Ockham and was soon joined by her mother, who found herself so excited at being among her 'birds' that she had to leave again. Dr King came in her stead, and was to stay until Christmas, when Ada received as her present the portrait of her father in Albanian costume.

Dr King had not proved as easily convinced as Annabella that Ada's marriage had changed her. He suggested she spend £10 of her pin money on religious and scientific books, as 'knowledge is evidence and evidence is absolutely necessary to settle the mind under doubts and temptations. Under a great temptation to do wrong, ignorant doubt of the Truth of Revelation might turn the balance of happiness for ever.'

Ada certainly continued her quest for mathematical knowledge, and the mood of harmony persisted into the autumn, when she announced that she had fulfilled her wifely duty and become pregnant. The tone of her mother's response to the news was, unlike her other recent effusions, far from elated. She told Ada to expect lots of discomfort, even depression. She then passed on to the latest bulletin concerning her own state of health, which, as always, was of more pressing concern than her daughter's.

Soon after, William and Ada stayed with Annabella at yet another rented house, this one in Southampton, where the exaltation of William reached new heights. Halley's comet had made a recent visit, and Annabella likened him to it. Both became more brilliant as they approached, she observed. 'Now,' she announced rapturously, 'I see the nucleus.'

In October 1835, after three months of marriage, Ada and William were separated for the first time. William, keen to assert his lordly presence in county matters, returned to Ockham to participate in manoeuvres with the Surrey Militia, and Ada stayed with her mother in Southampton. While she was there, she had her first experience of gambling – or at least the first to which she would admit in writing. She mentions the episode in a letter to William only to play it down, saying that any 'lurking propensity' she may have had was 'nipped in the bud' by the loss of four shillings. Keeping up the tone of flippancy, she also admits to a 'cheating exploit', when she took a walk on the pier without paying the twopence toll. She ends the letter by reporting she has not been allowed to read the letter the 'Hen' is sending him, arousing the suspicion in her mind that it portrays her as a 'bad bird' – and the suspicion in ours that she is only telling him about the gambling and 'cheating' episodes to pre-empt whatever he may hear from his mother-in-law.

These were the first of many scrapes that William would come to hear about, either from her or from others. He was, by and large, indulgent, and in any case, at the time he was more concerned about his relations with Annabella than with Ada.

As Ada was left to go about the business of being pregnant, and occupy her spare hours with mathematics and harp-playing (a growing enthusiasm), William formed the notion that he would create an experimental school at Ockham just like Annabella's at Ealing. No act of impersonation could have offered a sincerer form of flattery. Annabella was delighted to have her attention, currently preoccupied by an outbreak of the heart flutters that had once again confined her to bed, diverted by the project – indeed, far more delighted than to have it diverted by news of Ada's impending labour, the onset of which was not considered pressing enough to justify her being awoken during the night.

A little boy was born at William's London house in St James's Square on 12 May 1836 – coinciding, to Ada's brief annoyance, with an eclipse of the sun (one which proved to be of some scientific significance because it was among the first to be closely observed using the latest astronomical instruments, enabling Francis Baily for the first time to note the 'beads' of brilliant light – henceforward known as 'Baily's Beads' – around the rim of the moon caused by light bleeding through its cratered profile).

The boy born under the eerie gloom of a blinded sun would gleam in his mother's eye like one of those beads. But what to call him? There was one obvious name, but no one apparently dared suggest it until Annabella did: Byron. It was another instance of the perverse desire of Ada's mother to reinforce the link with the man she had done so much to distance herself from.

Annabella may have been uninterested in the pregnancy, but she was delighted with its result. Byron Junior was honoured with no less than two 'characters', one jocular version written within days of his birth, the other more serious after his first birthday, when she observed that he was physically precocious, able already to open and close cupboards, throw a ball, spin a thimble, play peepo, clap his hands in command and form the sweetest smile.

To begin with, Ada observed his development in more detached, almost scientific language. She noted that his attention was attracted by motion rather than colour. He responded to sounds, but only when they were discordant. He also had an experimental disposition, she observed, and reported how he would play with a chain belonging to his aunt Hester, watching the various positions into which it would fall. Nevertheless, she doted on her 'little treasure'. There were certain 'radicalisms & heresies', something of the 'devil' in him that she found enjoyable as well as challenging.

Little Byron's aunt Hester, William's sister, was to become of especial importance to Ada during this period. Ada had adopted her as a sister just as eagerly as Annabella adopted her sister-in-law Augusta when she married Byron – but not with such disastrous results. As Ada recovered from the birth, Hester pushed Ada around in a chair in the gardens at Ockham. Hester was also her companion when Ada went on rides in the carriage, such as on the occasion when they passed the Duchess of Kent driving a 'beautiful little phaeton with a pair of lovely piebald ponies'. Sitting by the Duchess's side that day was a young woman who Ada recognized to be Princess Victoria, then seventeen years old and a year away from her accession to the throne that would be hers for the rest of the century.

Victoria had already made her mark on Ada and particularly her mother. She had shown some sympathy with their reformist views during a visit to a children's school, which moved Annabella to write a sentimental verse pointing out that queens have in common with even the poorest children the need for paternal love . . . a curious sentiment, given Ada's circumstances.

By the end of 1836, Ada was pregnant again. It provoked her to speculations on her views about gender, writing in a letter to Mary Somerville that she preferred boys to girls – an observation she teasingly commanded Mary to read out to her daughters because Ada knew how much it would annoy them. Little Byron was staying with Mary at the time, and Ada missed him. She was also missing the chance to immerse herself in her studies, and wrote a succession of letters to Mary covering all the current ideas and disputes. She took a particular interest in a famous argument that had developed between Isaac Newton and John Flamsteed, the first Astronomer Royal, appointed to the post in 1675 by Charles II. The row, one of the most undignified and entertaining spats to erupt in scientific circles, did not come to light until 1835, more than a century after it took place, when Francis Baily (he of Baily Beads fame) published a biography of Flamsteed that contained the vitriolic letters the two eminent men had exchanged.

The argument arose over Newton's need for some lunar observations for his work on the moon's orbit. Flamsteed supplied the requested data, but felt that Newton was insufficiently appreciative. He suspected the reason for Newton's ingratitude to be the influence of another great scientist of those very exalted times, Edmund Halley, with whom Flamsteed was locked in a long-standing feud.

Francis Baily's book, which was broadly critical of Newton and favourable to the much less famous Flamsteed, appeared at a time when the idea of the great scientist – indeed, the very term – was still new, and applied to a profession that had yet to establish the authority it needed to drive home the great blows to cherished beliefs its theories were beginning to deliver. The idea that Newton, the father of modern science, could involve himself in such petty rivalries and jealousies represented a serious threat to the eminence his successors were desperate to acquire.

Ada had read Baily's book, and also seen some original manuscripts held by Sir David Brewster, a friend of Babbage's who was planning a biography of Newton. Ada believed that these documents would help restore Newton's reputation. She also believed they would help settle an argument then raging over Newton's theological beliefs. The Church of England was eager to co-opt him to its cause, as there was a growing awareness that science was encroaching on territory concerning the nature and organization of the universe that had previously been exclusively the Church's own. One theologian had triumphantly announced to Ada that when Newton's private papers were published, they would confirm his belief in the Trinity – a central tenet of orthodox Christian belief. This would strike a blow against the unorthodox (indeed, in the view of some, all but agnostic) Unitarians 'who had hitherto gloried in his authority, & appropriated him to themselves'. Ada, herself a Unitarian, angrily disputed this, maintaining that the argument was based on false assumptions about Unitarianism.

Her interests, theological and otherwise, were to be interrupted in late 1837 with the birth of a second child, a daughter she dutifully called Annabella in honour of her mother. Soon after she was struck down by a serious illness which she identified as cholera.

There is no medical evidence to support Ada's diagnosis, but historically speaking she was spot on. The first half of the nineteenth century saw two global outbreaks of the disease – terrible forebodings of epidemics to come. The final stages of the second and much more widespread of these coincided with the time she fell ill.

Cholera had its origins in India, and had traditionally been confined to the subcontinent. Then, during a pandemic of 1816, it escaped. Perhaps carried along by the new imperial and trade links that were being forged across the globe, it managed to make its way past the Caspian and Black Seas and through Russia and Turkey to the very doorstep of Europe. There it stopped in 1826, as though not yet ready to take on the seething industrial cities just over the horizon, and it receded.

Then another pandemic began in 1829. This one was stronger than the first, and was soon lapping at Europe's borders. It quickly broke through, and within three years had established a foothold not only in England but in America.

Cholera's effects were devastating and deadly, and it was untreatable. The symptoms were sickness and dizziness leading to violent vomiting and diarrhoea, followed by the most terrifying phase, when faeces emerged as a grey liquid, or, in the final stages, water contaminated by particles of the sufferer's own disintegrating intestinal tract – it was this material, produced copiously as the patient suffered from a raging thirst that could only be slaked with huge quantities of water, that spread the disease as it passed through inadequate sewerage systems into the water supply.

London's epidemic reached its height in 1832, when seven thousand were killed by it. Its devastating effect was largely the result of the sort of cramped, insanitary conditions that London's explosive economic growth had produced, and therefore it was the poor who inevitably suffered worst. However, the unstoppable bacillus was to find a way into homes at Ada's end of the social scale, too. With the introduction of piped water systems, one of the great marks of progress in the booming city, came the increasing use of water closets. These flushed waste not into cesspits, which would at least temporarily confine any infection, but into those other triumphs of Victorian engineering, underground sewers, which fed straight into the Thames. It was from this very same Thames that piped water was extracted between Chelsea and London Bridge. Thus a vast system for the incubation and circulation of infected sewage had, thanks to progress, been inadvertently created.

Ada presumably tapped into this system while she was at her husband's London house, which, thanks to William's passion for introducing new technological innovations to his homes, was undoubtedly connected into the new piped water system that private water companies were touting around the smarter parts of town. And so, just as the epidemic was about to die out in the poorer parts of the city it had so recently ravaged, she found herself suffering from its terrible symptoms.

Members of her class may not have been able to avoid the disease, but they stood a much better chance of surviving it. She was under the care of the best doctors, including Annabella's faithful physician Dr Herbert Mayo, and by the beginning of 1838 seems to have been on the way to recovery. She refers directly to the disease in only one undated letter to her mother. She was very thin, and still not eating, and Annabella, who retained through her life a belief in the miraculous effects of a good dose of 'divine' mutton, thought her undernourished. Unlike her mother, Ada did not like her illness being the subject of attention, and resisted interference from others, preferring instead, as she put it, to write about matters of the 'spirit' rather than the 'body'.

Ada's relationship with her body was never an easy one. She had almost abandoned it during her adolescent paralysis. And then, as soon as she had recovered power over it, it seemed to seize hold of her, overwhelming her with passions that nearly led to her ruin. The onset of cholera seemed to initiate a repetition of this cycle, with the disease confining her to her bed for months, leaving her weak and wretched, and recovery in the spring of 1838 producing a barely controllable surge of sexual longing. This had no outlet other than through her letters to William, whose insatiable interest for renovation had now been diverted from Ashley Combe to his London residence in St James's Square. As he worked away supervising the building and decorating work there, she placed herself alluringly before his mind's eye. She asked her 'mate to fondle her'. She was his little bird who 'wants to . . . nestle up to ou very much indeed. Ou won't hurt her I think, will ou? . . . I want my Cock at night to keep me warm.' She even suggested that she needed him on medical grounds, as her doctor, Charles Locock, considered an attempt at 'increasing the family' essential to recovery.

Such an increase duly followed, with Ada becoming pregnant in the autumn of 1838. The family also enjoyed an increase in status, with William appearing in the honours list issued to celebrate the accession of Victoria to the throne. This was a direct result of Annabella's family connection. William Lamb, the son of Annabella's aunt Lady Melbourne and the husband of Lady Caroline Lamb, had become Lord Melbourne on the death of his father in 1828, and Prime Minister of the Whig government which was in power at the time of Victoria's accession. As an acknowledgement of his rather distant family connection with Ada, he had decided to make William King an earl. William, in return, had chosen the name Lovelace for the peerage that had been created for him, as it was an extinct earldom he found buried in Annabella's family tree.

Having acquired such a rank, William did not rest on his laurels. He worked ceaselessly to promote his name and position. He took an active part in the work of the House of Lords, became a Fellow of the Royal Society and, in 1840, was appointed Lord Lieutenant of the County of Surrey, an honour which members of the land-owning classes regarded as one of the most exalted obtainable, carrying as it then did various important duties and entitlements.

All this only increased Annabella's attachment to the man she now called her 'dearest son'. The two were so hugger-mugger that Ada's only chance of getting a look-in was to start taking an interest in William's plans for Ockham School.

Her efforts were not wholly appreciated. William gave her the job of setting up the gymnasium, and it appears she overspent the budget, ordered the wrong supplies, misdirected the workmen, missed the deadline for completion and blamed the resulting mess on William's attempts to intervene.

She turned out to be far better at putting together a curriculum for the school's academic courses. She started looking around for suitable books, and was concerned to find that there were none on subjects like geography and domestic science suitable for young children. She also felt she needed a book for the parents who were to send their children to Ockham School. For them, mostly farming folk, the idea of education that was neither religious instruction nor craft apprenticeship would have been regarded as something of a novelty. They needed to be told what it was for.

Another subject area she wanted the school to embrace was one that many would have considered dangerously unorthodox for young minds: phrenology. Phrenology's champion, George Combe, had described the discipline as 'one of the most important discoveries ever communicated to mankind', nothing less than a 'system of Philosophy of the Human Mind, founded on the physiology of the brain'. This, too, the children must learn about, not least because Combe's books rivalled the Bible, in terms of popularity as well as philosophy.

*

The great French thinker Descartes was, as a nineteenth-century biographer put it, the first philosopher to 'enter the human mind with the torch of analysis in his hand'. The phrenologists, led by the field's founders Franz Joseph Gall and Johann Casper Spurzheim, followed behind with the measuring tape of empirical science.

Indeed, the phrenologists challenged the 'dualism' that underlaid Descartes' philosophy, the idea that the body belonged to the material realm while the mind existed only in the spiritual one. Descartes considered that this could be demonstrated merely through the act of self-contemplation – the famous 'I think, therefore I am', the ignition spark that could be used to light up the mental realm without having to resort to any knowledge of the physical one.

'But,' objected Combe,

> the mind, as it exists in this world, cannot *by itself* become an object of philosophical investigation. Placed in a material world, it cannot act or be acted upon, but through the medium of an organic apparatus. The soul sparkling in the eye of beauty trans-mits its sweet influence to a kindred spirit only through the fila-ments of an optic nerve; and even the bursts of eloquence which flow from the lips of an impassioned orator, when mind appears to transfuse itself almost directly into mind, emanate from, and are transmitted to, corporeal beings, through a voluminous apparatus of organs.

A moment of Descartes' vaunted introspection surely revealed the truth of this, he argued. Lowering clouds and a stormy sky can make us depressed or edgy, famine or disease can crush the spirit. The state of the world clearly influenced the state of the mind – no more so, as other scientists had already noted, than in this period of frantic industrial and social change. A veritable epidemic of nervous diseases was, according to the physician George Cheyne, released by modern humanity's habit of ransacking the globe for 'its whole Stock of Materials for Riot, Luxury, and to provoke Excess'. Ada agreed, at one point ascribing her own nervous problems to 'the high-pressure of the present age & epoch & state of society'.

In early nineteenth-century Britain, to suggest too strong a link between the mental and the physical was dangerous. You could be accused of supporting that most denigrated (and, worse, French) species of philosophy, 'materialism', which saw the world having no moral or spiritual dimension at all. To avoid this label, phrenologists claimed that their method shed no light on the nature of the connection of mind and matter. 'The phrenologist [only] regards man as he exists in this world,' Combe asserted, 'and desires to investigate the laws which regulate the connexion between the mind and its organs, but without attempting to discover the essence of either, or the manner in which they are united.'

The principal idea behind phrenology was that the brain was, like the rest of the body, made up of separate organs, and the mind arose from the activity of these organs. A variety of observations confirmed this, such as 'partial idiocy', where 'idiots' were shown to possess normal moral faculties despite having impaired intellectual ones. Each organ, it was further argued, is associated with a particular 'faculty' or behavioural tendency, and the larger the organ, the better developed the faculty. The relative size of organs could be measured simply by studying the shape of the skull, which would bulge over the bits of the brain containing the better-developed organs, and dip over the bits containing underdeveloped organs. Thus it was possible to get the measure of an individual's character simply by feeling his or her cranial bumps and comparing the result with a phrenological map showing the location of the mental organs. These maps, taking the form of porcelain heads with labelled regions drawn across the scalp, are now quaint antiques, but in the 1830s were regarded as important scientific instruments.

Like geologists and botanists, the phrenologists had a passion for classification, and Combe's books provide a complex system for dividing up the faculties. There are two orders (feelings and the intellect) which are subdivided into genera (such as 'propensities' and 'sentiments') and further subdivided into particular faculties, ranging from taste to 'Amativeness'.

The phrenological agenda was an extremely ambitious and ultimately influential one. It gave doctors the power to diagnose psychological as well as physiological disorders. For example, the phrenological term 'Amativeness' allowed for the first time the medical examination of perversion, as it replaced the traditional notion of an 'instinct for propagation' with the modern idea of sexual urges that could express themselves in any number of ways, such as through homosexuality and masturbation. Furthermore, because the organ of Amativeness was located in the cerebellum – the largest part of the brain and capable of interacting with all the others – sexuality was allowed to 'flood into all areas of life', as one historian has put it. When it was well developed, and when the 'moral and intellectual organs be weak', as Combe warned, the patient was liable to 'become a deceiver, destroyer, and sensual fiend of the most hideous description'.

*

It was Annabella who first introduced Ada's head to the phrenologist's fingerwork. There is no record of when she attended her first session, but it was likely to have been in 1831. The phrenologist was one Dr Deville. As might be guessed from his rather unconvincingly theatrical name, he was not the most respected practitioner of the science in medical circles. He supplied casts of interesting craniums to Phrenological Society set up by George Combe in 1822, but was regarded by at least one eminent member of the Society, the physician John Elliotson, as a 'vulgar being' who pandered to fashionable sensation-seekers.

Deville's procedure for a phrenological consultation would have been simple. The patient would be received in his consulting room, which was probably the drawing room of his house, and invited to sit in a chair. Deville would then examine the patient's head, at first simply by assessing its overall shape. His fingertips would then be deployed to explore the more subtle topography of the skull, looking for bumps and indentations at significant points.

When this procedure was performed on Ada, the results alarmed both Annabella and her friends. The girl's Imagination, Wonder, Constructiveness, Harmony and Intellectual faculties were found to be highly developed. In short, she appeared to have the head of a poet. As Sophia Frend observed, this simply did not correspond with what was known (or rather expected) of Ada's character. However, after some further analysis was performed on the results, a solution was eventually found to this 'difficulty'. 'Language was not so powerful as the others,' Sophia announced with relief. 'Music and mathematics were indicated in this head, and were both strong elements of character.'

Annabella had accompanied Ada on her first visit to Deville, and after the consultation was completed Ada was sent from the room and Annabella took her turn in the chair. She proved to be a far more legible subject. Deville found her organ of Sensitiveness to be so swelled as to call for immediate medical treatment – a diagnosis that exactly matched Annabella's feelings of distress following the recent publication of Thomas Moore's biography of Byron. To have the origin and intensity of her feelings so gratifying acknowledged by her bumps assured Annabella's complete acceptance of phrenology. Her letters subsequently became saturated with references to her faculties and organs, and she developed the habit of judging everyone she encountered by the shape of their head. For example, when she met Cardinal Newman, the leader of the conservative Oxford Movement promoting the return of Anglicanism to High Church traditions, she noted his dome to be strangely irregular, indicating a lack of Logic and Benevolence and an excess of Veneration. During a prison visit, she observed that the area of the skull above a convict's ears was stretched almost to translucency by the swelling of the organ beneath – the organ of Destructiveness, which apparently explained why the man had stabbed a woman without provocation.

Ada had an ambivalent attitude to phrenology. She accepted its scientific principles, but was not so sure about the elaborate theories that had grown up around them. Perhaps this was partly because she had a particular appreciation of the ironies of Combe discussing the faculty of 'Philoprogenitiveness' – the love of having children – with references to poetry by the father who had abandoned her.

In March 1841 she returned Dr Deville's consulting rooms, accompanied by her husband and a friend, the famous Egyptologist Sir John Gardner Wilkinson. This time it was her turn to be unsettled by the doctor's diagnosis. He observed she had an almost pathological fear of criticism, which she acknowledged to be accurate. But he also said that her Sentiments predominated over her Intellect.

This was not what Ada wanted to hear. Following the birth of her third child in July 1839, a son she called Ralph, Ada began to get restless again. She found the restraint of motherhood hard to bear, cramping her freedom just as her mother had done. She loved her children, but could not deal with them 'practically in details'. She was not exaggerating. She nearly succeeded in 'murdering', as she put it, her still infant daughter Annabella by overdosing her with calomel, a powerful mercury-based purgative. Certainly the children's nurses, no doubt enjoying a chance to play up their mistress's incompetence, were convinced the poor child would 'sink' under 'so dreadful a dose'.

She decided to immerse herself in her studies again – to develop her Intellect to control those Sentiments. She started a 'Mathematical Scrap Book' which she filled with formulae and problems, and started to look around for a tutor. Mary Somerville was no longer available, as she had moved to Italy in 1838. Ada therefore needed to find someone else in London who could give her some formal instruction, 'but,' she wrote to Charles Babbage, hinting heavily, 'the difficulty is to find the man. I have a peculiar way of learning, & I think it must be a peculiar man to teach me successfully.'

The issue of her peculiarity – something that nearly everyone commented upon – provoked her to add a paragraph about herself that must have taken Babbage by surprise, even shocked him a little. 'Do not reckon me conceited,' she wrote,

> but I believe I have the power of going just as far as I like in such pursuits, & where there is so very decided a taste, I should almost say a passion, as I have for them, I question if there is not always some portion of natural genius even. At any rate the taste is such that it must be gratified. I mention all this to you because I think

you are or may be in the way of meeting with the right sort of person, & I am sure you have at any rate the will to give me any assistance in your power.

This letter provides the first glimpse of a preoccupation that dominates her correspondence in the coming years – her peculiarity. She became increasingly aware of her uniqueness. It was as though she was the product of a cross-breeding experiment, an attempt to combine Sentiment and Intellect to produce a new type of human just as agricultural scientists were trying to combine fast growth and fertility to produce a new type of pig. She now needed to know how successful the experiment had been, whether it had resulted in a viable new breed, or a monster.

In her struggle to find an answer to this question, the word she tentatively, almost apologetically used to describe her attributes to Babbage – 'genius' – would become a familiar part of her vocabulary. In moments of mania, which now began to come with increasing frequency, she meant it in the familiar sense of some sort of super-human power, and her use of it in this way, as in clearer moments she acknowledged, did sound conceited. Like her father, she was susceptible to that 'no less imperious passion' Self-Love, as he described it. She described herself as poetical genius, a metaphysical genius, a musical and dramatic genius, even a skating genius.

At other times, however, she seemed to regard her genius as much an affliction as a gift, something that alienated her from everything that was familiar and comfortable, including her own family. When her preoccupation with her genius was at its most intense, she could hardly bear to be in the company of her husband or her children, and yearned to escape.

*

After some months of searching, a man was eventually found to help her develop her mathematical skills. He was the not particularly peculiar though very accomplished mathematician and logician Augustus De Morgan. The connection was made through his wife, Sophia, the daughter of William Frend, Annabella's now very aged former tutor. The link was not an altogether fortunate one, as it prolonged Ada's association with Sophia, who continued to accept Ada's confidences simply so she could offer them as supplications to the revered Annabella. Her husband Augustus, however, turned out to be a good tutor: patient and supportive.

With Augustus at her side, Ada was now ready to embark on a voyage of self- as well as scientific discovery, to find herself among the higher reaches of pure mathematics. But at the very moment she was to begin her ascent, she was pulled back. She could not go there until she had passed through the darker regions of her 'natural genius', until she had undertaken a journey that would take her into the depths of her past, and to the edge of sanity.

CHAPTER SIX

The Deformed Transformed

A BITTERLY COLD WINTER set in during the dying months of 1840. 'Mathematical weather', Ada called it, perfect for confining herself indoors and getting on with her studies. She did just that, spending most of December working away at her exercises in her Mathematical Scrapbook.

Outside, a strong, incessant east wind blasted the frozen Surrey landscape. Ada hated the east wind. It had an almost mystical power for her. Whenever it blew, she felt her life being disturbed. The situation was not helped by the 'plaguing' children. Shut into the house together for weeks, they were becoming unmanageable. Young Byron, Ada's older boy, developed a taste for terrifying his little sister Annabella, telling her that God made good little children come alive again if they died, but chopped naughty ones up into pieces.

There was a brief respite on Christmas Eve. A lake on the Ockham estate froze over, and Ada and William went out to learn how to skate on it. Ada became a great enthusiast, almost regretting the New Year bringing a thaw that 'injured' the ice.

The east wind subsided, but there was another source of disturbance blowing Ada's way, coming from the south. Annabella was in Paris, and there a story was unfolding whose profound consequences Annabella was beginning to spell out to her daughter.

Several months earlier, in July 1840, Annabella had received a begging letter from the lawyer Sir George Stephen. Sir George was after money not for himself but for one of his clients, a woman living in France – Elizabeth Medora Leigh, the daughter of Augusta Leigh who as an infant Annabella had observed arousing a particularly tender look in her husband's face.

It was not in Annabella's nature to grant anyone an ex gratia payment. Her purse had very long strings. However, she quickly realized that it might be fruitful to allow Medora to become entangled in them, so she set off for France to see what she could do.

In the French city of Tours Annabella found a twenty-six-year-old woman both destitute and ill, possibly with consumption. Medora must have presented a pathetic picture to Annabella, who was moved to take the young woman under her protection. She instructed her to take an assumed name and come with her to Paris, and Medora, a model of pliability and gratitude, did as she was told.

A young woman accepting and *appreciating* Annabella's help and guidance – it was such an unfamiliar experience that it aroused maternal feelings that almost overwhelmed her ... 'feelings that have long lain, like buried forests, beneath the moss of years, are called forth,' as she put it to Olivia Acheson in a rapturous letter dispatched during a stop-off at Fontainebleau. 'This is a little gleam in my life – it will not last, but its memory will be sweet. The object of my affection is marked for an early grave ...' These were sentiments she never came close to expressing about her own daughter – who, being a friend of Olivia Acheson's, may well have come to learn of them.

Annabella's outpouring was not aroused simply by Medora's pathetic state. It had, like Medora herself, a more questionable parentage. Annabella's relationship with Augusta, the woman she had so enthusiastically embraced as a sister during her marriage to Byron, had deteriorated badly following the Separation, and since 1835 had more or less collapsed. Augusta had made overtures, but, with her typical lack of tact, had only succeeded in making Annabella yet more hostile. Her worst mistake was to write to Annabella that she wished to be able to approach Ada in society, and would assume that Annabella consented to this if she did not reply. Forced to break a silence it indulged her sense of moral superiority to keep, Annabella sent a contemptuous refusal. Subsequent, more desperate entreaties for reconciliation were met with curt dismissals.

Now Annabella had within her complete control (or so she thought) the cursed woman's daughter. Not just that, but the daughter was ready to wreak revenge on her mother, and had the means of doing it – the conviction, ruinous to Augusta if ever revealed in public, that she was the product of Byron's incestuous union. In other words, to have a hand in this girl's salvation was to hold the means of Augusta's oppression.

Annabella's association with Medora went back to the late 1820s, when she lent Bifrons, her rented home near Canterbury, to Medora's sister Georgiana and Georgiana's new husband, Henry Trevanion.

The mercurial Henry had not been considered a good match for Georgiana, and everyone had tried to prevent their marriage – everyone except Augusta, Medora significantly noted. Despite such reservations, the union had gone ahead, and Georgiana soon became pregnant. She was due to have the baby while staying at Bifrons.

According to Medora's self-vindicating account of her life, which in the 1860s found its way into the hands of Dr Charles Mackay, the editor of the *Illustrated London News*, Georgiana and Henry got bored while they were at Bifrons, and took to inviting Medora down to keep them entertained – or, more accurately, to keep Henry entertained. Apparently Georgiana, whose jealousy was 'actuated' by her younger sister's 'superior beauty and talents', arranged for the innocent fifteen-year-old to be left alone with Henry.

The inevitable happened. 'Some months passed – I was ruined,' as Medora tersely put it.

Her ruination did not go unobserved. A clergyman, the Reverend William Eden, was also staying at Bifrons at the time with his wife Lady Grey (she had kept the title from a previous marriage). They soon guessed what the extraordinary ménage was up to in the other rooms, and lost no time in reporting their findings to George Anson Byron, the seventh lord and the poet's heir, who was handling Annabella's affairs during her absence abroad. It was decided, after consulting Annabella but not Augusta, that Georgiana, Henry and Medora should all be sent abroad, where their tangled affairs could be sorted out beyond society's gaze.

Medora had her child prematurely in Calais, and a Dr Louville took it into his care. Medora then went back to live with her mother at her St James's Palace apartment. Augusta remained totally unaware of what had passed, and set about launching her sixteen-year-old daughter into society.

Meanwhile, George Anson Byron sent an emissary to Calais,

apparently at Annabella's request, to search for Medora's baby. He was sent away empty-handed – though Medora, who never found out what happened to her child, at times deluded herself into believing that it had been secretly recovered and placed under Lady Byron's care.

Back in England, Medora claimed that she was once again thrown into the company of the detested Henry Trevanion, this time not just by her sister but by her *mother*.

It appears Augusta harboured powerful feelings for Henry. When Medora revealed to her that she had become pregnant by Henry twice, the second time under her mother's own roof, Augusta wrote two letters to Henry in which she declared quite openly to him that she had always loved him, and always would. But she also said that in her he would find 'the tenderness, the indulgence of a Mother' – not a lover. She also wrote to Medora, begging her to confide in a mother who doted on her. 'Let me implore of you to be comforted,' she wrote, 'to do your utmost to make the best of circumstances – to trust in my affection. That you are tried, SEVERELY tried, I feel – and I pray God to support you and comfort you and guide you.'

These frantic effusions are not those of a woman offering her daughter as bait to a sexual predator. They seem more like the incoherent outcries of a woman prone to histrionics who had been driven into a state of panic by, among other trials, the threat that her violent husband, Colonel Leigh, would discover what had happened and kill Henry.

Medora was sent away, again with Georgiana and Henry, this time to Bath. With her daughters and son-in-law safely out of harm's way, Augusta decided the time had come to tell her husband what had happened to Medora, his favourite. However, she did not have the nerve to do the job herself, so she delegated one of his most trusted friends, Colonel Henry Wyndham, to break the news on her behalf. Leigh flew into a predictable rage, and started to roam the streets of London in search of Medora with a pistol in his pocket.

He finally found out her whereabouts from Augusta, and raced off to Bath to fetch her back. There he managed to get into the house where she was staying and demanded she return with him to London. She eventually agreed, but not before plotting an escape with Trevanion. Georgiana was also there, and in a scene obviously saturated with uncontrollable emotions promised to give up her husband to her sister. What happened to the child Medora was then supposed to be carrying is unknown. It was probably stillborn before Colonel Leigh arrived.

Leigh dragged Medora back to London – not, as he had led her and Henry to believe, to her mother's apartment in St James's Palace, but to a secret address in Lisson Grove, west London, where she was to remain under house arrest under the surveillance of a Mrs Pollen.

Henry and Georgiana determined that they would save the girl they had jointly ruined from her protective father, and succeeded in tracking her down within a fortnight. With the connivance of Mrs Pollen, they got her out of the house. Henry then promptly eloped with Medora to the Continent, abandoning Georgiana and the three children she had had by him.

The couple spent the next two years living on the Normandy coast as brother and sister under the assumed names M. and Mlle Aubin until they ran out of money. Henry had expensive tastes and a meagre income, which largely comprised Georgiana's allowance from Augusta, which he still received since the law placed all of his wife's income under his control. A relation of Augusta's, Lord Chichester, had written to Henry, demanding that he surrender his wife's money, as Augusta could not afford to make up the deficit in Georgiana's now desperate finances. Henry refused, no doubt because he considered his situation even more desperate.

He embarked on a search for new sources of finances, visiting Jersey to see what he could get out of relatives who lived there. Medora, meanwhile, wrote to her mother asking for an allowance with promises of leaving Henry and becoming a nun. Augusta could only afford £60, a modest sum, but then Medora proposed living a modest lifestyle.

Medora found herself unequal to her reforming aspirations. She spent two months trying to admit herself to a convent, but was every time lured back to Henry's bed, which she found irresistible. Her plan was finally terminated by the discovery that she was pregnant again. She gave birth to a girl in May 1834, who she called Marie, and she and Henry lived together in a derelict chateau in Brittany – truly, Medora unconvincingly claimed, as brother and sister. She now devoted her time to her baby, while he turned his attention to shooting and, Medora reported without a hint of irony, religious studies.

After about a year of this life, Henry momentarily interrupted his theological pursuits to resume his financial ones. He went to England and pawned his marriage settlement. This ingenious transaction was made possible by the Reversionary Interest Society, an organization set up in 1823 which quickly thrived on the expansion of the land-owning classes by offering its impecunious heirs the means to realize their inheritances before they were due. In Trevanion's case, the Society was prepared to give him over £8,000 in return for signing over £11,000 worth of government securities, a portion of the £60,000 put in trust for Augusta's children by Byron. A now very flush Henry returned to Brittany, where Medora claimed the remains of her passionate attachment to him were once more aroused (whether by the money or his physical presence she does not specify). But now, she proudly proclaimed in her memoirs, she was twenty-one and no longer a child, so knew how to resist.

It hardly mattered. The magnificently monstrous Henry found a new mistress and relegated Medora to the role of a servant. By the spring of 1838, Medora, now alone and desperate, fell 'dangerously ill', according to her own diagnosis. She asked the doctor who attended her to help her get away from the man 'I had never loved & who every day convinced me more & more of his utter worthlessness'. The doctor helped her write to Augusta, who eventually sent her the £5 she needed to set up elsewhere, but not the £120 per year that was apparently the absolute minimum she needed to live 'in a very cheap spot'.

Medora now sought a little 'reversionary interest' for herself by offering to sell a deed which Augusta had drawn up in 1839 to provide £3,000 for Medora's child on Augusta's and Annabella's deaths. To do this, she needed to get hold of the original deed itself, which Augusta was refusing to give up. This is the moment she turned to Annabella, and when Ada found herself becoming irreversibly entangled in the whole grisly business.

*

References to the mysterious Medora, mentions of the unmentionable Aunt Augusta, sister of her even more unmentionable father – Ada soon realized that the family history that had lain so deeply buried since her birth was about rise, festering, from the grave. The ground had already been disturbed by Annabella, when she decided to give Ada Byron's inkstand and the portrait in Albanian dress. Now Ada was about to confront the ghost itself.

One of his emissaries came knocking at Annabella's door in 1838 and gave her a terrible fright. It was Fletcher, Byron's valet, who had served his master faithfully until his death. Byron had left Fletcher an allowance, but the maw of Augusta's monetary mishaps had swallowed it up, so he had been forced to take a job as the early nineteenth-century equivalent of a door-to-door salesman. Annabella was intrigued by his situation, and anxious to see if he had now formed the same opinion of Augusta that she had. Disappointingly, he had not. She had 'expensive' children to provide for, he said, and in any case was quite happy to give him a shilling or two in emergencies. This forbearance confounded Annabella, who assumed it to be a sign of simple-mindedness.

Annabella reported the whole episode to Ada, and, making one of the first open allusions to the darkness that still clouded memories of her marriage, confessed to the agitation the mere sight of Fletcher caused. 'No one so forcibly brings your father before me,' she wrote.

Byron was to be brought before Annabella again in France in 1840, when she embarked on her mercy mission to save Medora. The girl, Annabella observed to Sophia De Morgan, looked and moved very like Byron – more so, she claimed, than Ada. Not just that, but she was literally the embodiment (or so she assumed) of Byron's incest with Augusta – the pretext for Annabella making her estrangement from her husband permanent.

By this time, Ada was almost feverish with expectation, an absence of information sucking in a swarm of speculations. Now she was married, she had more or less free access to her father's poetry and the freedom to talk about it to others. She began to spot bio-graphical details hidden in the words, a pattern of allusions – a hidden story, even. It did not take the differential calculus she was still studying to put two and two together and work out what that story might be.

For example, what lay between the notorious lines found in the climactic moments in her father's great dramatic poem *Manfred*? The eponymous hero, driven to distraction by his memory of some 'half-maddening sin', stands by a cataract among the lower reaches of the Alps and summons up the Witch of the mountains. She appears and he confesses to her his love for one who . . .

> . . . was like me in lineaments – her eyes,
> Her hair, her features, all, to the very tone
> Even of her voice, they said were like to mine . . .

The lady was Astarte, named after an incestuous pagan goddess. Later in the poem one of the characters describes her as being Manfred's . . . what? At the moment the revelation is about to fall from his lips, he is interrupted. The truth remains just offstage, like Astarte (the name, incidentally, Ada's son Ralph took for his book about the Separation, which in adulthood he compiled to vindicate his grandmother's role).

There were those other references, that love in *The Giaour*, for example, which will find its way 'Through paths where wolves would fear to prey . . .' And in *The Bride of Abydos*, there were Zuleika's erotic feelings for Selim, the 'companion of her bower, the partner of her infancy' whose 'keen eye shone'

> With thoughts that long in darkness dwelt;
> With thoughts that burn – in rays that melt . . .

What thoughts? What darkness?

Medora herself was there, the wife of Conrad the villainous Corsair, who was redeemed only by his passion for her . . .

> Which only proved, all other virtues gone,
> Not guilt itself could quench this loveliest one!

No wonder, then, that Ada's first contact with Medora produced such an intense response – even though it was through the medium of a pincushion. It had been sent to her by her mother in early January 1841 from Paris, perhaps as a late Christmas or birthday gift. Annabella had probably suggested Medora make it for Ada to demonstrate her feminine virtues and domestic accomplishments. In Ada's hands, however, it took on an extraordinary symbolic power. She beheld it as a metaphor of Medora's life – perhaps her own too: the black background represented the hopelessness of Medora's situation, the red needlework and gold braid the brilliance of her future redemption. In Medora's case, the braid might not glitter in this world but in the next, since that was where Annabella – herself an expert on fatal diseases, since she survived a succession of them – still expected that the consumptive Medora would imminently be.

Ada wrote a letter of effusive thanks for the pincushion via her mother, and added a passionate endorsement of Annabella's behaviour. Annabella had shown herself to be a great judge of character in offering her support to so deserving a cause as Medora, a woman who had proved herself to be 'very remarkable & very exalted' in terrible circumstances.

Such words pleased Annabella no end, and she was now eager to reveal as much as she could about the real reasons for her interest in Medora – and why Ada should be interested too. For in this poor, wretched creature Annabella had at last found the sister Ada had for so long craved.

*

Ada and William were to go to Paris in the spring of 1841 to meet Medora, and in preparation Annabella wrote to them both setting out the background to Medora's story and revealing for the first time her belief that Byron was her father.

Annabella had clearly expected the revelation to come as a thunderbolt. In the event, Ada replied that it came as no surprise at all. She had already guessed at this 'most *strange* & *dreadful* history'. 'You merely *confirm* what I have for *years and years* felt scarcely a doubt about, but should have considered it most improper in me to hint to you that I in any way suspected', she wrote to her mother. She added that she had shared her thoughts with William the previous summer, when Medora had first approached Annabella, and at the time felt ashamed at voicing such '*monstrous* and *hideous*' suspicions. 'I only feel now upon it as I have long done, that the state of mind of both parties [Augusta and Byron] in which such a monstrosity could have originated, is indeed appalling'.

Annabella's response to this view was to deny hastily any responsibility for Ada having formed it. 'I know not, dearest Ada, what I ever said to you that could even suggest the idea of *vice* on [Augusta's] part. Of her want of veracity and artfulness I have spoken. When you alluded to the poems in which there was a remote reference to the fact [of incest], I have always avoided discussing them.' As for Byron, Annabella claimed she had left his 'aberrations' so 'indistinct' as to be accused by others of showing weakness, even of hiding the truth. Now Ada knew the truth, she could only hope that it would benefit both her and the father who was now 'in his purified state'.

One benefit was that mother and daughter could now be honest with one another. They had been kept apart, Annabella claimed, by an obligation on the mother's part to protect her daughter from having to take sides, to declare loyalties – the first duty of any parent following the breakdown of an insolvent marriage. Now Ada had so firmly taken her side, and on the basis of her own surmise rather than anything Annabella had said, the obligation was instantly lifted: 'I can make you the friend of my past as well as of my present life, *without reserve*', Annabella triumphantly declared.

Providing an immediate example of this new licence, Annabella invited Ada to 'ask what you please' about the marriage and her father. Ada duly did just that. A few days later she sent a letter to her mother wanting to know how she knew Medora was the 'result' of Byron's and Augusta's relationship. 'It might not have been easy to prove this, or even to feel any degree of certainty about it,' Ada reasonably argued, as Augusta was married at the time, which meant that Medora could well have been Colonel Leigh's child.

The sluice gates opened, the deluge was released. In a long letter, Annabella retold the story of Byron's behaviour during their marriage. There were the passages he read to her from Augusta's letters 'with great agitation on his part and hints of something fearful & mysterious'; the rage he flew into when she innocently (always innocently) remarked, as they stared at their joint reflection in a mirror, that similarities between them noted by Byron meant they could be brother and sister;

the references to William Godwin's novel *Caleb Williams* and the threat to 'persecute me for *ever*'; the times she was turned out of her own drawing room with 'bitter sarcasm' so that Byron and Augusta could 'amuse' each other alone. There had been the same 'remorseful expressions' even before their marriage, but Annabella had put them down to an overactive imagination, and throughout the whole grisly episode had given him the benefit of the doubt and kept her composure 'out of consideration for *the unborn*' – i.e., the young woman who now read these revelations. 'Poor thing!' Annabella concluded, 'you have a right to the excitability which appeared in you so early'.

*

Excitability was the word. Ada's response to the revelation of her father's sins was a powerful one – at first, outrage at his behaviour, then, quite unexpectedly, a dangerous glint of identification.

Her mother had effectively confirmed what she had guessed, that the poetry which her marriage had at last allowed her to read was biographical. She began to recognize the features of her father in Childe Harold and Conrad, Manfred and Don Juan, as one might see the features of a father in his daughter, or a brother in his sister.

In particular, she noted the theme of defiance. It was in the poetry. It was in the blood.

Defiance is one of the great themes of Byron's work, indeed of Romanticism. It was the quality shown by that icon of the Romantic movement, Prometheus, the Greek god who defied Zeus by stealing fire from the heavens and bringing it down to earth, whose

> . . . Godlike crime was to be kind,
>   To render with thy precepts less
>   The sum of human wretchedness
> And strengthen Man with his own mind . . .

as Byron himself expressed it in his poem celebrating the great Titan.

Ada had been strengthened with her own mind. She had a 'natural defiance of the law, of everything imposed', as she put it, and it made her tremble with excitement and terror to discover where it came from. All that prevented her from sharing the same fate as her 'unhappy parent' – Prometheus himself – was her scientific and mathematical understanding.

She resorted to phrenological language in an attempt to explain herself to her mother: she would be saved by her 'Causality', the 'reflective' faculty of abstracting laws from experience, and 'Hope', the 'superior' (i.e., unique to humans) sentiment of optimism, a belief in progress. 'Pray do not think the Bird is going mad; which you know has often been my own horror,' she told her mother – but something very like madness now seemed to grip her.

She announced that she was to embark on an extraordinary quest to save herself and redeem her father. 'I have an ambition to make a compensation to mankind for his misused genius,' she declared. 'If he has transmitted to me any portion of that genius, I would use it to bring out great truths & principles. I think he has bequeathed this task to me!' Through science she would overcome the profanities of art.

To her faithful confidant and lawyer Woronzow Greig she sent a letter outlining how she planned to accomplish this act of redemption. She had been divinely ordained for a 'peculiar intellectual-moral mission', she announced. Greig had been trying to persuade her to produce an original work, but it was, she felt, too early. Resorting to economic allusions, she saw the current phase of her life as an investment that would produce a proper return only in the long term. She was doing the intellectual equivalent of what all the railway investors were then doing, laying down the infrastructure for a system that might one day change the world.

She would fulfil this mission by deploying what she called her 'scientific Trinity' (a very Unitarian concept, calculated to tease Trinitarian Anglicans). This comprised intuition (which she also called 'tact', using the word in a now obsolete sense to mean an exceptionally sensitive form of perception), reasoning and concentration (in the phrenological sense, meaning not just to an ability to direct the mind to a particular subject, but to bring a lot of disparate ideas together, 'throw rays from every quarter of the universe into one vast focus', as Ada put it). She saw this Trinity as a huge 'engine' that Providence had put into her hands, and it was her responsibility to see what, in the next twenty years she saw allotted to her, she could achieve with it. It was a wonderful challenge, and a way of tapping the excess of energy she had which, as she now admitted to Annabella, 'has been the plague of your life and my own'.

Some of that energy she immediately directed towards analysing that great preoccupation of the Romantics and bugbear of her mother's, the imagination. All that remains of her effort is a fragment of an essay, made all the more intriguing by its provisional nature and truncated form.

As she pointed out in the opening sentence, imagination was usually discussed exclusively in terms of poetry and art. In this context it appeared ineffable, an occult power that, as Byron himself put it, welled up within the soul, threatening to erupt violently if it was not released through the pressure-valve of poetry. Ada saw this force as more benign, less like bubbling mental magma, more like the wings that during her childhood she had dreamed of strapping to her back so she could take flight and leave her earthly cares and constraints behind.

Ada identified imagination as having two jobs: combining disparate ideas and experiences into new combinations, and envisioning things that are otherwise invisible. In the latter respect, it is the key to religion, as it provides us with the means of seeing God, allowing us to 'live in the tone of the eternal'.

Ada, however, was not interested in religion. She was interested in what happens when the two very different aspects of imagination she had identified are wedded to one another. Then it becomes an instrument for reconciling poetry with physics, the artistic impulse with the scientific method ... her father with her mother, she implied. When this reconciliation was achieved, then it would reveal unseen worlds – not fantasy worlds, but scientific ones: the world stretching back millions of years revealed by geology, the world filled with a multiplicity of species revealed by zoology, the world of pure three-dimensional forms and imaginary numbers revealed by mathematics. Scientists stand at the threshold of these extraordinary, extrasensory landscapes, she claimed, and only if they allow themselves to be carried away on 'the fair white wings of Imagination' can they hope to 'soar further into the unexplored amidst which we live'.

This notion of the visible world being but a glimpse of a vaster, invisible one came to saturate her thoughts from now on. But, she gradually began to realize, the 'dark things of the world' that lay beyond her immediate perception were not just the geometric objects of some rational, scientific universe. They were demons and ghosts – the denizens of Bedlam.

*

On 1 March 1841 Ada wrote a letter to Sophia De Morgan following the death of Sophia's father, William Frend, Annabella's old tutor. Sophia did not like Ada, but nevertheless kept up a correspondence, if only to gather fresh ingredients for the witches' brew she and the other members of Annabella's coven of Furies continually stirred.

Ada's letter is written in a tone quite unfitted to a condolence. She begins appropriately enough with a few kind remarks about Sophia's father, but then suddenly lurches into a discussion on the theme of Death, as though Sophia needed a lecture on the subject. She proceeds with a discourse on how she sometimes feels *she* has died, and thinks that her 'remarkable tact' enables her to understand what it is like to make the transition.

This remarkably tactless letter provides one of the clearest signs that some sort of mania or hysteria – words that she was later to use when she came to recognize what was happening to her at this time – had now taken possession of her. The first signs of it became evident in the summer of 1840, coinciding with Annabella's departure to France to rescue Medora, and its escalation can be almost exactly plotted against the escalation of the Medora affair. Ada herself became aware around this time that something was wrong with her – some 'extraordinary illness', a 'Hydra-headed monster' that exposed her to 'no end of manias and whims' and which was 'no sooner vanquished in one shape, than it has sprung up in another'. She consulted her doctor, Locock, about it, but kept the details from her mother.

Her mother, however, must have guessed that something was wrong from the letters Ada was pouring out to her, letters which she wrote in such a state of excitement her face would swell up. She would tell her about her Promethean powers, the laboratory of her brain, passing from the night of her life into the dawn, her thoughts about the inhabitants of the moon . . .

She was, she seemed to be saying, deformed, deformed by her paternity – but it was something to celebrate, not lament, as her father had celebrated it in a poem he laboured on for years, *The Deformed Transformed*:

> . . . Deformity is daring.
> It is its essence to o'ertake mankind
> By heart and soul, and make itself the equal –
> Aye, the superior of the rest. There is
> A spur in its halt movements, to become
> All that the others cannot, in such things
> As still are free to both, to compensate
> For stepdàme Nature's avarice . . .

To become all that others cannot – this was her destiny, and she struck out on a search to the very edges of science in her efforts to fulfil it. She started to take an interest in occult ideas about the still novel, mysterious phenomenon of electricity. She was under its 'dominion', she confessed; it had taken over her mind and she was desperate to know more about it – 'Pray find out all you can for me, about everything *curious, mysterious, marvellous, electrical,*' she asked her mother. In particular, she wanted to explore its connection with that dark, dangerous but scintillating new scientific obsession of the early nineteenth century: mesmerism.

*

With the benefit of hindsight, it is easy to anoint mesmerism with the same snake oil that smeared other semi-scientific ideas of the time. In fact, traces of mesmerical concepts are to be found throughout modern psychoanalysis and psychotherapy. Hypnotism, for example, was a term coined by the mesmerist James Braid in 1842 to refer to what had previously been called animal magnetism and magnetic sleep. The whole idea of alternative states of consciousness owes its origins to mesmerism.

But mesmerism's influence runs even deeper than that. It established that there could be a connection between the mind and the body that was neither completely physical (and therefore the preserve of hard science) nor wholly metaphysical (and therefore the preserve of philosophy) – i.e., that was psychological. By attempting to open up this intermediary realm to scientific investigation, it made possible the sort of study of the mind and behaviour that Freud would eventually formalize at the end of the nineteenth century and bequeath to the twentieth.

So we cannot afford to be smug when we look back upon the herd of rich Victorian women who threw themselves into the embrace, literally, of charismatic young mesmerists. Nor can we lump all who showed an interest in mesmerism into the same credulous group. Ada for one approached the whole field with a degree of scepticism, even at the height of her mania. She demanded the same standard of evidence that other scientists, such as Michael Faraday, were then asking for. She openly attacked the most marginal forms of it (the 'new mesmerism') which embraced such ideas as 'spiritual telegraphy'. But she was also convinced that properly conducted mesmerical experiments were providing glimpses of something very important. It was a new mental terrain, perhaps the place where Coleridge's 'deep romantic chasm' could coexist with the 'unseen worlds' awaiting scientific discovery, where her peculiarities no longer seemed like madness and her 'natural genius' would finally discover its native habitat.

Mesmerism took its name and method from the work of the Austrian physician Franz Mesmer. In the 1760s he started to take an interest in Newton's idea of gravity as a force that held the whole universe together. It was widely believed at the time that the movement of the stars influenced human health, being capable even of causing disease. He wondered if gravity might be the reason for this, and postulated the existence of a special sort of 'universal gravity' that somehow connected the 'astral plane' with the human one, harmonizing the two like musical instruments (he was a keen musician, an accomplished player of the glass harmonica and a patron of Mozart).

In his efforts to understanding the nature of universal gravity, he started to experiment with magnets, which, like gravity, caused one object to be drawn towards another. He became convinced that there was a 'magnetic fluid', some invisible force that somehow could influence the material realm, which coursed through the human body. This fluid was the vital principle that distinguished living organisms from dead matter, the elusive 'quintessence' that animated life. It was also the cause of disease, because any disturbance in its flow produced a corresponding disturbance in the body.

His idea might have remained an academic curiosity had he not applied it to one of his patients. Franzl Oesterlin was a casualty of the epidemic of a particular nervous disease doctors then observed sweeping across Europe, and in particular its female inhabitants: hysteria. Franzl had all the usual symptoms – convulsions, stomach cramps, faintings, hallucinations and paralysis. Conventional treatments, which amounted to little more than prescribing ever larger doses of opium, had failed to relieve her condition, so, in desperation, Mesmer began to consider some less conventional ones.

The idea of female bodies, particularly nubile ones, being especially sensitive to the resonances of the music of the spheres was deeply embedded in popular lore – it was evident in the connections of menstruation and the moon. So it was but a short and reasonable step for Mesmer to apply his ideas of universal gravitation and magnetic fluid to a young woman in such a state of psychic disharmony.

If his theory was correct, the flow of magnetic fluid through Fräulein Oesterlin's body had somehow become disturbed in the region of her womb. So perhaps if he applied magnets around that area a more regular flow could be restored.

He managed to get a supply of powerful iron magnets from a Jesuit priest, and used these to 'magnetize' her body. He placed one magnet on her stomach and one on each of her legs and stood back to see what would happen. Initially, she reacted violently. The magnets seemed to be making her condition worse – which at least proved that they were having some sort of effect. Then, without warning, her condition began to improve. Symptoms that had assailed her for years disappeared for several hours. It was a triumph, if a temporary one, and Mesmer was convinced his magnets were responsible. He explained his patient's initial reaction as a 'beneficial crisis' that marked the body's magnetic 'tides' being restored to a state of equilibrium.

This astonishing result was widely reported in German newspapers, and both Mesmer and the new discipline named after him, Mesmerism, became an instant hit. Mesmer began to claim that he did not even need magnets to heal his patients, as he was himself magnetic, and could magnetize anyone or anything he touched – silk, leather, stone, glass, water, wood, even dogs. Everyone had this power – this 'animal' (as opposed to mineral) magnetism – but some had it more than others, and he had it in abundance. Not everyone could be magnetized, however; it only worked on those whose magnetic fluid was in some way blocked. Nor was every illness caused by such blockages; it seemed to be only those associated with the convulsive and hysterical symptoms found particularly in women.

Some remained unconvinced by his ideas. In a celebrated case involving a blind young female musician whose sight he claimed to have restored, several eminent physicians contested his results, and the girl's father withdrew her from Mesmer's care. He was forced to move his practice to Paris. There, however, more sceptics awaited him. A doctor disguised himself as a patient, and when Mesmer tried to treat symptoms the impostor had falsely described, denounced Mesmer as a fraud.

Despite these setbacks, the force of Mesmer's magnetism prevailed. As the denunciations continued, including one from a commission set up by the King of France and led by the celebrated scientist Benjamin Franklin, new recruits continued to be drawn into the field, each one trying out different methods of magnetizing and reporting miracle cures.

One of the most popular and influential of these disciples was the marquis de Puységur. In 1784, the marquis went to the aid of one of the peasants on his estate, who was suffering from a dangerous fever. Puységur tried to magnetize the young man and, after a few minutes, found that his patient had apparently fallen asleep. But this was no normal sleep. Still slumped in Puységur's arms, the peasant began to babble about his problems, and became increasingly agitated. Puységur attempted to soothe him by suggesting happier subjects, such as winning a shooting prize or dancing at a party, whereupon the man calmed down.

Puységur realized that he had encountered an important phenomenon: a new state of consciousness that lay somewhere between wakefulness and sleep. He subsequently found that he was able to induce this waking sleep in a variety of subjects, and noted certain other features about it: patients would become extremely suggestible, doing exactly what they were told to do by the mesmerist, and when they awoke they would instantly forget everything they had done while 'asleep'.

Puységur called this condition 'magnetic sleep', and it became one of the key features of mesmerical therapy. It also aroused increasing concern about the powers that mesmerists and magnetizers could exercise over their mostly female patients. In a secret annexe to the report Franklin's committee drew up for the King of France, the potential for mesmerism and magnetic sleep to be abused was spelt out in an account of a typical healing session:

> Ordinarily the magnetiser has the knees of the woman gripped between his own; the knees and all the parts below are therefore in contact. The hand is applied to the diaphragm area and sometimes lower, over the ovaries. So touch is being applied to many areas at once, and in the vicinity of some of the most sensitive parts of the body . . . All physical impressions are instantaneously shared and one would expect the mutual attraction of the sexes to be at its height. It is not surprising that the senses are inflamed. The imagination, which is active, fills the whole machine [body] with a kind of confusion. It suppresses judgement and diverts attention, so that women cannot give an account of what they are experiencing and are ignorant of their state.

The author of the report, Jean Sylvain Bailly, went on to recount in even more graphic detail the 'crisis' that the patient suffers as the intensity of the magnetic experience reaches its climax:

> When this kind of crisis is coming, the face reddens and the eyes become ardent – this is a sign from nature that desire is present. One notices that the woman lowers her head and puts her hand in front of her eyes to cover them (her habitual modesty is awakened and causes her to want to hide herself). Now the crisis continues

and the eye is troubled. This is an unequivocal sign of a total disorder of the senses. This disorder is not perceived by the woman but is obvious to the medical observer. When this sign occurs, the pupils become moist and breathing becomes shallow and uneven. Then convulsions occur, with sudden and short movements of the arms and legs or the whole body. With lively and sensitive women, a convulsion often occurs as the final degree and the termination of the sweetest emotions. This state is followed by languor, a weakness, and a sort of sleep of the senses which is the rest needed after a strong agitation.

Despite the growing concerns about the potential for abuse – concerns that Puységur himself voiced – mesmerism's spread across Europe seemed unstoppable by the turn of the century. The historian Fred Kaplan attributes its popularity to the way it 'anticipated the coming psychological crisis of Western man, the new illness of the nerves and the nervous system that was to become the characteristic disease of a society attempting to grapple with the problems of personal and public identity that the new age forced on many of Victoria's subjects'. It certainly succeeded in putting hysteria on the medical map, and indelibly linking it and many other female conditions with sex.

The one country that proved most resistant to mesmerism during its first flush of popularity in the late eighteenth century was England. There it did not really have any impact until the 1830s, when the chemist Richard Chevenix, who had been taught Puységur's methods, demonstrated them before an audience at St Thomas's Hospital in London that included among its number a well-known and popular doctor, and an acquaintance of Ada's, called John Elliotson.

*

John Elliotson was the man who set off the 'mesmerical mania' that gripped London in the 1830s and 1840s and for a time held Ada in its sway. He was a respected physician and medical academic: Professor of the Principles and Practices of Medicine at the new University College, and President of the Medical and Chirurgical Society of London, the precursor of the Royal Society of Medicine. Contemporary biographies hailed him for introducing clinical teaching in London hospitals, for his advocacy of the 'use of prussic acid, or iron in large doses, of creosote in nausea and vomiting' and for championing auscultation, the practice then regarded as eccentric of listening to the chest to establish the condition of the organs within – using a stethoscope, we would now call it.

He was also a great innovator and modernizer. As *Munk's Roll*, the standard work of medical biography, put it, he 'accepted nothing on the ground of authority or antiquity, and rejected nothing merely because it was new' . . . such as phrenology, of which he had been an enthusiastic supporter when George Combe set up the Phrenological Society in 1822.

This free-thinking bent (something he shared with Ada) expressed itself in a number of ways, not least in his liking for unconventional clothes (something else he shared with Ada). He insisted on wearing trousers at a time when knee-breeches and stockings were still considered to be the proper attire of medical men. He also sported rather alarming side-whiskers.

Such a character was bound to attract hostility as well as admiration, especially in a profession that is by nature conservative and cautious. His first serious run-in with the authorities came in 1837, when Elliotson invited the French mesmerist baron Du Potet to University College Hospital to demonstrate the medical uses of magnetic sleep, in particular to anaesthetize patients undergoing operations, and to treat hysterical and epileptic women. For a while, Du Potet's presence in the hospital was tolerated. However, the hospital's management board began to get anxious about the public attention this rather eccentric French baron was attracting, and eventually ordered him to practise elsewhere.

Elliotson was, typically, unconcerned by the controversy and started to conduct experiments of his own at the University College Hospital. There he was observed by Annabella's old physician, Herbert Mayo, then the very eminent Professor of Comparative Anatomy at the Royal College of Surgeons. Mayo was intrigued by what he saw, and said so, providing Elliotson not quite with an endorsement for his views, but certainly some encouragement to continue experimenting. Elliotson also managed to persuade the *Lancet* – then, as now, a leading medical journal – to publish an essay he had written expressing his enthusiasm for Du Potet's methods.

In 1838 Elliotson's association with mesmerism took a decisive turn when he encountered two sisters at University College Hospital, where they were being treated for severe epilepsy.

Elizabeth Okey, then about fourteen, and her sister Jane, twelve, were almost wild – small, dark creatures who jumped and danced around the room, gibbering away in a nonsensical language. Elizabeth, some male observers would subsequently note, was also strikingly beautiful.

Elliotson discovered that they were both exceptionally susceptible to being magnetized. A mere pass of the hand before their faces would put them into a deep trance, and in that state they were under his total control.

So impressive was this transformation that he decided to make them the centrepiece of a series of public demonstrations at University College Hospital, celebrated events that attracted the likes of Michael Faraday, who was already deep into his experiments linking electricity and magnetism, and Charles Dickens, who later indulged in his own mesmerical experiments, and appointed Elliotson as his family doctor. Elliotson also published a series of reports on his experiments with the Okeys in the *Lancet*.

The demonstrations began with him putting them into a deep magnetic sleep, instantly transforming them from dancing dervishes into compliant angels. While under, he would then submit them to a variety of sensations. He showed, for example, how it was possible to push a pin into Elizabeth's neck without her noticing. He also sought to establish a medical as well as literal connection between animal magnetism and electricity by wiring up the unfortunate Elizabeth to a generator to show how her body would convulse when exposed to shocks, without her apparently being conscious of any pain.

Then came the demonstrations of more theatrical and less medical mesmerical phenomena – Elizabeth's clairvoyance, which enabled her to 'see' bread through the back of her head when she was hungry,

Jane's susceptibility to being mesmerized merely by holding a shilling coin that Elliotson had previously magnetized.

It was apparently this shilling experiment that particularly caught the eye of Annabella when she attended one of the demonstrations at University College Hospital, which she reported to Ada in 1838. Ada was intrigued, and discussed the idea with various scientific friends, including Babbage and Faraday who both, at the time and like her, were suspending judgement on the mesmerical mania.

Ada also sought the opinion of Charles Locock, the doctor who had delivered her children. Locock was then on the threshold of becoming a celebrated accoucheur (what would later be called an obstetrician) and attended at some of Queen Victoria's confinements. He also exercised a great deal of influence over Ada, and she regularly referred to him about her own illnesses as well as new ways of treating them. Like Mayo, Locock was interested in mesmerism. He dismissed the more mystical claims such as clairvoyance, but believed that there must be some sort of physical phenomenon underlying animal magnetism. Ada agreed with this, and became increasingly convinced that the link was electricity.

She started to dabble with mesmerical experiments, trying out her own 'shilling experiment' on her sister-in-law Hester and herself. The coin itself seemed to take on a life of its own, oscillating violently in the glass in which it was contained, and sending tingling and throbbing sensations through her fingers. She reported feeling a 'gentle current' flowing through her head, a sensation similar to ones she had experienced during previous nervous attacks. This seemed to convince her of the potential of mesmerism to explore if not treat the 'derangements' she was now experiencing.

Elliotson, meanwhile, was having to experience some unpleasant sensations of his own. The Physiological Committee of the Royal Society, of which he was a member, decided to investigate some of his claims concerning the Okey sisters. At around the same time, an anonymous article appeared in the hitherto sympathetic *Lancet*, claiming to offer evidence that many mesmerical demonstrations were fraudulent. This elicited a series of sympathetic letters. The tide was turning.

To settle the issue, Thomas Wakely, editor of the *Lancet* and a friend of Elliotson's, arranged for Elliotson to perform a mesmerical session with the Okey sisters at Wakely's house in Bedford Square. There were to be ten witnesses, five nominated by Elliotson, the other five by his critics. One of the ten was Herbert Mayo.

Using a small piece of magnetized nickel, Elliotson managed to make Elizabeth become 'violently flushed, the eyes were convulsed into a startling squint, she fell back in the chair, her breathing was hurried, her limbs were rigid'. He then invited Wakely to have a go at using it himself. Wakely got an even stronger response, and Elliotson proclaimed himself vindicated.

Afterwards, Wakely privately told his friend that when he had touched the girl, the magnetized nickel had not been in his hand; he had secretly passed it to an associate who then stood some way off. Elliotson tried desperately to account for her reaction, suggesting some sort of transference between Wakely and the distant lump of metal, but it was hopeless. Mayo reported the results in the *Lancet* and Wakely wrote two editorials denouncing Elliotson's claims.

Despite the accusations, Elliotson was allowed to retain his post at University College Hospital on the understanding that he stopped holding his mesmerical demonstrations. He agreed, but started to take Elizabeth Okey into the male medical wards at night, where he would submit patients to her clairvoyant examination. One night, she claimed to see the Angel of Death standing next to one of the beds. Its occupant died soon after, setting off a furore through the rest of the ward. This was the final straw, and Elliotson was dismissed.

The indefatigable doctor soldiered on. He founded a London Mesmerical Infirmary and a journal on mesmerical science called the *Zoist*. He also continued to perform demonstrations with the Okeys at his elegant home in Conduit Street off Regent Street. There, his practices started to attract the interest of a variety of people, from sensation seekers to often desperate individuals in search of a cure for nervous disorders.

*

Ada started to make her visits to Elliotson in the spring of 1841, the time the Medora issue was erupting with its greatest intensity. During the initial consultations, Elliotson gave her a demonstration of his methods, on the first occasion using the Okey sisters, on the second a boy he had discovered in Belgium.

Such demonstrations aroused a great deal of public curiosity, which was reflected in the appearance of a lurid little pamphlet entitled *A Full Discovery of the Strange Practices of Dr. Elliotson*. This suggestively featured on its cover a blindfolded woman in a faint being held by three men and contained in its preamble the promise of sexual revelations:

> On the bodies of his FEMALE PATIENTS! At his house in Conduit Street, Hanover Square, with all the secret EXPER-IMENTS HE MAKES UPON THEM and the **Curious Postures they are put into while sitting or standing, when awake or asleep.** The whole as seen BY AN EYE-WITNESS and now fully divulged! &c. &c. &c.

This was followed by a few lines of doggerel:

> That which was call'd witchcraft, and the blackest of crimes,
> Is admired, and sought after, in these latter times,
> Ah 'Mesmer!' once defam'd – could thou again but rise
> Whole multitudes would pay thee, each their sacrifice!

The following report, though unequal to its sensationalist billing, was saturated with latent sexuality, focusing on Jane. 'The first experiment performed was something really startling,' the eye-witness reported.

> I had asked for a glass of water, which the doctor ordered Jane to bring into the room. She had just set down the water bottle and glass on a side table, when Dr Elliotson at the distance of about twenty feet, and unseen by her, by a wave of his open hand transfixed her in the attitude in which she happened to be at the moment.

Later, the visitor decided to have a go at mesmerizing the obliging Jane himself.

> I had, unperceived by Jane, waved my hand behind her, and it
> uniformly and constantly fixed her into rigidity. Other gentlemen
> present took similar opportunities of magnetising her by a pass of
> the hand, and always with the same result, for the power seems to
> reside in anyone . . . Before my departure, I took out my watch and
> held it towards her, as a person holds a watch before a child's face
> to engage its attention. I asked her to kiss the watch, and doing so
> she was instantly fixed in a stupor in a bending attitude . . . She
> recovered by my blowing in her face . . .

It is easy to view such passages as nothing more than examples of
Victorian male domination fantasies, which to a large extent they
probably were. But in Elliotson's time, many doctors and scientists
would have seen a serious purpose in what he was doing. With their
crises and convulsions, waking sleep and magnetic attractions, such
mesmerical experiments supplied a language to start a tentative exploration of female sexuality.

In the early nineteenth century women were not generally regarded,
at least from a medical perspective, as sexual creatures. They did not
have 'needs' like men, just obligations – to have and nurture children,
a commodity that was essential at a time when the economic survival
of families, to a great extent, still relied on the number of offspring it
produced.

However, with the rise of the industrial bourgeoisie came a class of
women who were no longer required to devote their bodies completely
to childbearing, and who were, unlike the vast bulk of the population,
rich enough to have their illnesses treated by a physicians. The medical
profession, which was obligingly expanding both in number and in
expertise to meet this demand, began to observe in these women a
whole new set of symptoms – notably hysteria – that they assumed
must be caused by pent-up erotic passions that would normally find
their release or elimination through childbirth. The startling (to them)
conclusion was that these passions were overwhelmingly powerful.
Indeed these women – perhaps *all* women – were positive steam
engines of sexual energy.

The evidence for this conclusion came from observing what happened when the steam engine was left to run free: nymphomania.

Philippe Pinel, the French pioneer of modern psychiatry and the man who introduced the concept of 'alienation' to the study of madness (which resulted in early psychiatrists being called alienists), described in detail the symptoms of this terrifying disorder:

> Nyphomania is most frequently caused by lascivious reading, by severe restraint and secluded life, by the habit of masturbation, an extreme sensitivity of the uterus, and a skin eruption upon the genital organs ... In the beginning the imagination is constantly obsessed by lascivious or obscure matters. The patient is in a state of sadness and restlessness; she becomes taciturn, seeks solitude, loses sleep and appetite, conducts a private battle between sentiments of modesty and the impulse towards frantic desires. In the second phase she abandons herself to her voluptuous leanings, she stops fighting them, she forgets all the rules of modesty and propriety; her looks and actions are provocative, her gestures indecent; she begins to solicit at the moment of the approach of the first man, she makes efforts to throw herself in his arms. She threatens and flares up if the man tries to resist her. In the third phase her mental alienation is complete, her obscenity disgusting, her fury blind with the only desire to wound and to revile. She is often on fire though without fever, and finally, she manifests all the different symptoms of maniacal condition.

Ada was no nymphomaniac. But she was sexual, unashamedly sexual – sufficiently to be in the eyes of some doctors pathologically so. Could she be an hysteric? She had read the books, such as Marshall Hall's *Lectures on the Nervous System and its Diseases*, and they plotted a terrifying trajectory from the early symptoms of mania – 'an expression of the eye and of the countenance, a manner, a demeanour, a loquacity, which denote the utmost excitement' – to suicide, homicide or nymphomania. She had herself described to Mary Somerville the 'agitated look & manner' that overcame her when she suffered a nervous attack. She also suffered from stomach pains – 'gastritis', she diagnosed it – which were regarded as a common symptom of hysteria (she half-jokingly referred to her stomach as the 'seat of "*original sin*"'). She probably knew of the case studies, too, which described circumstances disturbingly similar to her own, such as a celebrated example cited by Pinel of a young girl 'only just past the age of puberty' who had suffered an hysterical breakdown following an elopement 'with a young man of lowly origin'.

How could she avoid a similar fate? She had already tried small amounts of opium (in the form of laudanum) and it had failed to have any effect. Perhaps mesmerism would work, since its most celebrated successes involved the treatment of hysterical women. Perhaps this would provide her with one last chance to slay the 'Hydra-headed monster' that afflicted her before it destroyed her.

*

We can only imagine how Ada became entangled with mesmerical experiments, as she leaves no personal account of her experiences. However, our job is made a lot easier by a journal kept in July of 1841 by the physician and educationalist Dr James Phillips Kay. Kay – like, as we shall see, so many of the men Ada encountered – quickly developed an overpowering infatuation for her. It drove him to sending her letters which, by the standards of the time, were almost indecent in their candour, especially as they were sent by a bachelor to a married countess.

She was his 'Will o' the Wisp', he wrote in one passionate missive, a 'spirit of the wild', a 'delusive & beautiful light flickering with wayward course over every dangerous pitfall, deep morass and miry slough'. He was married the following year to the heiress of the Shuttleworths of Gawthorpe Hall in Lancashire, and added his bride's name to his own. He later all but admitted that only by submitting to the bonds of matrimony could he restrain himself from committing some terrible indiscretion with Ada.

Kay probably first met Ada through Annabella. In 1839, as a result of his pioneering work on the conditions endured by the working classes in Manchester cotton mills, he was appointed a secretary of a new committee of the Privy Council set up to study the idea of the government undertaking the job of educating the children of the poor – a radical idea, particularly as until then it was considered to be part of the Church's work. Despite ecclesiastical resistance, he succeeded in using his post to lay the foundations of a secular state education system, an enormous achievement that inevitably brought him to the attention of the prophetess of Pestalozzi and industrial schools, Lady Byron.

When he met Annabella, Kay was living in Battersea, a village on the bank of the Thames opposite Chelsea. It was formerly famous for its market gardens but was already being transformed by the railway revolution into an industrial conurbation. By July 1841, he was, as his journal attests, a regular visitor not only to Ealing but the Lovelace home in St James's Square. There are few direct references to Ada in his journal, for reasons of discretion (it was written to be read by others, including members of his family). Nevertheless, it was at her home in St James's Square that he was to have his own first experiences of mesmerical methods.

The earliest mentions of mesmerism are in connection with some correspondence he was having with a Greek curate called Mr Calliphonas, who would soon become the husband of William's sister Charlotte. The first experiment he attended at Ada's home took place on the evening of 5 July 1841. Kay had been in a state of agitation for most of that week, partly due to the ongoing general election that would sweep aside the collapsing Whig regime and replace it with Robert Peel's Tory administration, partly because of some unspecified business that had left him feeling like a man 'unjustly suspected . . . under the strongest circumstantial evidence, for a crime of which he has the greatest abhorrence'. He leaves no evidence as to what that crime might have been.

At around lunchtime he went to a bar called Dubouys on the Haymarket, which is just a short step from St James's Square, where he drank some wine (a reflection of his state of mind, since it is the only occasion when he reported taking any form of alcoholic drink). He turned up at Ada's house feeling 'exceptionally fatigued, nearly asleep, and only kept from actually drowsing by the twinges of a severe headache'. There he found not just members of the household, but Charles Wheatstone, the famous co-inventor of the electric telegraph. There was also a mesmerist in attendance, but Kay does not identify who it was.

William's sister Charlotte was the first and not very successful subject for experimentation, followed by one of the housemaids, who proved more susceptible. The mesmerist or 'operator' quickly put her in a trance, whereupon he invited other members of the audience to speak to her. Wheatstone obliged, but she seemed unable to hear him. She only responded to the voice of the 'operator', who then demonstrated his influence over her by drawing her out of her chair and round the table in the middle of the room with nothing more than a few gestures of the hand. He then invited everyone in the room to stand behind her, and she managed to identify who was who while still having her back to them – she could 'feel' who they were, she claimed.

Kay was not completely convinced by this demonstration, but attended two others, in the following week, this time in the company of Mr Calliphonas, who was obviously using these meetings as an opportunity to get to know his prospective bride. The second occasion did not take place at St James's Square but at another address. It was attended by Hester and Charlotte King and some anonymous others (probably including Ada) together with a *Who's Who* of mesmerism: John Elliotson himself was in attendance, along with Chauncy Hare Townshend, a clergyman and the author of an influential and very popular book on the subject. But the star was Charles Lafontaine, a Parisian who had experimented with Elliotson on using magnetic sleep to anaesthetize patients, and who drew enormous crowds with his flamboyant public demonstrations of his magnetic powers.

With a sense of showmanship that distinguished so many great mesmerists, Lafontaine began his demonstration without actually being present in the room. Rather, to demonstrate the reach of his animal magnetism, he had chosen to sit in a neighbouring apartment from where he placed the 'patient' now sat before his august audience into a 'cataleptic slumber'. He then entered the room – a 'porcine' man with a thick beard and moustache, Kay observed – and proceeded to show the extent of his influence over his unfortunate patient by making him contort his body and suffer violent spasms. He then stuck a pin into the young man's forehead, and connected it to a large 'galvanic' battery. The charge held by this battery was so great that none of the members of the audience could touch its terminals for more than a second. Yet, despite being connected up to it for a while, the young man registered no pain.

The first patient was awoken and excused and replaced by a young woman who, the audience was told, had been deaf and dumb since birth. After he had put her in a trance, Lafontaine claimed she could now hear, and members of the audience were invited to say monosyllabic words into her ear, to which she responded. It was, Kay wrote, a 'pleasing' experiment, especially compared to the one with the young man, which had disgusted him.

It was during just such an experiment, probably conducted at her home, that Ada felt something strange happen to her. She was obviously standing near the mesmerist's subject, and felt herself being touched by the overwhelming magnetic miasma that filled the room. The effect was profound, causing, she believed 'unnatural feeling, mental & bodily' to erupt within. Such feelings would assail her for years to come, and she became convinced that they were responsible for some of the strange behaviour she exhibited. She was also sure that her resulting condition might therefore be curable, if only she could discover the mechanism of mesmerical influence.

*

Presumably because her mother's involvement prohibited it, Ada never seemed to recognize the most obvious catalyst of her mental problems: the issue of Medora. As she was drawn into the morass of revelations, so her mania intensified, and the two seemed to be approaching a crisis when Annabella asked her to visit Paris. There she was to meet for the first time the young, sickly woman who everyone now assumed to be her biological half-sister, and who her mother wanted to be her spiritual whole sister.

Ada dropped everything to fulfil her mother's wishes, including arrangements for a concert to raise money for an impoverished prodigy she had taken under her wing. She departed on the ship to Boulogne on 6 April 1841 without William, who was confined to his bed by illness. She arrived in Paris the next day, where she was received by a mother so transfused with satisfaction and excitement she was barely recognizable.

In the warmth of a Parisian spring, Ada found herself swamped in maternal affection. She had never experienced anything like it before, and it seemed to puzzle her. Its cause was clearly something to do with Medora, which she happily accepted, indeed indulged by showing a sisterly interest in the young woman she was about to meet.

Annabella had rented a house in the Place Vendôme (where Mesmer had set up his Paris consulting rooms), and installed Medora in a self-contained apartment in one wing, where she could be left in peace and privacy to recuperate from her illness. It was in that apartment that Ada was to get to know her, and to learn the gruesome details of her terrible story.

Ada spent some time with Medora, and, with Annabella's coaxing, came to the view that she was a woman of principle who was the innocent victim of a most terrible history.

Annabella would often declare in the most self-righteous terms that she did not spread rumours. This was either a lie or, more likely since Annabella sincerely regarded herself incapable of lying, a delusion. For example, many years later she would tell Harriet Beecher Stowe, the American author of *Uncle Tom's Cabin*, that she had not mentioned the incest allegations against Byron and Augusta to anyone, not even her closest relatives. In fact she had written about them in detail to Ada and William, and at least alluded to them in the company of her circle of Furies.

Annabella made an even less convincing claim to discretion in a narrative she wrote at the time she took custody of Medora. She claimed Medora had 'remained [in Paris] with me unknown'. Unknown by whom? Certainly not Ada or William, nor by Augusta – nor, according to the great Byron authority Doris Langley Moore, by George Anson Byron, Mary Montgomery, Lady Wilmot Horton, Mrs George Lamb, Selina Doyle and the entire Court of Chancery, which was to hear a case she was helping Medora to bring against her mother.

Nor did the terrible allegation about Augusta lining up her daughters for Henry's gratification remain unknown with her. Ada had barely unpacked her bags before she was told it, with further embellishments added for her benefit. It now appeared Medora had been forced to conform to Augusta's dreadful scheme 'by means of *drugging* the victim, who found herself ruined on coming to her senses'. And now Henry had tired of Medora, Augusta had 'lined up' Emily, her youngest daughter, to lure him back to her side.

To compound this outrage, Augusta was still refusing to release the original of the deed that she had drawn up to provide for Medora's illegitimate baby by Henry, Marie. Annabella had now pledged to help Medora get hold of this document, which the same Reversionary Interest Society that Henry had used to mortgage Georgiana's inheritance demanded if it was to allow Medora to do the same with Marie's. A suit had been lodged with the Court of Chancery, paid for out of Annabella's funds, and Augusta faced the prospect of her incestuous past being made public if she did not relent – which she ultimately would.

No one dared challenge this increasingly grotesque farrago, least of all Ada. If she even questioned what she had been told, the maternal affection now so bounteously showered upon her would have dried up immediately. No truth, least of all one concerning Augusta, someone Ada had spent a lifetime learning to loathe, was worth that. Support for Medora had become a test of loyalty, loyalty to Annabella's side of the Separation saga that still informed every aspect of her life and thinking. Ada had to take the test, and pass it.

This she did. She never showed the slightest speck of suspicion or jealousy towards the woman who had become the recipient of all the maternal feelings previously denied to her. She was meticulous in her dealings with Medora, being unfailingly kind and helpful, even offering to let her and her daughter stay at Ockham should she come to England – an invitation which, unfortunately for all parties, she would later accept.

Ada also accepted William being co-opted by Annabella to act as Medora's champion in the affair. He was now the man in Annabella's life, there to defend her should she be threatened, such as when she melodramatically claimed she would be sued for libel by Augusta for threatening to tell the truth.

But Ada never showed any real affection for Medora – she was not embraced as the longed-for sibling, a position that only her second cousin George (son of the seventh baron) and William's sister Hester could ever occupy.

*

Annabella eventually brought Medora and her daughter back to England, where she was to resume the name she had adopted during her years with Henry, Aubin.

At first Medora treated Annabella with the gratitude and affection she so keenly sought, even referring to her in her letters as 'Pip', the pet name given to Annabella by Byron. She also obliged Annabella with a continuing campaign of hostility towards her mother. In one long, entertaining letter, written, she admits, after drinking 'quantities of wine', she reported to her 'dearest Pip' how she had spotted Augusta while accompanying William in a carriage back to the Lovelace home in St James's Square. 'I instantly recognized her – she is unchanged in face – & turned my head as if waiting for William who was ringing at the door . . .' Augusta glanced over at the carriage, which she must have known was waiting outside Ada's house, but she did not recognize her daughter, who had her veil down. This gave Medora the opportunity to contemplate at leisure through the 'little back window' in the rear of the carriage the mother she had not seen since she her last elopement with Henry. 'Her large eyes are ever & indeed *unchanged*, her walk is most altered – she shuffles along as if she tried to carry the ground she walks on with her & she looks WICKED.'

Fortified by further slurps of wine, Medora went on to fill page after page with more on the same theme, lurching between maudlin self-pity and rambling incoherence:

> Oh how dearly fondly I loved her, & had she only stifled the existence her sin gave me – but God *is* there – I will do my best to bear as I have ever done but it is so long, so constant – God forgive her. Oh how horrible she looked – so wicked – so hyena-like – That I could have loved her so!

This drunken lament was music to Annabella's ears. It confirmed Medora's place in the legions Annabella had lined up to fight her cause and vindicate her actions in the great Byronic battle.

However, Medora soon tired of the subservient role allotted to her. She hated having to defer to her benefactor and the friends and relations with whom she found herself deposited like an unwanted pet. In no time at all, she was transformed from a lapdog into a trapped cat, lashing out at anyone who tried to help her. Poor Anna Jameson, perhaps Annabella's only really likeable friend and the one who had the misfortune of being appointed Medora's principal guardian while she was in England, had offered the girl real devotion, only to be lacerated with insults in return. Even when the deed she had so desperately wanted from her mother was finally procured for her, she complained that she had not been properly consulted.

Annabella, too, found herself on the receiving end. 'You cannot imagine the scenes of fury that she exhibited,' she wrote in exasperation to Sophia De Morgan, 'I had *never* seen anything like it. She has told me I was her bitterest enemy and threatened every kind of revenge, I not having from first to last ever uttered an unkind word to her.' Annabella confessed herself at a loss, but could not sign off without making one plaintive admission, perhaps to demonstrate her forbearance, perhaps to express a genuine feeling: 'But I loved her still . . .'

Annabella, a woman who aroused awe if not actual terror in just about everyone she met, even found herself becoming intimidated by Medora. On one occasion it was announced that Medora was coming to Annabella's house in Esher to ask for more money. Rather than face yet another demand, and the rages that would accompany its refusal, Annabella reportedly climbed out of the drawing room window, jumped into a carriage and did not stop until she reached Brighton.

*

Medora eventually secured enough money from Annabella to travel back to France, but the affair by no means ended there. Soon after, yet another begging letter arrived, one even more brazen than all the others, for Medora had got her nine-year-old daughter Marie to write it for her, explaining how poor her mama had become and how much she looked forward to a letter from her former benefactor.

Annabella, showing that her strange infatuation for Medora was as strong as ever, found herself unable to refuse. However, this time she did not dispense further largesse without taking the precaution of sending the money via the Beaurepaires, a married couple she imagined to be under her control, as they had been appointed Medora's manservant and maid on her behalf by Ada just before Medora's return to France.

The demands continued to arrive, and Annabella, whose medical treatments made her more familiar than most to the tactics of bloodsuckers, now decided she must staunch the flow once and for all. After an inevitable deluge of protests and pleas, she agreed to send 1,000 francs as an absolutely final payment to cover the Beaurepaires' unpaid wages.

The foreclosing sum was sent after some delay, and the Beaurepaires were forced to spend most of it paying off the debts that had accrued in the meantime. It was a foolish move, as the Beaurepaires now found themselves dependent on Medora rather than Annabella, and duly switched allegiances. They offered to use the remainder of the money they had received from Annabella – around 400 francs – to pay for Medora and themselves to go to Paris to consult lawyers.

Medora, Marie and the Beaurepaires arrived at the luxurious Hôtel du Rhin, possibly pursued by creditors. They managed to inveigle their way in by promising that their bill would be settled by Annabella, the famous wife of Byron whose credit, the hotel manager would have known, was beyond reproach.

It so happened that Selina Doyle was in Paris at the time and somehow managed to find out what was going on (as Annabella's Furies were wont to do). She loaned the little nest of cuckoos now comfortably ensconced in the Hôtel du Rhin's finest rooms 100 francs on Annabella's behalf, on the understanding that they would move on to cheaper accommodation as soon as possible. Annabella sent some more money by messenger, and a note to the hotel manager instructing him to extend no further credit.

Medora reacted by sending Annabella a letter in which she poured out all that remained in her very considerable reserves of bitterness and resentment. On being shown the letter, Ada sprang to her mother's defence, and wrote a long, passionate, indignant reply demanding Medora's gratitude for the 'benefactress' who had saved her from ruin and even death. The appeal, as Ada probably expected, went unheeded.

Having been insulted, leeched, traduced and duped, Annabella was forced to resort to the remedy she had used when she thought another young woman was taking her to the brink of ruination. That time it had been Ada herself, and the man who had redeemed her was that doctor of morals and medicine, William King.

Dr King, still heavily involved in the Co-operative Movement, was also now running a lunatic asylum in Sussex. Annabella asked him to go to Paris to assess whether or not Medora was mad.

He duly went and returned to give his diagnosis – but not before he, like everyone else who came into contact with Medora, had been relieved of a generous donation from his own pocket. She was, he pronounced, sane. As with Byron, Annabella took this to mean that there could be no mitigation for Medora's behaviour, so she must be punished by being completely isolated from Annabella's benefaction.

This, however, was to be a tale with a very tattered ending. Because Ada had employed the Beaurepaires on Annabella's behalf, and as Annabella had stopped paying their wages, they had effectively been dismissed without notice. Worse, when Mme Beaurepaire sought a reference to get a new job, Annabella accused her of being an extortioner.

It was the first and last time that Annabella was provoked into an act of such ill-judged impulsiveness. Byron had failed, Augusta had failed, ultimately even Ada and William would fail; but there was something about Medora that moved Annabella beyond the realm of reason. She had now exposed herself not just to criticism for abandoning a woman she had taken responsibility for, but to the very real risk of legal action, which the resourceful and independent Beaurepaires duly threatened, along with public denunciations of Annabella and, since she had been involved in hiring them, Ada.

Ada was once again called upon to help sort out the whole sorry mess, which she did by visiting the French embassy after Mme Beaurepaire had gone there to make her case.

The lawyers who had been so deeply involved in sorting out the Separation had the responsibility of brokering an agreement, and one was eventually reached, though not before William, who could be hot-blooded on such occasions, intervened at a particularly delicate moment with an attempt to get the Beaurepaires arrested for writing threatening letters.

Medora finally disappeared from Ada's life in 1844, when she carried the precious deed Annabella had helped secure from her mother back to France, and lived off loans raised against it until she married a soldier in 1847. She died in 1849 from smallpox.

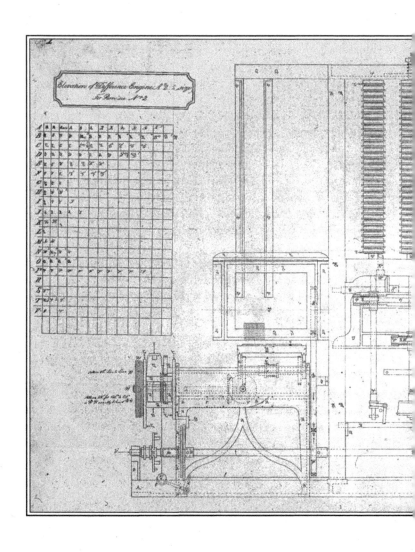

Elevation of Difference Engine No 2, 1847,
for Review No 2

CHAPTER SEVEN

A Completely Professional Person

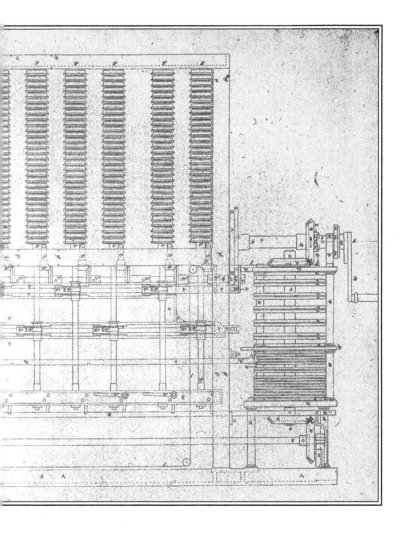

THE NOISE THAT SOUNDED across England in the early 1840s was like the scream Catherine the Great emitted as she died, observed the French writer Hippolyte Taine. It was the whistle of trains.

This was the era of 'railway mania', yet another craze in a crazed age barely recovered from Byromania and still in the thrall of the mesmerical mania. Like the magnetizers, the screaming trains both hypnotized and terrified those who witnessed them – justifiably so, as they were to change every area of life and set the pattern for the technological revolutions to come, from the introduction of mains electricity to the spread of the Internet.

Alexander Somerville, no relation of Mary but the author of the *Autobiography of a Working Man*, called the opening of the Liverpool and Manchester line an 'epoch in the history of the world', and he was barely exaggerating.

> All sights which I had seen in London or elsewhere – the beautiful, the grand, the wonderful – shrunk into comparative nothingness, when I saw . . . the white steam shooting through the landscape of trees, meadows and villages, and the long train, loaded with merchandise, men and women, and human enterprise, rolling along under the steam. I had seen no sight like that; I have seen nothing to excel it since. In beauty and grandeur the world has nothing beyond it.

Charles Dickens expressed another view of the railway in *Dombey and Son*, Ada's favourite book. There it is a thing not of beauty and grandeur but of terror and destruction, the iron horseman of the Apocalypse: 'The power that forced itself upon its iron way – its own – defiant of all paths and roads, piercing through the heart of every obstacle, and dragging living creatures of all classes, ages and degrees behind it, was a type of triumphant monster, Death.' That shriek, the roar and rattle, drilling

> through the damp earth, booming on in darkness and heavy air . . . through the hollow, on the height, by the heath, where the factory is smoking, where the stream is running . . . breasting the wind and light, the shower and sunshine . . . with a shrill yell of exultation, roaring, rattling, tearing on, spurning everything with its dark breath.

The death rattle disturbed William Wordsworth, who had passed into a more conservative, curmudgeonly mood since he put his walks with Coleridge across the Quantock Hills behind him and headed for the Lakes.

> Now, for your shame, a Power, the Thirst of Gold,
> That rules o'er Britain like a baneful star,
> Wills that your peace, your beauty, shall be sold,
> And clear way made for her triumphal car
> Through the beloved retreats your arms enfold!
> Heard YE that Whistle? As her long-linked Train
> Swept onwards, did the vision cross your view?

Ada heard that whistle – but ran towards, not away from it. In 1841 it cut through the peace of the Surrey landscape, when the South Western line that was to link London to the port of Southampton reached Weybridge, four miles from the Lovelace estate at Ockham. One of the first people it was to convey from London was another rail enthusiast, Charles Babbage, who braved an open carriage in the middle of a freezing winter to take up an invitation to visit Ada at her home.

Ada was desperate to see Babbage again, to talk about the latest ideas and technologies, and pointed out that, thanks to the train, he was now less than an hour from Weybridge, where she could have a carriage waiting for him to whisk him off to Ockham. What a contrast that was to the arrangements she had to make to get visitors down from London a few years before. In 1836, for example, she had to direct Mary Somerville to take a stage coach called the Accommodation which left from the Ship Inn at London's Charing Cross. It departed daily just after lunch and passed by Ockham in time for dinner.

Her interest in railways, however, went beyond their convenience. She knew they were changing the world, distorting time and space as, a century and a half later, it was predicted that computers would.

*

The great philosopher Diderot once described love as 'the voluptuous rubbing of two intestines'. This was the sort of dry-bone Enlightenment thinking the Romantics were reacting against, and its legacy was that many came to regard a display of enthusiasm for science and technology as a defect of character.

This antagonism was sharpened as the pace of technological change accelerated in the mid-nineteenth century. Modern machines and systems now insinuated themselves into the most personal aspects of life in a way they never had done before. There was not just Catherine's death scream echoing through the valleys and across the hills, but the more subtle chirrup of telegraphic communications and rustle of the penny-stamp post. It was too much for some. 'The probabilities of fabulous tales are left far behind,' a Reverend Eagles mournfully observed in a newspaper column, which reported a walk he had taken across the Quantocks with the scientist Andrew Crosse, a future friend of Ada's. 'Really the world, with all its exciting busy doings, is too much for us. There is no leisure, no slow movement, it is all railway pace, or, infinitely more, telegraphic.' Crosse had shocked the poor Reverend Eagles by suddenly breaking into a recital of poetry as he strode through the gorse. 'I did not think he had possessed such a poetic vein, addicted, as he has ever been, to science,' the bemused vicar observed. 'What is the cause of this? Is it that the magnitude of realities of science is a great and overwhelming poetry?'

It was a terrifying thought. What if poetry was a part of the scientific world, rather than an escape from it? This was certainly Ada's view. The machines taking over the world were not to her the dull, dehumanizing materializations of the mathematical mindset. They were the inventions of creative minds, artefacts of the same 'unseen world' where the poets had discovered their Corsairs and their deep romantic chasms. They were complicated, of course they were, but no more complicated than the machinations of a Medora or her mother.

Coleridge captured the view in his essay *Hints Towards the Formation of a more comprehensive Theory of Life*, in which he breathlessly surveyed the success of science, where 'not only all things in external nature, but the subtlest mysteries of life and organization, and even of the moral being, were conjured within the magic circle of mathematical formula'.

Ada was now drawn into this magic circle, and became increasingly fascinated by the technology it conjured up. She studied it in what she called the laboratory of her mind, a strange place, crackling with the excess energy of her mania, populated with odd inventions and freakish philosophies. There could be found such innovations as a 'reading machine', an educational device she had apparently designed for Dr Kay, who instructed a cabinetmaker to build it, presumably for use in his school for poor children at Battersea. There were ideas like mesmerism and a growing interest in electricity, that still mysterious, invisible and deadly power that Faraday had begun to master through his work on electromagnetism.

Railways provided a particularly rich supply of curiosities for Ada's laboratory. There were new services, such as Thomas Cook's cheap group excursions, which began in 1841 with day trips to temperance clubs – the unlikely precursor of the packaged holiday that would later convey less sober bands of fun-seekers to Las Vegas and Ibiza. And there were new forms of locomotion like Henry Pinkus's patent 'gaso-pneumatic' railway system, and Parkin's proposal to have carriages tugged along with cables worked by windmills stationed along the line. These were not regarded as mere mechanical oddities. As Thomas Macaulay put it,

> Every improvement of the means of locomotion benefits mankind morally as well as materially ... [It] not only facilitates the interchange of the various productions, of nature and art, but tends to remove national and provincial antipathies, and to bind all the branches of the great human family.

One improvement that promised exceptional benefits for mankind's morals, and captured Ada's particular interest, was being put to the test on a segment of the West London line which ran across Wormwood Scrubs, an area of empty wasteland to the west of London that was then, and remains, as unglamorous as its name.

The West London line was a typical casualty of the technological and financial chaos that became a feature of the railway mania – and of just about every technological mania that was to follow. The industrial cities in the north of England were linked to each other and to London via a railway system developed by George Stephenson, the inventor of the steam locomotive. This used rails set apart the same distance as they were in collieries, 4′ 8″. Isambard Kingdom Brunel, the engineer responsible for building the Great Western Railway, had chosen to base his line linking London to Bristol on a system developed by his father Marc. It had been designed to transport timber around the Royal Dockyard at Chatham, and, perhaps reflecting its use in a more spacious setting, had been given the much wider gauge of 7′. Thus began what became known as the 'battle of the gauges', a struggle to establish a new technical standard that was to tear the railway industry apart for decades to come.

The combative Charles Babbage charged straight into the mêlée by publicly championing the broad gauge adopted by his friend Brunel. To prove how much better it was, he devised a special train carriage in which he would clatter across the Great Western's network, using a sort of giant seismometer of his own devising to record every jiggle and jolt on strips of wallpaper. Having covered hundreds of miles of track and about two miles of wallpaper, the seismometer's tracings were used to decorate the walls of the London Tavern in 1839, where a meeting was held by the GWR's directors to thrash out the gauge issue. The event was a triumph for Babbage, who gave a presentation that persuaded even a delegation of railway proprietors from the narrow-gauge north of the broad gauge's merits. As he proudly proclaimed in his autobiography, without him, Brunel's system would not have survived – which it ultimately did not as the narrow gauge later came to be adopted as the national standard.

The West London line crossing Wormwood Scrubs was caught in

the middle of this battle, as it straddled the broad-gauge GWR and the narrow-gauge lines leading north and was supposed to connect the two. This uncomfortable position was made worse by the GWR's imperious view of attempts to link its own network with a lowly branch line of any gauge. The only connection it would allow between the West London network and its own was a turntable for the transfer of single wagons and carriages. Thus marooned by technical incompatibilities and corporate haughtiness, the West London line found it had nowhere to go, and became the subject of merciless lampooning by *Punch* magazine, for which it provided a convenient stick to beat bumptious engineers and vulgar speculators. So when two brothers, Joseph D'Aguila and Jacob Samuda, approached the directors of the West London Railway with a proposal to lease a stretch of line to conduct a strange experiment they were only too happy to oblige.

Their scheme was to build a railway based on the principle of Atmospheric Propulsion. The idea had already attracted a great deal of attention when a working scale model appeared at the Adelaide Gallery of Practical Science. This is where Ada would have first seen it, as she was one of the Gallery's keener patrons.

The atmospheric principle was first proposed by George Medhurst as far back as 1810, when he had implausibly suggested that carriages could be propelled through huge thirty-foot-diameter tubes by compressed air like peas through the barrel of a shooter. The Samuda system was more realistic, comprising a tube less than a foot in diameter running between the rails of a conventional line. A vacuum would be created in the tube by pumping stations along its length, and this would draw a piston from one end to the other. The problem was finding a way of connecting the piston to the rolling stock it was supposed to pull. The answer was a simple metal plate linking one to the other passing through a gap in the top of the pipe. To preserve the vacuum in the pipe, the gap was covered by an ingenious leather flap impregnated with beeswax and tallow, which opened up as the plate passed along it, but formed an airtight seal either side.

The Samuda brothers laid a half-mile length of this 'traction piping' along a stretch of the West London line, terminating at Wormwood Scrubs, where a state-of-the-art sixteen-horsepower steam pump was installed to create the vacuum. The brothers commenced testing the system in the summer and autumn of 1840. By December, they were able to perform regular runs, and achieve breathtaking (literally so, some technophobic commentators feared) speeds of forty-five miles per hour.

The following year they started to hold public demonstrations, and they were a sensation. Ada attended one on 11 March, and was so thrilled she took two trips. She quickly grasped the system's technological benefits, which were principally related to its ability to traverse slopes, something locomotive-powered trains found difficult. She also considered its prospects as an investment, perhaps on behalf of her mother, who may have been planning to give the idea some financial backing. But what really excited her was the speed: twenty-five miles per hour, she recorded.

The Samudas' 'atmospheric railway' technology was eventually deployed in Ireland, linking Dublin to Kingston, the south-west of England, where Brunel used it to build a line through the hills of Devon, and on the London to Croydon line. The latter was the most ambitious scheme. The London and Croydon was one of the capital's most spectacular lines, with the first four miles out of its London Bridge terminus being entirely supported on a brick viaduct. With its 878 arches, the viaduct had become one of south London's most prominent man-made landmarks, a demonstration of the ability of engineers to create rather than merely follow physical geography.

The atmospheric railway was laid alongside the conventional one, and the first section opened in 1845 to great acclaim. It demonstrated not just the power of the atmospheric principle, but two other innovations.

The first was architectural: the four pumping stations that lined the route, each one containing an enormous hundred-horsepower engine to maintain the vacuum in the fifteen-inch-diameter traction pipe, were designed by the architect W. H. Brakespear to look like churches, an early example of neo-Gothic design being applied to such modern facilities. Brakespear's aim, the *Pictorial Times* reported in August 1845, was to show that 'the most uncouth forms may be so decorated as to become ornaments of the landscape'. In particular, that most uncouth form of all, the chimney stack, had been transformed into a soaring spire – an attempt to reverse a process of spires becoming stacks that had blighted the London skyline since the beginnings of the industrial revolution.

The other innovation adopted by the London to Croydon atmospheric line, one that was already appearing elsewhere on the rail network, was a sophisticated electric telegraphic system. Ada's friend and co-mesmerical experimenter Charles Wheatstone had invented the first practical application of the technology in Europe together with Charles Cook. Ada had gone to see it exhibited at Exeter Hall, and was intensely interested by what she saw – despite another spectator showing a similarly intense interest in her (his attentions ended up forcing her to leave).

Telegraphy and trains were made for one another. The railway lines that cut so cleanly across the countryside provided a perfect route for the wires needed to carry telegraphic signals, and the need to coordinate an increasingly complex network demanded a form of communication that went much further and faster than conventional post. Thus the railway owners let the telegraphic companies lay down the wires in return for being allowed to use them to coordinate their railway services. This financed the telegraphic network's rapid spread, and opening it up to public use delivered the profits.

Just as with the Internet and email a hundred and fifty years later, the public was at first wary of the idea of telegraphic communications. There were particular concerns about security, as lines could be tapped and messages intercepted without anyone knowing. Experts started to consider the use of encrypting messages to ensure they could only be read by their intended recipients. Once again, the name that popped up with the most radical proposals for this idea was the indefatigable Babbage, who had spent some time studying the mathematics of ciphers.

Ada was not concerned about security. As with the atmospheric railway, the feature of the telegraph that interested her was its speed. In a letter to William brimming with excitement, she imagined the possibilities the technology created. The previous week she had been returning to London from their Somerset home at Ashley Combe by rail from Bridgwater, then the nearest mainline station. She had missed her intended train, and as a result arrived at Paddington two hours late. If only there had been a telegraph service along that route as there now was along 'her' line, the South Western linking London to Weybridge, she could have 'sent words in two seconds to London' of her delay. And that was not all. Wheatstone had told her it was possible to hold conversations between stations. She could, for example, discuss an order with a tradesman in London while staying at Ockham – an early example of what might now be called on-line shopping. 'Wonderful agent and invention!'

Ada's exultation at the pace and impact of technological change became so intense, she began to wonder if it might overpower her. She started to look for other outlets for her energies that might divert her '*hysteria*', as she now openly referred to it, 'from all its mischievous & irritating channels'. Drama and music became one new outlet. She started to practise singing several hours a day, performing operatic arias in the library at Ockham, and while in London, braving the 'filthy & odious' atmosphere produced by the city's thickening pollution to make regular visits to Drury Lane.

More alarming to William was her growing interest in poetry and literature. Barely any of her work in this area survives. In a letter to William, for example, she mentions a reworking of a ballad by the great German poet Schiller that she sent him to demonstrate her skills. There is no longer any sign of it among her papers. All that is to be found is a review of *Morley Ernstein*, one of George Payne Rainsford James's popular historical novels. On the face of it, her choice of subject was a poor one. Dr Kay, whom Ada consulted on literary matters, declared James to be a 'dull, undramatic dunce', reflecting the judgement of contemporary and future opinion that he was the author of so many (far too many) undistinguished potboilers. In the judgement of Robert Harrison, the librarian at Thomas Carlyle's newly instituted London Library, James's 'want of originality, of insight into character, of vivid portraiture, is redeemed, in the opinion of the uncritical portion of mankind, by a regular story, elaborate descriptions, rigid poetic justice in the catastrophe, and an even, unobjectionable style of writing'.

In her review of *Morley Ernstein*, Ada clearly lumped herself squarely in 'the uncritical portion of mankind'. She could not praise James too highly. His 'enlightened' view of penal reform, for example, stood in such shining contrast to that of Wordsworth's, which had recently become as reactionary as his attitude to trains.

Just as television has more recently been dismissed as a superficial and addictive form of entertainment, so then was the novel, and Ada thought that the term should be disassociated from such an important book. Indeed, it was a work of metaphysics, as was obvious even in its opening paragraphs, which dealt with the distinction between the 'spirit of the soul' and the 'spirit of the flesh', a theme that ran through the work.

Kay was flabbergasted by Ada's opinion. Her review was so favourable, anyone reading it would imagine that James was a friend of hers, he told her. Perhaps this was evidence that her ever widening range of interests had declined into dilettantism.

However, it is obvious from reading the book that there was more to Ada's assessment than a lapse of critical judgement. *Morley Ernstein*'s central concern was hers, namely the split between art and science, passion and intellect, the 'spirit of the soul' and the 'spirit of the flesh'.

One passage in the novel deals very directly with the challenge of the science against the arts. 'Great discoveries have been made,' the narrator observes, mulling over the character of the era in which the story takes place. 'We have had Herschels, Laplaces, Faradays . . . we have invented steam-engines, rail-roads, electric telegraphs.' These were now considered to represent the summit of human ability. But, argued the narrator, they were no such thing. Scientific research was a cumulative process, one generation handing on a legacy of knowledge to the next. This meant that the achievements of the Herschels, Laplaces and Faradays depended on the achievements of the Newtons, Halleys and Copernicuses. 'If I have seen further, it is by standing on the shoulders of giants,' as Newton himself put it. Art, on the other hand, was reinvented with each generation. Every artist worthy of the title must confront the human condition afresh, as innocently as a baby beholding the world, and produce an original response to it. That accounted for the lack of great poets, and the abundance of famous scientists.

Putting aside the validity of this view, its importance for Ada was that it squarely confronted the issue now dominating her thoughts: what was the relationship between science and art, between the realms of reason and passion?

In many of the letters she sent in 1841 and 1842 to William, who was away performing ever more of his official duties as Lord Lieutenant of the county, she struggled with this dichotomy. She was stoked up, had built up a head of steam, was ready to roar into the future with a yell of exultation – but could not decide in which direction she was supposed to go. The sciences were not enough, the arts were not enough, the demands of domestic life were not enough. 'There seems to be a vast mass of useless & irritating POWER OF EXPRESSION which longs to have full scope,' she wrote to William.

'I am a d—d ODD animal,' she admitted to a worried Woronzow Greig, resorting to language that would have given the somewhat straight-laced lawyer a frisson. Either an angel or a devil watched over her, she added, but could not decide which.

The feeling that she was something of a misfit was willingly confirmed by those around her, Woronzow among them, who remembered in his memoirs how the anecdotes she told him were all more or less eccentric. Her letters are filled with spicy allusions designed to shock their readers, references to being 'naughty', to being the devil, to shocking the neighbours.

Her style of dress was what in a more recent age might have been called 'arty', as Doris Langley Moore, the curator of a costume museum as well as Ada's biographer, put it. Ada experimented with the design of swimming costumes, choosing to wear a body suit made of black camlet (a mixture of wool and goat's hair) rather than the more conventional petticoats, which she observed not only impeded swimming but trapped air around the midriff and threatened to topple the swimmer headfirst into the water.

Her unconventionality was the focus of much attention when she was in society – something that often made her a reluctant guest.

She was known for sending a tremor through a dinner party with the uninhibited expression of some extraordinary idea, such as when she congratulated the novelist Eliot Warburton on the arrival of his first son with some observations about overpopulation (remarks which Warburton, to his credit, took in the satirical spirit in which they were presumably offered). Byron's old friend Hobhouse found her conversation far too rich for his tastes, though he would have heard much richer from her father. 'I sat next to Lady Lovelace at dinner,' he reported in his journal, adding primly, 'she spoke to me very freely on subjects few men & no women venture to touch upon'. The topic on that occasion was the future, a word which was then frequently used to refer to the afterlife. She expressed the very advanced opinion that a belief in the future was really a form of wish-fulfilment.

The future, in all senses of the word, weighed heavily on Ada's mind during this period. Sometimes, perhaps helped along by one of the doses of laudanum that she was now almost habitually prescribed by Dr Kay or Dr Locock, it appeared to her in a dazzling light: 'I have now gone thro' the night of my life . . . and . . . am now approaching the *Dawn*,' she wrote to her mother.

On other occasions, she looked forward, and beheld only darkness. She yearned to *achieve*, she despairingly told William, but could not find a focus for her interests that she could be confident of yielding a result.

Her friends agreed. It was hard finding an area that matched her unique sensibilities. But she must do so. Woronzow became concerned that she was overreaching herself by darting around in so many different directions. He had even, he hinted to her, begun to question whether she had the moral qualities needed to achieve her potential. However, the way she had dealt with the Medora affair, the 'deep feeling and total abandonment of self' she ha0d shown, proved to him that she did have those qualities, and it was now only a matter of time before they would enable her to attain fulfilment. '*Festina lente*', he counselled, quoting the punning motto of the venerable house of Onslow: make haste slowly. 'Your immediate prospect is uninviting, the self denial required is great, and the sacrifice enormous – but the end is glorious and an approach to immortality will be your reward . . .'

However, a glimpse of the glory, of those first beams of sunlight announcing the dawn, was to come sooner than Woronzow or Ada expected. It arrived in the form of a French paper published in a Swiss journal about a scientific meeting in Italy. This apparently obscure article would change her life.

*

In 1840, Italy's foremost scientists gathered together in the city of Turin to discuss the state of their nation's scientific achievements. It was an occasion of some significance, since despite this being the country that gave the world the Renaissance, Italy's scientific fraternity had yet to achieve the sense of identity and confidence that was so well developed in neighbouring France. The organizers were anxious to make an impact, and so decided to invite a foreign guest of honour: Charles Babbage.

Babbage had visited Italy in 1828, and made a deep impression on the Grand Duke of Tuscany, one of Italy's most important scientific patrons. Babbage had turned down the invitation he had received the year before to the first scientific gathering, held in Pisa, but found the invitation to the second irresistible.

It came from the mathematician Giovanni Plana. He had been making enquiries about Babbage's latest project, a successor to the (still unbuilt) Difference Engine, to be called the Analytical Engine. From all he had heard, Plana had concluded the following: 'Hitherto the *legislative* department of our analysis has been all-powerful – the *executive* all feeble. Your engine seems to give us the same control over the executive which we have hitherto only possessed over the legislative department.' In other words, great progress had been made in discovering the laws of mathematics, but very little in actually applying them – music to Babbage's ears, which by now had tired of the discordant chorus of doubts his plans had received in England. The invitation was duly accepted.

Thus, laden with his usual accoutrement of gifts and exhibits – plans of the Analytical Engine, an album of experimental calotypes from the pioneer of photography William Fox Talbot, a silk portrait of Jacquard made using one of his looms programmed by no fewer than twenty-four thousand cards – Babbage arrived in Turin in the autumn of

1840, eager to tell anyone who would listen about his latest invention. He came with a friend to act as interpreter, Fortunato Prandi. Prandi, who would later play a mysterious role in Ada's life, was a friend and presumed accomplice of Giuseppe Mazzini, then exiled in London. Mazzini was a radical figurehead of the Risorgimento, the movement for Italian political unification, and the presence of one of his followers in Turin attracted the attention of the secret police. Eventually, a royal dispensation had to be sought by Babbage to allow Prandi and himself to go about their business unhampered.

Babbage soon established himself in an apartment, stuck his plans and diagrams for the Analytical Engine to the walls and proceeded to receive a procession of the 'most eminent geometers and engineers of Italy', including Plana himself and Captain Luigi Menabrea, a young military engineer who would go on to become Prime Minister of a newly unified Italy.

Babbage excitedly explained his ideas to these gentlemen, and Plana started to take notes. However, due to certain 'laborious pursuits' he needed to attend to, Plana handed over the task to the young captain.

Babbage quickly ran into difficulties demonstrating the significance of the Analytical Engine to his audience, even though it did include some of Italy's foremost mathematicians and engineers. His engine would take the principle of mechanical calculation to a different level altogether, he told them. Whereas the Difference Engine used one particular method or formula to solve mathematical problems, this machine could use any, and apparently decide which depending the results of earlier calculations. One baffled professor asked how it could do such a thing – for example determine which of several different possible approaches to a given mathematical problem to apply – when such a decision seemed to demand an act of judgement?

The professor's question went to the very heart of the issue, and indicated the chief innovation that the Analytical Engine embodied: the ability to perform different mathematical functions using the same mechanism.

In the age of silicon chips, the technology is sufficiently inscrutable to make the idea of a machine somehow managing to 'rewire' itself to perform different functions in different circumstances at least seem feasible. We also have the distinction between hardware and software to help, the idea that information is in some way separate from the machinery that processes and displays it. But in Babbage's time, in the age of cogs and ratchets, this was a far tougher proposition to grasp. Menabrea put it thus in the paper he wrote based on the notes he had taken during Babbage's visit:

> Mr Babbage has devoted some years to the realization of a gigantic idea. He proposed to himself nothing less than the construction of a machine capable of executing not merely arithmetical calculations, but even all those of analysis [i.e., any mathematical formula], if their laws are known. The imagination is at first astounded at the idea of such an undertaking; but the more calm reflection we bestow on it, the less impossible does success appear . . .

A gigantic idea that astounds the imagination – perhaps it was this awed reaction to Babbage's proposal that made Charles Wheatstone think of a particular name when he first saw Menabrea's paper published in French in the October 1842 edition of the *Bibliothèque Universelle de Genève*. *Taylor's Scientific Memoirs* had recently been founded to publish translations of interesting papers appearing in foreign journals, and Wheatstone had been retained to come up with suggestions. Menabrea's article was an obvious candidate, and Wheatstone knew of just the person to translate it: Augusta Ada Lovelace.

When Wheatstone approached Ada with his proposal, she evidently snatched at it. This was surely what she had been waiting for, something that she could focus on that embraced her 'peculiar combination' of qualities. She immediately commenced work on the project, and a few months after the original's publication delivered the result of her efforts to Wheatstone for submission to the *Scientific Memoirs*, anxious to know whether or not it would be accepted.

At around the same time, probably early 1843, she told Babbage what she had done. He had been ill, and knew nothing about her labours. 'I asked why she had not herself written an original paper on a subject with which she was so intimately acquainted?' he recorded in his memoirs. 'To this Lady Lovelace replied that the thought had not occurred to her. I then suggested that she should add some notes to Menabrea's memoir.'

It was a startling proposal. Women rarely wrote papers for scientific journals. Mary Somerville was an exceptional case. The few women who wrote about scientific subjects (for example Maria Edgeworth, the novelist who was so admired by Annabella, or Jane Marcet, whose *Conversations on Chemistry* inspired the self-taught Faraday to take up chemistry) did so only on the basis that they were making ideas discovered by men available to a mostly female lay readership. The sort of notes being proposed by Babbage would be for male readers – expert, scientific male readers. How could a woman, a countess whose only claim to fame thus far was that her father was a poet, possibly be the best candidate for such a task? In more contemporary terms, it would be like nominating Lisa-Marie Presley to annotate a study of quantum computation.

Ada, however, felt herself to be well prepared for the task. She had resumed her mathematical studies in earnest in 1842 under the guidance of Augustus De Morgan, and was making steady progress. De Morgan, the University of London's first Professor of Mathematics, was like many academics of the time an enthusiastic contributor to popularizing scientific texts such as the *Penny Cyclopaedia*. He was good at steering Ada through the obscurities of functional equations and trigonometry, and at gently tugging her back to earth when her speculative bent got the better of her.

She was eager to get ahead – so eager that De Morgan at one point issued the same warning as Woronzow Greig: '*festina lente*', make haste slowly. He quoted Laplace's wise words: '*Ce que nous connaissons est peu de chose; ce que nous ignorons est immense*' – what we know is little; what we do not know is vast. She attempted to heed the advice. Her letters, which were in turns overflowing with gratitude for his assistance and saturated with apologies for her shortcomings, show her grappling with the dull details of differential calculus.

One of her main problems was her irrepressible curiosity. When she was supposed to be concentrating on one subject, she would find herself being drawn towards another, apparently more exciting one.

For example, she was fascinated by the meaning of one of those mathematical curiosities that can send non-mathematical minds spinning: the square root of a negative number. In conventional arithmetic, a negative number does not have a square root. This is because if you square a negative number, you end up with a positive one (–2 × –2 = 4, not –4). Thus, all positive numbers (such as 4) in fact have two square roots (2 and –2). Which poses the question, what number could there be that, when squared, results in a negative number? We know that it is neither a negative nor a positive one. The answer, according to, among others, the sixteenth-century Italian mathematician Rafaello Bombelli who Ada herself had studied, was that it was a different type of number altogether, an *imaginary* one.

For obvious reasons, Ada found the concept of imaginary numbers fascinating. They behaved slightly differently from real (i.e., normal) numbers, but could be manipulated using the same basic arithmetic. And, as various mathematicians had shown at the end of the eighteenth century and the beginning of the nineteenth, they could be combined with real numbers, yielding 'complex numbers', the properties of which could be explored by plotting a graph, with one axis representing the imaginary component of the complex number, the other axis the real part. This created a new form of two-dimensional geometry (a geometry that a hundred and fifty years later would be shown to have as one of its distinctive shapes that icon of chaos theory, the Mandelbrot Set).

Ada's response to this was to ask what sounded like a simple, technical question but which was in fact a deeply profound one: could you have a third set of numbers in addition to real and imaginary ones which would yield a *three*-dimensional geometry – a whole new, previously unexplored mathematical space, in other words? De Morgan had already grappled with this question, and he was defeated by it. It was the Irish mathematician Sir William Rowan Hamilton who came up with the answer a few months after Ada posed the question. He had been struggling with the issue for years, and discovered that you had to go up a further dimension to four to come up with a workable solution. He called his new numbers quaternions, and they were to prove extremely useful in understanding the bizarre realm revealed by modern physics.

Imaginary numbers were not the only arithmetical phenomenon to lure Ada away from mathematics into metaphysics. Her struggles with certain aspects of algebra produced the same response, but more with an aim of escaping the drudgery of substitution and calculation than probing new intellectual frontiers. The way one formula could be derived from another, for example, continued to perplex her, and she began to observe half-seriously how much algebraic expressions were like 'sprites and fairies', 'deceptive, troublesome & tantalizing' little creatures that could adopt any form they chose.

The fairies, in fact, fluttered into her mathematical work with increasing frequency. In her now chummy letters to Babbage, she even described herself as being one. 'Science has thrown its net over me, & has fairly ensnared the fairy, or whatever she is,' she wrote on one occasion. On another, following a visit to Babbage's house at Dorset Street, she chided,

> think of my having to walk! (or rather run), to the Station, in half
> an *hour* last evening; while I suppose *you* were feasting & flirting in
> luxury & ease at your dinner. It must be a very pleasant merry sort
> of thing to have a Fairy in one's service, mind & limbs! – I envy
> you! – I, poor little Fairy, can only get dull heavy *mortals*, to wait on
> me!

When it came to Babbage, this 'fairyism', as she dubbed her obsession, was evidently deployed to contrast her own more speculative approach to mathematics with his more down-to-earth one. It was her chance to indulge in 'a little *play* & *scope*', as she put it. But it also reflected a much wider issue.

*

By the mid-nineteenth century fairies were everywhere, breeding like flies in the compost of medieval escapism and Gothic romanticism that made up the Victorian artistic imagination. During the eighteenth-century Enlightenment they had become an endangered species, but now they were back with a vengeance, and had somehow managed to make their way into the impregnable edifice of reason and materialism.

It was a growing interest in fairy tales that had reactivated them.

Walter Scott's 1802 essay on 'The Fairies of Border Suspicion', which appeared in a collection of traditional ballads called *The Minstrelsy of the Scottish Border*, had been particularly influential. In 1828, Thomas Keightley published an entire book, *The Fairy Mythology*, designed to preserve the best stories, and in the process powerfully reinvoked the nostalgic nether world of fairyland:

> It is pleasing to us, now in the autumn of our life, to return in imagination to where we passed its spring – its most happy spring. As we read and meditate, its mountains and its vales, its verdant fields and lucid streams, objects upon which we probably shall never again gaze, rise up in their primal freshness and beauty before us, and we are once more present, buoyant with youth in the scene where we first heard of fairy legends of which we are now to treat.

That spring, that most happy spring seemed to be bursting out all over the Victorian arts. It could be found in the revival of interest in Shakespeare after years of neglect. His were the plays that Ada insisted on seeing at Drury Lane when her interest in drama was at its height. One of the most popular was *A Midsummer Night's Dream*. In the eighteenth century the story of Bottom, Puck, Titania and Oberon was just the sort of silly, mystical nonsense that provoked Dr Johnson to criticize Shakespeare for his failure to observe the classical unities of time, space and action. Now the kingdom of the fairies that Shakespeare so playfully invoked captured the Victorian imagination. It inspired not just new productions and favourable reappraisals, but a whole genre of art, the fairy painting.

One of the most telling fairy paintings was by J. M. W. Turner. *Queen Mab's Cave* was finished a few years after Ada started to work with Babbage, and it was fitting that an artist who so brilliantly navigated the line between Romanticism and science should have painted it. The grotto he created was apparently his invention. There is no such cave in either Shakespeare's *Romeo and Juliet* or in Shelley's *Queen Mab*. Nor is there anything quite like it in other contemporary works – except his own. The mouth of *Queen Mab's Cave* recalls the opening to a kiln in Coalbrookdale, the centre of iron production, which he painted around 1797. Or perhaps it is the interior of a forge depicted in *The Hero of a Hundred Flights*, which shows the casting of the huge equestrian statue of the Duke of Wellington. Or perhaps, with its straight sides and gently arching roof, the mouth of Queen Mab's cave is like the opening of a railway tunnel, with the body of tunnel receding along its perpendicular course into the depths of the fantastical landscape behind. All these openings lead into a darkness blasted with a shimmering light, only in the case of Queen Mab's Cave what emerges is not a shining iron ingot, or a heroic statue of the Iron Duke or a fire-breathing iron horse, but a flock of diaphanous spirit forms dancing across the canvas, a 'daylight dream in all the wantonness of gorgeous, bright and positive colour, not painted but apparently flung at the canvas in kaleidoscopic confusion,' as the journal *Art-Union* put it.

Those same wanton forms danced across the floor at Covent Garden. In 1832 the Italian dancer Filippo Taglioni put on his own production of *La Sylphide*, the story of the doomed love between a mortal man and a fairy, featuring his daughter Marie. The result was the creation of one of the most enduring images of Romanticism: the ethereal ballerina dancing on air. As one of Marie's balletic successors, Margot Fonteyn, put it, 'she and her father, between them, were to forge the image of the ballerina that we now take for granted, standing on one toe in an airy arabesque'. Marie popularized the method of dancing on the point of the toe and wearing a tutu that were to become the signatures of what is now, perhaps inappropriately, known as 'classical' ballet.

The features all these fairies shared were ephemerality, capriciousness, even puckishness, and it was with these characteristics that Ada was identifying when she flirted with fairyism. It expressed her own ambivalence towards science, the feeling that there was a part of her that was too romantic for Babbage's solid, permanent, consistent scientific world. For, as Charles Dickens was later to show in an outburst against scientists entitled 'Fairyland in Fifty-four', fairies were on the front line when it came to the divide between the arts and the sciences:

What have I done that all the gold and jewels and flowers of Fairyland should have been ground up in a base mechanical mill and kneaded by you – ruthless mechanical philosophers – into Household Bread of Useful Knowledge administered to me in tough slices at lectures and forced down my throat by convincing experiments? Are the Good People ... to give up the ghost; and am I to be deprived of all the delicious imaginings of my childhood and have nothing in their stead?

Now, however, the time had come to decide whether she wanted to remain a fairy, or turn herself into something more substantial. Faced with Babbage's invitation to write a series of detailed notes about his Analytical Engine – a work that would become the definitive (indeed, the only) detailed account of its design and applications – Ada had to decide whether she was really up to such a task, whether this was the departure point for the '*intellectual-moral* mission' that she had been awaiting, and that she considered to be her destiny.

It was not an easy decision to take. To begin with, Babbage's Analytical Engine would not be an easy machine to explain. Hardly anyone at the time really grasped what it was about. Even the distinguished Italian men of science to whom Babbage had tried to explain it had difficulties. It involved ideas about machines and mathematics, the relationship of calculators and, as we would now call them, computers that most people would still find confusing over a century later. Indeed, in some respects they were ideas that would only make sense a century later, when the electronics were available to make the implementation of them practicable.

Furthermore, Ada had to make a leap of faith. She had to believe that Babbage's machine was not just some absurd technological chimera dreamed up by deranged engineering fantasist. She had every reason to be suspicious. Why were British scientists and engineers – including her own tutor, Augustus De Morgan – so uninterested in his ideas? Why did he have to go all the way to Italy before he could find anyone to listen to him? Why should she, or anyone, regard it as being any more significant than Pinkus's patent gaso-pneumatic railway system? That was certainly the view of Robert Peel's Tory administration, which was then in power. 'What shall we do to get rid of Babbage's calculating machine?' Peel wrote to William Buckland, the geologist and future Dean of Westminster. Peel thought it a worthless device, its only conceivable use being to work out exactly how little benefit it would be to science.

The machine was not the only unknown quantity, not the only entity that might turn out to be a marvel or a mere oddity. There was herself. To grasp the true significance of the Analytical Engine required science combined with imagination, an ability to go beyond the 'threshold of the unknown' and into the unseen world that lay hidden within the miasma of mathematics. Ada possessed this 'singular combination of qualities', she was sure of it. This was her make-up, the 'stratification of poetry and mathematics' that was, as she later told a friend, her distinctive geology.

But what if she failed? What if she, a woman, a countess, a public figure, already expected by her intellectual as well as aristocratic peers to make a name for herself, were to prove unequal to this task? Would it not be easier to leave expectations unfulfilled rather than prove that they were too exalted? More to the point, would she be able to accept what failure said about herself?

She had emerged from the Medora episode with an overwhelming belief that she had inherited the 'flower' of her father's characteristics and, she implied, her mother's. But if she failed, it would prove that she had inherited only the weeds. She would have to confront the fact that this particular breeding experiment had produced not a new type of superhuman, which in moments of mania she believed herself to be, but a freak.

Despite such risks, she decided to accept the challenge, to take on the 'great object' of writing about Babbage's extraordinary invention. Indeed, she largely did so in order to distract herself from her mania, and the constant self-analysis it provoked. 'I verily believe I should be in a bad way but for my *great objects*,' she wrote.

> If I were in circumstances that permitted of my dwelling much or long on *myself* & my *sensations*, it would be a very unpromising business. But I have happily got fairly *entraînée* [carried away] now, into a course which leaves me little opportunity for more *self* study & speculation & which moreover gives me passing motives to desire

not only a mere continuance, but a highly active & efficient continuance, of life in this world for many more years to come.

*

Ada set about writing the notes in the spring of 1843. She had decided upon a strange format for the work, perhaps because she hurtled headlong into the project without first thinking through how best to approach it. She would write a series of notes to be appended to her original translation of Menabrea's paper. However, it quickly became apparent that the notes would end up being substantially longer than the paper they were supposed to annotate.

Despite this structural problem, she persisted, with gusto, egged on by an increasingly excited Babbage. He sent her his own designs and formulae – already a large collection that would amount to thirty volumes by the time of his death. He also began to anticipate who might read the finished work. Perhaps they should send a copy to Prince Albert, he suggested. The Prince Consort had already displayed an interest in engineering and scientific developments that was unusual for a royal, and had a sympathy for many of the ideas Babbage's engines were designed to embody, such as the division of labour, which the prince had considered the 'moving power of civilization'.

Ada thought it too early to start thinking about such matters, and was far more concerned with getting to grips with what was emerging as a much bigger task than she had at first imagined.

She started by examining the most important and difficult issue: the feature that made the Analytical Engine different from the Difference Engine. This she discussed in the first note, identified 'A', which is the least mathematical and most metaphysical. It also appears to be the one to which Babbage contributed least, because when she first sent it to him for his views, he wrote back to her saying he did not want to return it because he did not want her to alter it. It demonstrated, he felt, a depth of understanding that even he had not anticipated.

He was not merely trying to be encouraging. Note A is, arguably, the most important in the entire work, as it grapples with explaining the true significance of the Analytical Engine – an extremely difficult task as nothing like it had ever been imagined let alone described before.

She reached the core of the issue when she discussed the Engine's architecture. As Menabrea's paper itself explains, it is divided into two sections. Drawing on the analogy of a cotton mill, one part was called the 'store', the other the 'mill'. The store held the numbers to be processed and the mill processed them. The mechanisms used in both sections were in some respects similar to those developed for the Difference Engine. For example, numbers were stored in columns of cogs, with each cog representing a single digit. However, in one of his boldest technological innovations, Babbage had proposed a new method for controlling these cogs: using cards of the sort Jacquard had invented to 'program' looms to weave particular patterns. These would be strung together in a way that allowed the Engine to perform such programming acrobatics as loops, where a sequence of instructions is repeated over and over again, or conditional branching, where one series of cards is skipped and another read if certain conditions are met (for example, if the result of the calculation currently being executed is 0).

It was when considering these cards that Ada started to explore the essence of the Analytical Engine – when she began to show that what she was writing about was not a mere calculator, a machine for performing particular arithmetical operations, but what we would now recognize as a computer: 'The bounds of *arithmetic* were . . . outstepped the moment the idea of applying the cards had occurred,' she wrote. It opened up a whole new technology for designing machines that manipulated symbols rather than just numbers, which had the potential to bring the entire abstract realm of mathematics into the physical realm of machines in a way never before possible.

> Thus not only the mental and the material, but also the theoretical and the practical in the mathematical world, are brought into more intimate and effective connexion with each other. We are not aware of its being on record that anything partaking of the nature of what is so well designated the *Analytical* Engine has been hitherto proposed, or even thought of, as a practical possibility, any more than the idea of a thinking or of a reasoning machine.

She gave several practical examples of what she meant by this in the last of her notes, designated 'G'. The best known of these is a demonstration of how the machine would work out the Bernoulli numbers. This is an endless sequence of numbers that stretches off in an haphazard fashion towards infinity.

The Bernoulli numbers provided a convenient way of demonstrating how the Analytical Engine would be programmed. She showed this by breaking down the equation normally used to calculate them into a series of much simpler formulae. By means of a table, she then showed how each formula would be entered into the machine.

It is this table which is used to justify the claim that Ada was the world's first computer programmer. Indeed, most modern-day programmers would recognize it as a program, containing a numbered list of instructions showing what operations (addition, subtraction, multiplication and division) are performed on which variables (the initial values and intermediary results used in the calculation).

In fact, Ada created the table out of Babbage's own formulae and data, and it was by no means the first program that he had drawn up (though it was the first to be published). He had experimented with several others during his earlier work on the design of the Analytical Engine. Nevertheless, Ada did not simply reproduce what Babbage gave her. That much is evident in the letters she sent to him as she continued to work away at the notes. 'I am doggedly attacking & sifting to the very bottom, all the ways of deducing the Bernoulli Numbers. In the manner I am grappling with this subject, & connecting it with others, I shall be some days upon it,' she wrote to him in early July 1843, adding in her typically mischievous manner that it was as well she was not living at a time when she might have directed her efforts towards 'the sword, poison, & intrigue, in the place of x, y, & z.'

Over the coming weeks, she started to find the abstruse Bernoullis getting on top of her. 'I am in much dismay at having got into so amazing a quagmire & botheration with these *Numbers*,' she confessed on one occasion. She spent days labouring on the illustrations, only to report once they had been completed, 'Think of my horror . . . at just discovering that the Table & Diagram, (over which I have been spending infinite patience & pains) are seriously wrong in one or two points.' Despite the setback, she could not resist observing in a tone of impish self-admiration, 'I have done them however in a beautiful manner'.

In the end, with publishing deadlines pressing, she had to redo the table in pencil and, having checked it over several times, conscript her husband to ink it over. No sooner had the final version been delivered to the printers than it was lost, and it turned out it was Babbage's fault. He had forgotten that he had taken it home for one last check. Once it was recovered, Ada wrote a furious letter to him, but one that cannot be read without sensing a hint of an indulgent smile: 'I do not think you possess half *my* forethought, & power of seeing all *possible* contingencies . . . How *very* careless of you to forget that Note,' she scolded, '& how much *waiting on* & *service* you owe me, to compensate.'

The same tone, though slightly moderated, is to be found in a number of her letters around this time, most of them concerning Babbage's carelessness or interventions. She chided him for getting various notes mixed up: 'I have always fancied you were a little harumscarum & inaccurate now & then about the exact *order* & *arrangement* of sheets, pages, & paragraphs &c; (witness that paragraph which you so carelessly *pasted over*!).' She berated him for making unauthorized changes: 'I am much annoyed at your having altered my Note. You know I am always willing to make any required alterations myself, but that I cannot endure another person to meddle with my sentences.' Babbage, despite his reputation for irascibility, acted the role of whipped dog impeccably. He always accepted her admonitions and followed her instructions in good humour. For her part, she would soften her swipes with a little satire. She also happily accepted his alterations when she thought them appropriate. It was, in short, a fruitful, fun if occasionally fraught working relationship – like all the best ones are.

However, as the moment of publication approached, she clearly suffered another severe attack of the nervous disorder that had been dogging her since 1841. This is revealed both in reports about her health in her letters to Babbage and her mother, and by the intrusion of the same sort of exultant tone that afflicted the letters she wrote following the Medora revelations. She started to dub the notes 'this first child of mine'. 'I cannot refrain from expressing my amazement at my own child,' she wrote at the end of July, as the last proofs were being corrected. 'The pithy & vigorous nature of the style seem to me to be most striking; and there is at times a half-satirical & humorous dryness, which would I suspect make me a most formidable reviewer. I am quite thunder-struck at the *power* of the writing. It is especially unlike a woman's style surely; but neither can I compare it with any man's exactly.' She announced her intention to turn her critical powers upon the Reverend William Whewell, one of the country's leading intellectuals, who introduced the term 'scientist'. 'Do you not pity him?' she asked Babbage rhetorically, 'and yet the beauty of it would be that I should never say one single *ostensibly* cutting, or harsh, or condemnatory thing'.

*

Ada's mania-charged assessment of her own work stands in contrast to the far more complex reception it would later receive. At the beginning of note 'G', the one containing the table that had caused her so much grief, she wrote:

> It is desirable to guard against the possibility of exaggerated ideas that might arise as to the powers of the Analytical Engine. In considering any new subject, there is frequently a tendency, first, to *overrate* what we find to be already interesting or remarkable; and, secondly, by a sort of natural reaction, to *undervalue* the true state of the case, when we do discover that our notions have surpassed those that were really tenable.

The same oscillating reaction was to mark the reception she and her Notes have since received. She, too, was a 'new subject', an invention of the early nineteenth century that anticipated the twentieth, a strong woman with ambitions to succeed in a male world and a deep interest in science and technology – there had never been anything like her before. Nobody knew what such a creature was capable of doing, so it was inevitable that some would claim her to be capable of just about anything – such as inventing computer programming. This belief emerged particularly strongly in the 1980s, after the US Department of Defense decided to name its standard programming language in honour of her (the Department even acknowledged the year of her birth in its choice of military standard specification number: MIL-STD-1815).

Various books and articles followed which hailed her as an icon of feminism, generally basing this assessment on her refusal to accept limits set by men. For example, De Morgan, her own tutor, reflected the prevailing view of the time when he told Annabella that he felt that the 'very great tension of mind' demanded by mathematical problems was beyond a woman's physical capabilities. Despite this, Ada pushed on with her mathematical work – and did so when her health was in a particularly precarious state. She thus showed the same sort of mettle as later generations of women athletes and business executives, who proved through dogged determination that they could compete in their chosen fields just as well as men.

Like many of those sports- and businesswomen, however, she was no feminist. She did not challenge the system, and (unlike her mother) showed very little interest in political or social affairs. She did what she did on behalf of herself, not her sex.

Nevertheless, she instinctively rebelled against the oppressions the rest of her gender were forced to endure. She refused to accept her status as her husband's chattel. She yearned for financial independence and, above all, a 'profession'. It was her proudest boast following the publication of her notes about Analytical Engine that she had become 'a completely *professional* person'. 'I really have become as much tied to a profession as *you* are,' she wrote to Woronzow Greig in 1844, 'And so much the better for me. I always required this.'

She also has a claim to making at least some contribution to the invention of the idea of the computer. Note A, the first she wrote and the one over which Babbage had the least influence, contains a sophisticated analysis of the idea and implications of mechanical computation. In this sense she is as important to the idea of computing as, say, Thomas Huxley would later be to the idea of Darwinian evolution. There is, however, one fundamental difference that separates her from Huxley: Darwinism arrived at just the right time, when the world had the conceptual equipment to grapple with if not fully accept it; Babbage's engines, on the other hand, were marooned on an island far away from any familiar intellectual territory.

Babbage only ever succeeded in building the demonstration model of his Difference Engine that he had on show at his London home. The full version, and all subsequent designs, including the Analytical Engine, remained on the drawing board. The argument goes that this was because they demanded a level of engineering sophistication that was not available when he designed them. It is, on the face of it, a plausible theory. The Analytical Engine required the fabrication of tens of thousands of parts machined to the finest tolerances to be intricately assembled in a frame 15' high, 6' wide and up to 20' long – the size of a small locomotive and probably requiring about as much steam power to drive its enormous bulk.

In 1991, to mark the bicentenary of Babbage's birth, the Science Museum in London undertook an ambitious experiment to test this theory. Using the sorts of materials and tools available in the mid-nineteenth century, it attempted to build a complete working model of the Difference Engine No. 2, the more advanced version Babbage designed in the late 1840s to take advantage of the many innovations he had developed for the Analytical Engine. The resulting mechanism worked, proving that the reasons for his failure to start the computer revolution a century early probably had little to do with engineering.

A more likely explanation is that the world was ready for evolution but not for computing. In 1990, the science fiction authors William Gibson and Bruce Sterling published a counter-factual or 'virtual' history of Victorian technology. In their version, 1850s London is run by steam-powered Analytical Engines. A Central Statistics Bureau sits at the centre of metropolis, government and empire, monitoring and controlling the seething industrial landscape beyond:

> The whole vast pile was riddled top to bottom with thick black telegraph-lines, as though individual streams of the Empire's information had bored through solid stone. A dense growth of wiring swooped down, from conduits and brackets, to telegraph-poles crowded thick as the rigging in a busy harbour . . .
>
> Behind the glass loomed a vast hall of towering Engines . . . It was like some carnival deception, meant to trick the eye – the giant

identical Engines, clock-like constructions of intricately interlocking brass, big as railcars set on end, each on its foot-thick padded blocks. The whitewashed ceiling, thirty feet overhead, was alive with spinning pulley-belts, the lesser gears drawing power from tremendous spoked flywheels on socketed iron columns. White-coated clackers [machine operators], dwarfed by their machines, paced the spotless aisles. Their hair was swaddled in wrinkled white berets, their mouths and noses hidden behind squares of white gauze.

It is a striking vision, and the product of its authors taking some imaginative liberties with the facts, not the least of which is assuming that Byron and Annabella stayed together rather than separated, which meant he did not feel the need to exile himself from his native country, which meant he could escape his untimely death in Missolonghi and follow a political career that would culminate in him becoming Prime Minister. Ada, too, is allowed to cheat history, surviving as a spinster, chaperoned by one 'Lady' Mary Somerville. Ada becomes nationally celebrated as the Queen of Engines, the Enchantress of Numbers (a title Babbage really did bestow upon her) and an adept of the Society of Light, the secret inner body and international propaganda arm of the Industrial Radical Party.

None of these alterations to the course of history is completely implausible. They are not the reasons why *The Difference Engine* is a novel rather than a model of what might have been. The real reason concerns the wider environment. At the time Ada was writing her paper about the Analytical Engine, the most basic ideas that underpin the modern state were still hot from the forge of development: income tax was barely forty years old, professional government administration was still more than a decade away. Banking, limited liability companies, currency exchanges, stock markets – many of these were still embryonic and would not be recognizable to modern eyes for years to come.

Even time as we know it now was then a novelty. In 1839, George Bradshaw, a Lancastrian printer and map maker, published the first national railway timetable. When he was compiling it, one railway company director he approached thought the idea ridiculous, since it implied punctuality to be some sort of 'obligation'. It would soon become one, because within a decade the railway network was so large and complex it was impossible to operate without what Bradshaw went on to produce: a timetable synchronized to a single temporal standard, Greenwich Mean Time.

Until the arrival of the railway and Bradshaw's timetables, most towns and villages observed only local time, often set by the church clock, which ticked to its own, idiosyncratic beat. When the railway companies built their stations, they wanted to impose the new national standard, because they used it to coordinate their services. So the new, ultra-accurate clocks they installed became the centrepieces of their stations. These, rather than the church clock, determined the times of arrivals and departures, deliveries and dispatches, the pulse of commerce and communication.

The station clock also came to serve a symbolic purpose. Where the church clock oversaw communal events, a gathering of people who lived with each other and knew each other's affairs, the station clock was the meeting point for strangers, people trying to escape their localities – for brief encounters of a sort that Ada herself would soon experience.

With so many of the most basic aspects of modernity still taking shape, it is hardly surprising that the Analytical Engine, and the steam-powered information age Gibson and Sterling imagined it might initiate, did not happen. All the areas of life, government and industry that the computer has since revolutionized – telecommunications, administration, automation – barely existed at the time Ada translated and annotated the Menabrea notes. The Analytical Engine was no more viable in that setting than a human would have been had he or she, through some freakish act of mutation, emerged seventy million years earlier in the age of the dinosaurs.

For this reason, to imagine that Ada somehow had a hand in the invention of the computer is an example of the sort of overrating that, as she wrote, accompanies the emergence of any new idea. When the electronic computer emerged a hundred years later, its inventors knew very little about Babbage and Ada. Indeed, it was only as a result of curiosity aroused by the emergence of computers that historians started to examine their work in detail.

Dorothy Stein, who wrote a biography of Ada that examined her mathematical skills, provided some stark evidence that Ada herself has been somewhat overrated. Ada's command of mathematics was, Stein demonstrated, quite limited. Ada struggled with relatively elementary concepts in the months leading up to her starting work on the Notes, such as when she complained about her inability to cope with substitution in algebra. Stein also noted that Ada translated a printer's error in the original French edition of Menabrea's paper, which resulted in her reproducing a formula that even an amateur mathematician should have recognized to be nonsensical (the French '*le cas* n $= \infty$' was printed as '*le cos.* n $= \infty$', which Ada translated as 'when the cos of $n = \infty$', i.e. the cosine of n is equal to infinity; the line should have read simply 'in the case of $n = \infty \ldots$').

Stein's discoveries provided a useful and sobering corrective. There are, though, factors to be considered that at least soften their impact, and which begin to swing the equation for assessing her work away from overrating and towards undervaluing. The most obvious mitigation concerning the printing error is that not only did Ada miss it, but so did Babbage, who actually participated in the publication of Menabrea's original paper, and proof-read Ada's translation. Everyone else apparently missed it, too, until a century later.

But this is quibbling. Through detailed scholarship, Dorothy Stein showed that Ada probably did not have the mathematical knowledge to write the Menabrea notes without Babbage's help. Ada was not a great mathematician (though she was a much better one than her mother). As De Morgan implied in his own assessment of her capabilities, unlike Mary Somerville, she really was not sufficiently engaged with the details to be one.

The point is, Ada's notes are not to be assessed as a work of mathematics, but as a work of a more speculative, experimental nature. She was trying to apply her 'singular combination' of qualities, to add some imagination as well as understanding to the study of a scientific subject, to see what together they could achieve that separated they could not. It was thus not just a test of her approach but of herself, as the embodiment of these two polarizing sensibilities. She wanted to see if, as she had argued in the essay she wrote at the time of the Medora revelations, imagination could help reveal the 'unseen worlds' science was attempting to penetrate.

One might consider this a less exacting test, an early and very successful example of the old undergraduate trick of trying to use broad generalizations to disguise a shallow understanding of the subject. Except that Ada's notes are no student essay. They are detailed and thorough. And *still* they are metaphysical, meaningfully so.

In short, Ada managed to rise above the technical minutiae of Babbage's extraordinary invention to reveal its true grandeur. When she observed that 'the Analytical Engine weaves algebraical patterns just as the Jacquard-loom weaves flowers and leaves', she showed what imagination could reveal that mathematics alone could not.

This is what 'poetical science' had enabled her to do: see something that would remain invisible to the rest of the world for a century more.

*

As the moment of publication approached, Ada seemed almost to hold her breath: this could be the moment of truth, the approaching dawn of her life. Her sense of self-esteem had been made brittle by the fluctuating temperature of her moods. Now it would have something of substance to rest upon, a foundation for her future.

Then Babbage threw a large spanner into the whole intricate works.

At the time, Ada was in the middle of proof-reading the final revises of the translation and notes, or 'memoir', as they came to be collectively known. Babbage presented her with a statement he had written himself which set out his trenchant criticisms concerning the way the government had handled the development of his engines. She read it, and suggested a few alterations, but had no objection to its publication with the memoir.

It had just been decided, after some debate between Ada and William, that each of her notes would be signed with her initials, 'A. A. L.'. She did not want to advertise her authorship, but nor was she prepared to accept the pose of shrinking violet that female authors were expected to adopt. Because of this, she made the stipulation that

Babbage's statement should be clearly distinguished from her and Menabrea's work, lest anyone should be confused about who was saying what.

Little did she realize that confusion was precisely Babbage's intention. As Ada beavered away in the summer heat at Ockham, secret meetings were taking place in London to discuss the decision announced by Babbage that his statement should appear as a preface to Ada's memoir, and should be unsigned.

Taylor's Scientific Memoirs was by now desperate to publish the memoir, its star contribution, in time for the next meeting of the British Association, due to take place in Cork in September 1843. It was early August, and time was running out. However, William Francis, *Taylor's* co-editor, was worried about publishing Babbage's statement unsigned without the agreement of Taylor himself, who was out of the country and not due back before the publication deadline would be passed.

Sir Charles Lyell, the by now very distinguished geologist who in his younger days had enjoyed attending Babbage's soirées to admire the women, was called in to arbitrate. He had a meeting with Francis together with Wheatstone, who had remained attached to the project since he first suggested Ada translate Menabrea's paper. Babbage was absent. So was Ada, who had apparently been kept in complete ignorance of the proceedings that were deciding the fate of her 'first child'.

The august gathering of gentlemen discussed various options. One was to publish the statement in *Taylor's Scientific Memoirs*, and Ada's memoir in the *Philosophical Magazine*, a sister publication. Lyell thought the *Philosophical Magazine*, which contained a miscellany of letters and notes, an insufficiently prestigious organ for the debut of such a promising talent. It was eventually decided that the statement should appear as a pamphlet accompanying the edition of *Taylor's* containing the memoir. To prevent any risk of its authorship being attributed to Ada, Wheatstone generously offered to sign the statement if Babbage would not.

Lyell wrote to Babbage to report what he evidently considered a satisfactory compromise, and suggested that Babbage accept it, and that he 'manfully' sign the statement with his own name rather than wimpishly leave it to Wheatstone.

Babbage reacted with a self-destructive stubbornness that would characterize many of his dealings with government. He refused to accept the terms, and wrote to Ada asking her to withdraw her work from *Taylor's*.

This was apparently the first Ada knew of his intentions, and she was thunderstruck. It was a betrayal. Ada's mother, who was in the midst of developing one of her implacable dislikes for Babbage, melodramatically described it as an act of murder – and, given the damage it was liable to do to Babbage's reputation, suicide. He had revealed himself a Herod, threatening to slaughter this the treasured 'first child' of Ada's intellectual career for his own benefit. Ada wrote back to him flatly refusing to do what Babbage evidently felt it was his prerogative to demand. She was not another clockwork Silver Lady for him to pirouette before his public.

Babbage was first furious, then bewildered. Why was she behaving like this? He had not a clue.

He could have found one on the opening page of the memoirs he wrote twenty years later: 'The inheritance of a celebrated name is not ... without its disadvantages.' His actions suggested that he had so enthusiastically sought Ada's involvement, and so happily indulged all her requests, commands and scoldings, not because he had perceived in her any special talent, but because of her 'celebrated name'. He knew it would attract attention to his invention, and wanted to use that attention as a platform to air his grievances. She and all her work was thus reduced to nothing more than an instrument of publicity.

On 14 August 1843 Ada wrote Babbage an extraordinary sixteen-page letter. A whole climate of moods passed through it. It was, in turns, respectful, conciliatory, bitter, perceptive, delusional, affectionate, practical, arrogant and submissive. It began with a few words about how hard she had been working and her satisfaction with the project's progress, and how she now wished she had started writing it differently (presumably as an account to be published in its own right, rather than as a series of annotations to someone else's). She then considered the 'note' in which he had set out his demand that she withdraw her memoir, which she would have 'thrown aside with a smile of contempt' were he not 'so old & so esteemed a friend, *& one whose genius I not only so highly appreciate myself, but wish to see fairly appreciated by others.*'

Showing how well she understood him, she recognized that her failure to fulfil his wishes would become yet another entry in the lengthening list of 'disappointments and mis-comprehensions' that made up the ledger of his life. Nevertheless, he must understand that on this occasion it was he who was at fault, through his *'double-dealing'*. Her engagement had been 'unconditional', and could not be made subject to terms dreamed up in retrospect.

She then passed on to the basis of any collaboration they might have in the future. First, she insisted that if she were to continue working on his calculating engines, he must leave all practical matters to her, particularly when it came to dealing with others. This was a sensible suggestion, as he was even now proving.

Her second condition sounds less reasonable, indeed barely rational: that he gave his mind *'wholly* & *undividedly* . . . to those matters in which I shall at times require your intellectual *assistance* & *supervision'*, and do so in a more organized fashion than he had done during the writing of her notes.

Finally, she wanted to take over the management of the Analytical Engine project, developing a business plan which she would subject to the approval of trustees to be nominated from among his friends. This last proposal probably seems the most presumptuous. However, it anticipated the manner in which many high-technology companies would be set up in the late twentieth century, when inventors would often find themselves forced by investors to surrender executive powers over the companies they set up to market their ideas. Ada, admittedly, had no experience of management, but she could reasonably argue that she had better access to investment capital than Babbage, and she hinted in the letter that she had already found someone who might back a project to build the Engine.

Her tone now changed character once more, dissolving into a smothering cloud of piety and misty visions of grandeur. She boasted that she prized truth and God above fame and glory, and sanctimoniously claimed that in Babbage these priorities were reversed. She was as ambitious as anyone, she admitted, but only in order to fulfil her destiny. 'I wish to add my mite towards *expounding* & *interpreting* the Almighty, & his laws & works, for the most effective use of mankind,' she wrote, 'and certainly I should feel it no small glory if I were enabled to be one of his most noted prophets (using this word in my own peculiar sense) in this world.' 'Every year adds to the unlimited nature of my trust & hope in the Creator,' she proclaimed, in a tone totally out of character with the mischievous tone of her other letters to Babbage. It was almost as though she had become monetarily possessed by the ghost of one of her mother's cronies, as if Dr King had come to her shoulder and laid a hand upon her pen.

Then, as suddenly as she had puffed herself up, she deflated. She almost apologized for the sentences she had just written. 'My dear friend, if you knew what sad and direful experience I have had . . .' she wrote, referring to the continuing crisis regarding Medora.

At the very end of this exhausting journey through the mountain range of her feelings she apologized for the blots that covered the increasingly messy pages – blots of not just of ink but of tears – or perhaps the laudanum and claret prescribed for her stomach complaints and nerves.

She soon sobered up. The following morning she sent some more of her revisions to him, together with a brief note. 'You will have had my long letter this morning. Perhaps you will not choose to have anything more to do with me. But I hope [for] the best, & so I just write & send to you as if nothing had passed.'

Babbage would have more, much more, to do with Ada.

*

The Menabrea translation and notes appeared a few days later in *Taylor's Scientific Memoirs*. Babbage's statement appeared anonymously in the *Philosophical Magazine*. Ada's work was well received. De Morgan sent her a letter saying how good he thought it was. Friends were impressed. William was proud. Ada seemed to be happy.

A month later, Ada and Babbage exchanged letters suffused with mutual admiration and affection. He described her as the 'Enchantress of Numbers' and his 'dear and much admired Interpreter', she congratulated him on surrendering himself to her 'Fairy-Guidance!', adding in her familiar tone of mischievous flirtatiousness, 'I advise you to allow yourself to be unresistingly bewitched, neck, & crop, out & out, whole seas over, &c, &c, &c, by that curious little being!'

However, neither of them completely recovered from the argument. The publication of his statement and Ada's memoir was supposed to be the moment of truth, when the scientific establishment would fall to its knees, the public would clamour, the government would relent, the money would be found and the Analytical Engine would be built. Instead, it all remained as intangible as the machine itself, a pattern woven in air. When Cornelia Crosse, the widow of the electrician Andrew, visited Babbage at his Dorset Street home in the 1860s, he was still bitter about the circumstances surrounding the publication of the memoir. Sitting with him in the 'dreary, ghost-haunted' drawing room that had once been the setting for his glittering soirées, each time he mentioned Ada, he could not prevent himself from returning again and again to that hot, stormy August of 1843.

Ada, too, seemed to have been changed by the experience. The publication of the paper did mark the start of a new phase of her life, but not the one she had anticipated. Having used mathematics to make sense of one 'thinking' machine, she now set her sights on using it to master another, far more complex, mysterious and perplexing one: herself.

CHAPTER EIGHT

The Death of Romance

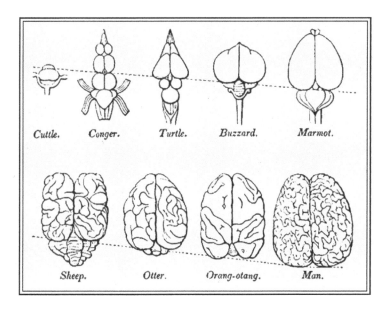

Cuttle.   Conger.   Turtle.   Buzzard.   Marmot.

Sheep.   Otter.   Orang-otang.   Man.

ONE OF THE FIRST PAPERS written by Dr William Benjamin Carpenter, before he became a famous physiologist, was entitled *On the Voluntary and Instinctive Actions of Living Things*. After meeting Ada in the winter of 1843, a few months following the publication of her notes about the Analytical Engine, the balance between these two sides of human behaviour was to become, for him at least, precarious to the point of ruinous.

The Dr Carpenter who arrived at Ashley Combe for his first encounter with Ada was a tall, wiry, earnest man two years older than she, just turned thirty. He had already made a name for himself as the author of *Principles of General and Comparative Physiology*, which he had written while at Edinburgh University. There he had been studying the physiology of invertebrates, and had shown a particular interest in the nervous system.

In 1841 he moved back to his home town of Bristol with his wife Louisa to take up a position teaching at the city's medical school. He soon attracted the attention of Annabella, who was staying in Clifton, the genteel Georgian suburb of Bristol that would become famous for its suspension bridge (another of Isambard Kingdom Brunel's engineering marvels).

Annabella and her son-in-law William had formed one of their pacts concerning the raising of Ada's three children. William despaired of their unruliness, and the failure of their tutor, a German called Kraemer, to correct it. The problem was getting worse, William complained to Annabella, because Ada would not intervene, nor would she let him. One solution, of course, was the customary aristocratic one of sending the children away to boarding school, at least the boys Byron and Ralph. However Ada apparently objected to the compulsory religious instruction and worship they would receive at an institution like Eton, William's alma mater.

So Annabella, whose opportunities for intervening in the lives of others had recently been drastically reduced by Medora's disappearance back to the Continent, took it upon herself to find a new tutor for Ada's children.

The young scientist who presented himself to her at Clifton seemed to fit the role perfectly. He was a nice, polite young man, clever but deferential. He had ambitions, but tempered by good sense. He had thought deeply about the great scientific controversies of the day, and come to conclusions that were largely in accord with Annabella's own. He was a Unitarian, was obviously very moral, and did not drink (he would later write a paper, entitled *The Use and Abuse of Alcoholic Liquors*, which recommended total abstinence). He had a nice wife called Louisa, upon whom he publicly bestowed many gestures of affection, which were reciprocated. He was, in short, the perfect candidate to nurture not just her grandchildren's intellectual and moral development, but that of her still unruly child.

Thus Annabella recommended to William that he appoint Dr Carpenter as the new tutor of the Lovelace children. What she did not tell her doting son-in-law was that she had also instructed Dr Carpenter to see what he could do about Ada, a mission that the doctor embarked upon with enthusiasm as soon as the opportunity presented itself.

*

When Ada first met Dr Carpenter, they were supposed to discuss what he could do for the education of her children. However, it was not long before Carpenter, following Annabella's orders, had turned the subject of the conversation around to Ada herself. He discovered a woman in a state of frustration and confusion. She, like him, had notched up a scientific publication (though a more modest one than his), and appeared to have a bright future before her as a scientific writer and researcher. But something was holding her back.

Her health was one obstacle. It was still precarious. She was prone to bouts of depression, as well as her chronic stomach condition. But worse even than this was an anxiety about her 'moral' – we would now call it mental – state. She had spent the best part of two years immersed in mathematics, and still they did not extinguish those 'objectionable thoughts'.

She confided much of this to Dr Carpenter as a patient to a doctor, a relationship which, initially at least, Dr Carpenter, cultivated by promising to compare notes with her personal physician, Charles Locock. Nevertheless, he also felt bound to tell Annabella what he had learned (something that Locock would never do), and wrote to her setting out his views a few days before he was due to go and stay with Ada at Ashley Combe.

He felt that, far from helping her moral development, her concentration on 'intellectual' matters had hindered it. Focusing on ideas tends to cause the mind to 'feed' on itself, which promotes egotism, he reported. 'We ought not to forget that God is Love, as well as Power and Wisdom,' Carpenter wrote, meaning that it was the development of Ada's feelings that should now take centre stage. She needed a course of 'mental training', which he himself could provide while attending to her children.

Dr Carpenter's diagnosis stood in complete contradiction to everything that had been proposed before. That hardly mattered to Annabella – the mention of morality and God was all she needed to approve of Carpenter's scheme. He had, however, been unspecific about one issue. How, exactly, did he propose to develop Ada's feelings?

Such matters were obviously regarded by Annabella as mere technicalities, and Dr Carpenter set off for Ashley with her blessing. There he was interviewed by William, who naturally endorsed his mother-in-law's selection and wrote immediately to congratulate her on proposing a man of such exemplary moral character and philosophical views.

Having secured his employment prospects, Dr Carpenter spent some time alone with Ada. While at Ashley, Ada loved to take walks along the precarious coastal paths, or rides up to Exmoor and the Quantock Hills, along the paths and tracks that Coleridge and Wordsworth had wandered fifty years before. Dr Carpenter no doubt accompanied her, and they talked for hours about science and, increasingly, the part of herself that had become so heavily entangled with it. Carpenter even suggested that they collaborate on a philosophical work, together exploring the metaphysical themes arising out of her first scientific child, the notes on Babbage's marvellous machine.

*

In some respects, Carpenter's philosophy was similar to Babbage's. He, too, believed that the clockwork universe had been wound up by God at the moment of its creation, and set off to run its course. He, too, thought that miracles were just the product of the activity of physical laws yet to be discovered. All are derived from one law, God's law, he argued, the 'Almighty *fiat* which created matter out of nothing'.

In his own first child, *Principles of General and Comparative Physiology*, Carpenter had even dared to argue that the human body itself fell within the jurisdiction of this Almighty *fiat*. The same science was involved in the study of the human body as the study of the movement of the planets, or the behaviour of chemical elements, or the layers of rock in a cliff face, or – though no one had yet gone *this* far – the evolution of species and the architecture of the human mind. 'The boundaries which at present divide the sciences disappear,' Carpenter had boldly asserted, 'just as the aeronaut, in enlarging his horizon, successively loses sight of the divisions which the art of man or the hand of nature has interposed to separate from each other, estates provinces and kingdoms.' As we shall see, this idea left a deep impression on Ada, and the first glimmering of what she thought would be the culminating project of her life.

Also like Babbage, Carpenter had found that his ideas brought him in contention with the establishment – but here the similarities between the two abruptly end. Where Babbage was confrontational, Carpenter was deftly conciliatory. He prepared the ground carefully for the appearance of his *Principles*. He consulted the Great and the Good of religion as well as science before publishing, and sent them presentation copies of the work afterwards.

When the inevitable critical backlash began, he responded immediately. Thanks to the work he had already put in, it was comparatively muted; amounting to a few accusations of materialism. What had begun as an exercise in 'fancy or in skepticism' had ended up as 'dogmatism or nonsense', said the critic of the *Medico-Chirurgical Review*.

The *Edinburgh Medical and Surgical Journal* attacked him for reducing the miracle of Creation to a machine that 'runs its allotted course without requiring the superintendence of the Creator'.

His response was not aggressive, as Babbage's would have been, but defensive. He emphasized the non-controversial elements of his argument, tried to show that it was 'safe'. The main contention had already appeared in a prize-winning student essay Carpenter had written at Edinburgh some years before, he pointed out, implying that it was not only old news, but carried the endorsement of that worthy institution's professors.

He also published an appendix in a sympathetic journal which included numerous testimonials from a wide range of established names: Babbage's friend, the astronomer Sir John Herschel; Peter Mark Roget, then secretary to the Royal Society and now better remembered for his thesaurus; Henry Holland, the London physician to Prince Albert; and James Cowles Prichard, the ethnologist who abolished, as a Victorian biographical dictionary put it, the 'physiological and ethnological distinction between Teutons and Celts, between the Hindoo and his English conqueror' – in other words established that they were all from the same human species.

This eagerness to please the establishment underlay Carpenter's relationship with Ada. He loved the idea of being in charge of her development as well as that of her children, of exploring her mind and, more exciting still, her morals. He was a member of the professional classes – which had not yet quite acquired the confidence or status they would enjoy in such full measure in coming decades. The idea that a man of such a station could become a trusted adviser and even intimate friend of an aristocrat – a *famous* aristocrat – was irresistible.

Following his first visit to Ashley, he worked hard at cultivating his new friendship, and Ada responded positively. She never much liked formalities, and in her dealings with Carpenter she quickly dispensed with them, addressing him frankly as an equal rather than a prospective employee. She did not even bother to sign the letters she sent him, which he found both astonishing and scintillating.

But more scintillating still were the confidences that she began to entrust to him, which were contained in letters she wrote to him in early December 1843, just as the negotiations for his post as her children's tutor were about to be finalized.

Having agreed to keep everything she told him secret, even from her mother, he stood open-mouthed as she began to unravel herself before him.

She now felt estranged from the people she was supposed to love, she told him; her husband, her children, perhaps even her mother. After years of study, she had become a 'walking intellect' who had become frightened of her own feelings.

She explained to Carpenter how this had come to be: about her teenage elopement, and perhaps of other lapses and indiscretions that followed; about Dr King's patent remedy: mathematics, which would, the good Doctor had so earnestly promised, banish the 'objectionable thoughts' that she even now knew to be gestating within her; about the Furies, and the fears that she was ungoverned and ungovernable. She was both electrified by the power of the feelings within her, and petrified of what they would do if they were discharged.

At times, her condition had plunged her into the blackest depressions, she told him, 'periods of mental slumber or death', as though she were in a dungeon, like Byron's Prisoner of Chillon . . .

> For all was blank, and bleak, and grey,
> It was not night – it was not day,
> It was not even the dungeon-light,
> So hateful to my heavy sight,
> But vacancy absorbing space,
> And fixedness – without a place;
> There were no stars – no earth – no time –
> No check – no change – no good – no crime
> But silence, and a stirless breath
> Which neither was of life nor death;
> A sea of stagnant idleness,
> Blind, boundless, mute, and motionless!

She may have even revealed to Carpenter that she once contemplated suicide. Two years before she had asked Woronzow Greig to send her a copy of Robert Christison's 1829 *Treatise on Poisons*, which she later explained to him had not proved suitable for her researches – a cover, perhaps, for the darker purpose behind her request for it. If she had contemplated poisoning herself, she seemed to abandon the idea quickly. She felt that her death was likely to be premature without her intervention, thanks to her chronic ill-health.

Carpenter's response to such revelations was not the sort of straightforward, clinical one that might have been expected of a man originally commissioned by Annabella to study her daughter's mental health. He wrote a long letter to Ada, addressing in the frankest terms the issues she had raised. He wanted to cherish and nurture her feelings. He could safely do this, he reassured her, because his own powers of self-control would ensure that nothing untoward happened.

It was a very dangerous letter, given that Ada's husband had no idea Carpenter was to act as Ada's moral tutor as well as their children's scientific one. This makes all the more extraordinary Carpenter's decision to enclose it, clearly marked 'Private' and addressed to Ada, as a postscript to a letter to William about his employment terms. It is hard to understand why he would have exposed both himself and Ada to such risk, unless he was trying to ensure a level of deniability – he could claim that, since he had sent the letter to the lady's husband, he had obviously not meant anything improper by it.

Ada, ingenuous Ada, did not mind nor care. Indeed, she reacted with excitement at finding a man who seemed to understand and sympathize with her. She identified with his diagnosis. She was indeed filled with the 'latent' feelings of 'affection' he warned her about. She agreed that she had been working too hard at her studies, and needed to loosen up a little. That might mean surrendering a portion of her intellectual powers, but it was a price worth paying.

Carpenter realized that she needed someone to become the object of these pent-up affections. Her husband could not fulfil that role, so he offered himself as a substitute. He knew she could not form any 'unbecoming' attachment to him, and she had his promise that he would never go 'too far'.

Ada took him at his word. She sent letter after letter brimming with feelings. In response, his deferential tone was quickly replaced by a more daring one. When she wrote to him about how her passions were burning her up, he jokingly conjured up a picture of her having spontaneously combusted into a heap of ashes, and of William commanding a servant to come and 'sweep away your mistress, and bring me another pipe'. It was an image rich with the sorts of domestic and scientific allusions that would have tickled Ada's fancy (as it did Dickens's, who famously assigned the same fate to Krook in *Bleak House*, basing his account on material he had received from, among others, the mesmerist John Elliotson).

Carpenter thrilled at her requests for more of his 'kind lines' – she hungered for them. They must see each other, he proposed. He suggested Bristol station as a venue, since she would be passing through on the train from Ashley Combe to London. She resisted, he insisted, and eventually it was arranged.

They met a few days before Christmas – underneath the station clock, no doubt. It was an awkward as well as brief encounter. Carpenter encouraged her to talk more about her problems, and she did so, at first without restraint, apparently at one point using language so colourful that it shocked him (though he would later claim that, of course, he was unshockable). Then she suddenly became aware that an elderly gentleman might have overheard them. She was terrified he would be able to identify who she was, and fled on the express to Paddington.

A few days later she wrote to Carpenter, berating him for his naivety, accusing him of saying things that had exposed her to danger.

Carpenter understood nothing about Ada's lifelong history of being an object of public curiosity and gossip, and took her rebuke as a sign of aristocratic imperiousness. In tones of mock reassurance, he dismissed her concerns, all but accusing her of having delusions of grandeur. He had only mentioned Ockham once, he claimed, and not one in ten thousand passengers on the Great Western Railway would know who lived there.

He also suggested that the 'revelations' she had shared with him might be delusions too, or at least imaginary, made up to shock a simple provincial doctor and his simple provincial wife. It was all just attention-seeking, he seemed to be saying, which, like delusions of grandeur, was a well-known symptom of hysteria.

Adopting the 'no, but seriously' tone so beloved of people who are always serious, he cast aside these speculations as flippant japes and started to be pious. She must learn to protect her mind, he warned, against the temptations that would inevitably lead her into evil. This would require great strength, and he would pray to God to provide her with enough.

Despite the misunderstanding at Bristol station, they were to meet alone several times more. The first occasion was at Ockham, soon after Christmas. Here, in the privacy of her home, with William away no doubt fulfilling his duties as the county's Lord Lieutenant, they came closer to each other than ever. She saw in him, as she saw in any man who treated her with such sympathy and understanding, who promised to care for her and watch over her, not a father – she never suggested any man could fulfil that role – but the makings of a brother. She felt safe with him. They were both anchored from drifting into danger by marriage, which in his case, he repeatedly reassured Ada, was such a happy one there was no possibility of him getting out of his depth. He even showed Ada letters from his wife Louisa, in which she expressed her approval of his increasingly intimate relationship with his prospective employer.

Thus Ada felt they had a licence to indulge in certain 'familiarities' which would not otherwise be permitted, a position that Carpenter was only too happy to affirm.

During his stay at Ockham, Carpenter beheld what he later described as the 'feminine parts of her character'. It thrilled him, and for the first time, they indulged in exchanging expressions of tenderness and even a few kisses.

After staying at Ockham for two days, Carpenter went up to London, and stayed there alone, probably making arrangements for his prospective appointment as Fullerian Professor of Physiology at the Royal Institution and a lectureship at the London Hospital. Then he received a letter from Ada inviting him to meet her at her London home in St James's Square.

Carpenter seems to have arrived for the meeting in a state of great excitement, the balance between voluntary and instinctive feelings he had written about as a student tilted decisively in favour of the latter.

Having spent a week in the city alone, away from friends, immersed in work, he yearned for her company – for more of the femininity he had discovered at Ockham, for more of the kisses that they had begun to exchange there.

Ada, however, became concerned that his endearments were not brotherly gestures, but of a quite different nature. He was being 'naughty', she scolded him, trying to cool things down, and jokingly said she would tell his wife about it. Carpenter did not take the hint, and imagined their strange half-carnal, half-filial relationship to be deepening. A few days later, he sent Ada a letter in which he gave full vent to the feelings he had now developed for her.

Ada replied immediately. Her tone was gentle, but her message stern. There had been a misunderstanding. Things were getting out of hand. They needed to put a 'check' on their relationship.

The sudden intrusion of a controlling hand, and the unexpected shock of it being Ada's, sent Carpenter into a panic. He realized that he was dangerously exposed. He had been developing a relationship with this famous aristocrat which he evidently assumed to be the preliminary of some sort of affair. Perhaps he thought that was how aristocrats behaved. Whatever his feelings, he could see his very handsomely paid job as the tutor of the Lovelace children, the house at Ockham that came with the job, his future as a physician and a physiologist, the reputation he had so assiduously cultivated with the eminences of the scientific and religious establishment, all of this turning to dust.

Cornered and desperate, he decided to deploy the same pre-emptive strategy he had adopted with his book on human physiology. He would write directly to Lord Lovelace, providing an account of his relationship with the noble lord's wife before she did the job for him. He would also get his wife to write to Ada, to set out exactly how intimately she was acquainted with her husband's intimacies, and how much she approved of them, at least on the basis he imagined they had been conducted – i.e., as some sort of medical treatment for her mental condition.

Obviously written while he was still lost in a red mist of terror, Carpenter's letter to William succeeded in achieving the exact opposite of what was intended: rather than allaying fears, it provoked them; rather than soothing his future employer, it insulted both him and his wife. He told William about the Ockham visit, how he had discovered the femininity of a woman who had described herself as becoming a 'walking intellect'. He explained how his feelings for her had intensified, but how he had felt they were still unobjectionable because they sprang from a simple desire to help her. He also claimed that he had always managed to keep a grip on his feelings, even before he had the additional 'safeguard' of a 'beloved wife' to rein them in. He added that he had been led to believe that her feelings towards him were equally wholesome, as they were those a sister might have for a brother (he was probably genuinely unaware of the special significance that such a statement might have when applied to the daughter of Byron).

He had no doubts of his usefulness to Ada, and looked forward to being able to 'watch over her mind' in the future. By helping her to control it, he argued, he would enable her to become healthier and happier, which would benefit her family as much as herself.

Carpenter had walked off the cliff edge, but a total lack of awareness of his position somehow kept him aloft. He now looked down, and beheld the yawning chasm about to engulf him. He tried to step back. He softened his tone, voiced his sincere regrets, begged not to appear presumptuous, sought retrospective permission to address his noble lordship as an equal because both William and Ada had always treated him as one.

It was too late. William replied immediately, his fury all but bursting the buttons of his lordly dignity. He regretted he had ever had anything to do with Carpenter. Neither he nor Ada had ever imagined him 'watching over' his wife's mind, just her children's education. She had not entrusted her confidences in him so he could in some way treat her; she was merely being friendly.

Carpenter was mortified by the response, and replied in terms that only made a bad business infinitely worse. He suggested that William could not see the merit of Carpenter's motives because of his class, and implied that if William's relationship with his wife had been in a better state the misunderstanding would never have arisen. As the ultimate gesture of his sense of righteousness, he announced he wanted to resign his post immediately . . . but could not do as he had already given up his job in Bristol for the sake of the Lovelaces. Instead, he asked to be given a few days to reconsider his position.

Carpenter did not tell Ada about his letter to her husband until two days after he had sent it, when he sent her a much longer one in which he sought to blame his own behaviour on hers. All the encouragement had come from her, he said. His feelings simply reflected hers, he had been a mirror of her morals. She had chastised him for assuming that he enjoyed so close a relationship with her when they had known each other for so short a time, but the intensity of their relationship meant they knew each other more intimately than old friends. He trailed off into non-sequiturs: she had hardly been in his thoughts . . . he would not forget his position again . . . unlike other men, he knew how to control his passions.

One of the themes that kept cropping up in Carpenter's letters was class. He took every opportunity to remind his antagonists, as he now saw Ada and William, of his social status relative to theirs, implying that they were to some degree exploiting their position in order to attack his. In this sense, the confrontation was a telling one, an early example of the professional classes testing their strength, beginning to assert a place in the social order that would soon make them the strongest challengers to, and ultimately the successors of, aristocratic hegemony.

William chose not to deploy any feudal droit de seigneur in his subsequent treatment of Carpenter. This was because there was an even loftier droit to consider, that of his mother-in-law, who had originally proposed Carpenter, and now apparently stood by him. Despite his earlier pledge to Ada to keep her confidences secret, Carpenter had done his duty and given Annabella his own account of Ada's behaviour during the affair. Annabella had evidently formed a very different opinion from William as to who was at fault.

*

Carpenter started teaching the three Lovelace children in early 1844, and he seemed to do a good job of it, as both the junior Byron and Annabella told their grandmother they were enjoying his lessons so much they were prepared to relinquish a trip to Ashley Combe to be with their mother in order to continue them.

The fly in the ointment was the property the Carpenters had been promised on the Ockham estate as one of the conditions of their employment. William kept saying he would attend to the matter, but failed to do so. This was partly because he was far more interested in a new project: the renovation of East Horsley Place, an estate neighbouring Ockham.

William had bought the property in 1840 from William Currie, a London banker. In 1845, he began an ambitious scheme to turn it into his principal seat.

William loved tunnels and towers. He had added both to Ashley Combe, the tunnels allowing tradesmen to approach the house without interrupting its view of the Bristol Channel. He now planned to implement similar schemes at East Horsley. The house had been designed by Charles Barry, the architect of the Houses of Parliament currently under construction in London. It was an undistinguished half-timbered Tudor revival mansion, and William set about turning it into the sort of mock-medieval castle so beloved of Victorian aristocrats. His first additions were a huge round tower on the west wing, and a great banqueting hall. His plans for the hall were particularly ambitious, ingeniously combining the modern with the medieval. It was to have a vaulted ceiling made up of wooden ribs that were so large they had to be bent using steam presses that William had himself helped to develop (he would give a lecture on the subject to the Institute of Civil Engineers in 1849, and was praised for his ingenuity by Isambard Kingdom Brunel, no less).

Such schemes were of little use to Carpenter, who was becoming progressively frustrated with William's evasions about the promised residence. The situation was made worse by the gap that seemed to be opening ever wider between William and Ada. By now, they rarely communicated, or even cohabited, Ada spending more and more time in London or Somerset. When Carpenter suggested to Ada a day trip for young Byron to alleviate the strains of a very punishing timetable, Ada told him to refer the matter to William, who accused Carpenter of trying to avoid his duties.

William took out his frustrations by dreaming up ever more grandiose plans for East Horsley, and ever stricter disciplinary regimes for his children. Carpenter found his employer to be increasingly inflexible, and unlike the wooden beams in the ceiling of his new great hall, no amount of pressure would bend him. He was particularly tough on the eldest, Byron, who had now become Viscount Ockham, the courtesy title of his father, acquired when he became earl. Young Byron was not interested in academic pursuits, and far preferred to spend his time labouring with the estate workers than learning from tutors. William considered this was inappropriate behaviour in the heir to an earldom, and felt the boy needed to spend more time exercising his mind. On one occasion, he made Ockham describe to his grandmother a riding accident he had just suffered, and do so in Latin. On another, he threatened to put horse's blinkers on the boy if he did not concentrate more on his studies.

In the midst of such domestic issues, Carpenter was the last person William felt like favouring with his attention, and when he did finally yield to perpetual badgering and produce a cottage for the tutor, it was in such a dilapidated state Carpenter could only refuse it.

Now realizing that he was being edged out of his job by a vindictive employer, Carpenter appealed to Annabella to intervene. Annabella's patronage did not extend to championing his case in favour of her beloved son-in-law's, and he was rebuffed. As before, Carpenter conscripted his wife Louisa to write an appeal on his behalf, but Annabella now resorted to her old tactic of returning correspondence unopened, which left Carpenter with no alternative but to sort out matters through Woronzow Greig, the Lovelaces' lawyer.

Carpenter was eventually forced into a humiliating retreat. He was offered £300 by Greig as a payoff, which he accepted. However, in one last desperate attempt to recover a little dignity, he said he would only take the money on condition that Annabella accept the letters from his wife which had been returned unopened. He must have calculated that Annabella was bound to accept such a undemanding stipulation to put an end to the whole ugly business. He calculated wrong. Annabella refused, but offered to pay a visit to Louisa as a compromise, which she conveniently did at a time when Mrs Carpenter was away.

One of the characteristics of aristocrats, at least the ones involved in this story, is that their *oblige* was never allowed to compromise their *noblesse*. Rank, in the end, always prevailed over reason and decency. Servants, employees and, as Carpenter so humiliatingly discovered, even professional advisers were immediately dispensed with when their interests threatened to impinge on those of their employers or patrons. Clearly the power of the aristocracy was still something to be reckoned with.

However, the lower orders did have one method for getting if not even then at least a little less uneven. With rank went not so much responsibility as reputation. A judicious rumour circulated in the right quarters could do great damage to aristocratic hauteur, and Carpenter set about spreading stories about both Annabella and the Lovelaces. However, even this endeavour backfired, as within a few months of his departure William could gleefully report that Carpenter had once again got into trouble over a young lady. 'The sex will ruin him,' William observed.

*

Carpenter eventually managed to rise above such shortcomings, and went on to have a distinguished and very prosperous career as a physiologist and writer.

Ada, however, now confronted the realization that she was never to enjoy a similar escape. The irony of the Carpenter episode is that while he may have been resentful of her rank she would probably have happily surrendered it for the chance to have his career. The £300 he had been offered as a payoff was exactly the same as the amount of pin money she received from her husband as her own annual income to cover all her expenses. It was a large sum by the standards of a servant or a manual worker, but small compared to the earnings of men like Carpenter, which could easily amount to three or four times as much (as part of the negotiations for Carpenter's original employment, his potential earnings for the year he was to work for the Lovelaces were calculated at £1,000).

Ada felt she was trapped, inadequate, wasting away. Her mind and feelings were paralysed in adulthood just as her body had been when she was a child. It all came flooding out one February evening in 1845 she spent with Woronzow Greig. Woronzow mentioned that for all the problems she may have had with her health, she at least enjoyed the blessing of a good marriage. Ada's response was a hysterical laugh. 'It seemed to me so cruel & dreadful a *satire*,' she wrote to him a few days later, apologizing for her behaviour. The brutal truth was that William meant nothing to her. Despite his best efforts to help her, despite him being a 'good & just man', despite her best efforts to summon up more intense feelings for him, she could not love him as a husband, if anything only as a son.

Having failed as a wife, she also felt she was failing as a mother. She could not but regard her children as anything other than 'irksome duties'. 'I am sorry for them,' she bitterly remarked. The only consolation she could offer was that they would find her a 'harmless and inoffensive parent, if nothing more'. 'My existence is one continuous & unbroken series of small disappointments, & has long been so,' she confessed, and she could see no means of redeeming it.

Her problems were not helped by her chronic health problems, her growing dependence on drugs and her repeated scrapes with scandal.

In the winter of 1844, Dr Locock called at St James's Square to see her. He found her suffering from an extraordinary swelling of the face that gave her a 'mad look' so peculiar and horrible he said he would never forget it. He immediately prescribed twenty-five drops – more than double the standard dose – of laudanum. It was, Ada later reported, 'like a release from Hell-fire'.

This happened at a time when Ada found her weight to be fluctuating wildly – a condition that her father had also suffered. She could go from feeling fat to thin in the space of a week. This was partly because she was subject to raging thirsts, and would slake them with bottle after bottle of 'thin fluids', just as Byron had done the night before she was born when, according to Hobhouse, he kept Annabella awake by knocking off the tops of soda bottles with a poker.

Dr Locock recommended that she should undergo what he called the 'opium system'. Annabella disapproved of the idea, but Locock remained the one person in Ada's medical entourage whom Annabella could not influence, and the system was duly adopted.

Locock was a great believer in the therapeutic effects of narcotics such as drugs and alcohol. He had already tried the claret system on her – prescribing judicious quantities of the wine to regulate her nerves. She had obviously found it of some benefit, and at one point proposed trying gin as an alternative.

The opium system required her to take a dose of opium either in the form of laudanum or the more recently discovered morphine every few days. On these, her 'laudanum days', there were to be no excitements or stimulations.

It was obviously on such a laudanum day that she had a vision of herself as the sun within her own planetary system. In a rambling letter to her mother, she conjured up this solar system, and began to populate it with planets. Locock would be one, she said, probably knowing how much it would annoy her mother. Another would be the Reverend Gamlen, the vicar who married her to William. He was a warm, carefree man who would sweep her up in his arms and tell her that she was 'quite my own little girl'. Neither Babbage nor Dr King was to be allowed into her orbit, but Mary Somerville, now in Italy but still very close to Ada's heart, was to be included. There would also be several comets, which she had yet to identify. Annabella herself, the 'Hen' as Ada affectionately referred to her, would have to decide for herself if she wanted to fall into Ada's gravitational pull. One name she did not mention at all was that of her husband William.

She had drawn up one of these lists before, in a letter to William. This time she called it her colony of friends and, besides Locock and Gamlen, it was also to include Frederick Knight and Dr Kay. Poor William, she teasingly allowed, might be admitted, but she was not sure what use she had for such an 'old Crow'.

Frederick Knight was a neighbour of the Lovelaces in Somerset, and would become well known in the area for his involvement in the enclosure of Exmoor for agriculture. Ada enjoyed his company simply because, like all her favourite men, he was a bit 'naughty' – but, unlike Carpenter, did not take the naughtiness too far or too seriously. She mentioned him in several of her letters to Babbage. 'I am more his ladye-love than ever. He is an excellent creature, & deserves to have a nice ladye-love,' she wrote on one occasion. On another she forwarded a note Knight had sent her. 'It will show you the tone of writing between us; & the sort of footing we are on,' she observed. 'Are you not amused at "your Ladyship" which just means, by the way that I am *anything* but My *Ladyship* to him!' She wrote in a similar fashion about other planetary bodies that came into her orbit, such as Sir William Molesworth, the owner and editor of the very influential *Westminster Review*. She went for a ride with him, and, though impressed with his intellect, was very disappointed by his gentlemanly behaviour.

Inevitably, such a free and easy manner with men would get her into yet more trouble. London society's busy telegraph of whispers continually buzzed with rumours about Ada, and occasionally these would become sufficiently noisy to surface in correspondence. 'That old gossip Mrs Park had made out that "a gentleman was with me" & that the *marito* was away,' she complained to Woronzow. 'I said quite coolly "Oh yes! And the gentleman's wife was in yr hands today."' On another occasion, William was forced to go through the visitors' book at Ockham Park to identify a 'traitor' who had passed on news about some indiscretion that Ada had committed there (it may have been the encounter with Carpenter). On yet another, Woronzow reported a rumour he had heard about her being seen with an unidentified man. An unrepentant Ada blithely remarked that without a name it could have been any of several.

She also still suffered the occasional attentions of sensation- and money-seekers trying to exploit her Byronic legacy. One 'G. Byron' wrote to Byron's publisher John Murray in 1843, claiming to be the illegitimate son of a liaison between Byron and the unconvincingly named aristocrat 'Countess De Luna'. It was the first of a number of melodramatic missives that he circulated among Byron's remaining relatives, and it was not long before he got to the Lovelaces. Demanding money with menaces of publicizing his kinship, it seems that William eventually agreed to meet him. William tried to lay a trap, but the 'extortioner' spotted it, and made his escape.

Such a combination of celebrity, notoriety and philosophical accomplishment made Ada the source of ever more intense public interest.

With no work having appeared from her following the publication of her notes on the Analytical Engine, people began to wonder what she was up to. Some even seemed to imagine she was some sort of intellectual Pimpernel, her occasional appearance in society merely being a cover for some heroic, secret mission into the heart of the scientific revolution. Then, in late 1844, a rumour began to circulate as to what that mission might have been. She, it was whispered, must be the anonymous author of a book just out that seemed to be regarded by reverends and even some scientists as the most dangerous publication to appear since the Marquis de Sade's *Justine*. It was called *Vestiges of the Natural History of Creation*.

*

In 1844 Charles Darwin, now deep in the development of his theory of natural selection, told a friend that publicly declaring that species could evolve was so dangerous it was like confessing to a murder. He could not yet bring himself to make such an admission, but someone could – the author of *Vestiges*. It was a brave move, but made anonymously. When *Vestiges* first appeared with no author's name attached to it, the Reverend Adam Sedgwick, an eminent geologist and keen collector of fossils, predicted 'ruin and confusion in such a creed'. 'It will undermine the whole moral and social fabric,' he predicted in a tone of escalating panic, and bring 'discord and deadly mischief in its train'. Sedgwick was one of those who suspected Ada to be the author of this scandalous work, mentioning his suspicions in a letter to Mcvey Napier, editor of the *Edinburgh Review*.

Considering the ideas and circumstances that inspired it, Sedgwick's nomination of Ada was an astute one. Like Ada, the author of *Vestiges* had a strong interest in the phrenological ideas of George Combe, and wanted to publish a work that made Combe's ideas accessible to school children. Like Ada, the author subscribed to the so-called 'nebulous hypothesis', which explained the formation of the solar system in terms of a gradual process of coalescence from a chaotic cloud – the 'inexhaustible womb of the future', as one writer called it – to an ordered cosmos. Like Ada, the author's assumptions about the application of natural laws to animal and human physiology were based on an intimate understanding of the work of one William Benjamin Carpenter.

Vestiges was the first popular British work to portray humans as part of a universe governed by natural laws. Without demur, it proposed that species evolved from primitive to more complex forms – indeed, the idea of progress that so excited the Victorian mind was here revealed to be a process at work throughout the universe. This was Darwin's confession of murder, not whispered to a scientific sect via an article in an obscure learned journal but bellowed to the world through the pages of a highly readable book that sold in its thousands, running to at least thirteen editions in Britain and over twenty in America.

It was immediately obvious that *Vestiges* was not written by a scientist. It contained a number of quite elementary errors (many of which were corrected in successive editions, with the help of William Carpenter), which provided an excuse for scientific critics to dismiss it without having to address the more dangerous matter of its evolutionary ideas. It was also very clearly written, containing none of the dense language that scientists now preferred to use in their self-conscious efforts to appear professional. It also criticized the scientific establishment for becoming paternalistic and conservative. For these reasons, some critics assumed it must have been the work of a woman.

Charles Babbage obviously heard the rumour that Ada might be the author, recommending that she must certainly read it if she had not written it.

She had not written it. The author was Robert Chambers, the Scottish writer and publisher who, with his brother, would later publish *Chambers' Encyclopaedia*. Chambers had a strong interest in both science and anonymity. His first book, about Walter Scott's equally sensational anonymous authorship of *Waverley*, had as its frontispiece a portrait of the mysterious writer, his face titillatingly obscured by a half-drawn curtain.

When he was asked years later what reasons he had for not putting his name to *Vestiges*, Chambers pointed to his house, in which lived his eleven children, and said, 'I have eleven reasons.' That was a measure of how frightened he was of the response the work aroused. Another was the decision to destroy all his papers relating to it. For this reason, it was possible, with a great deal of imagination, to implicate Ada in its authorship, given the parallels with her own philosophical development (she even suffered a mental breakdown at the same time as Chambers, and apparently for the same reason – overwork of the intellect). But even though she did not write the work, she certainly would have a role in helping to shape a more sympathetic critical reception for it.

On 15 November Ada wrote to Woronzow Greig, thanking him for sending her a copy of *Vestiges*. In the same letter she gives the first hint of her plans for the future, which were still being interrupted by her 'but too common *Gastritis*', as she called her various nervous complaints. 'I have my hopes, & very distinct ones too, of one day getting *cerebral* phenomena such that I can put them into mathematical equations; in short a law or laws for the mutual actions of the molecules of brain,' she wrote. 'It does not appear to me that cerebral matter need be more unmanageable to mathematicians than sidereal & planetary matters & movements, if they would but inspect it from the right point of view.'

Despite the matter-of-fact tone of its delivery, it was an altogether extraordinary proposal: she wanted to explore the possibility of developing a 'calculus of the nervous system', as she called it, a mathematical model that would show how the brain gives rise to thought, and how nerves give rise to feelings – an idea that would be revived over a century later with the study of artificial intelligence and neural networks.

If Darwin was dabbling with intellectual murder, then this was genocide, the murder of the entire human race. What was left of humanity if thoughts and feelings were the reduced to numbers and formulae? Nothing but a few spinning cogs and Jacquard cards in a biological calculating engine.

To respectable early Victorian opinion, this was an outrageous idea. As Harriet Beecher Stowe put it twenty-five years later: 'Physiological considerations of the influence of the body on the soul, of the power of brain and nerve over moral development, had then not even entered the general thought of society.' She was actually writing in connection with Ada's father Byron, whose own moral development, Beecher Stowe felt, had been fatally compromised by his physiology. But the same applied in Ada's time. For most people, morality never would be, never could be brought into the scientific orbit.

So what made Ada entertain such an idea? Why did she want to push it to such a frightening conclusion? Because she was not frightened. Indeed, she was comforted. To think of herself as a machine gave her some reassurance. It was a plea of diminished responsibility in the days before any such plea was permitted. It gave her a way of escaping the growing sense of guilt she felt for the disappointments, the disobedience, the 'weight of all my past iniquities'. She had felt she was 'condemned to liberty', she wrote to her mother in 1844. 'Would that I were a slave.' Well, if she could prove the existence of this calculus of the nervous system, she would prove she was a slave. The life sentence of freedom would be revoked.

In short, science was her salvation. 'Religion to me is science, and science is religion,' she wrote. 'I am more than ever now the bride of science.'

The sort of science, however, that would achieve this salvation was not the conventional sort. It would require what she called a 'poetical science', a type that unites reason with imagination. Who embodied such a unity? It is, perhaps, a measure of how far her mania was now diminished that she did not assume it would be herself.

Nevertheless, she did begin to make notes and consider strategies for having a go. The few documents that remain show that the first issue she wanted to address concerned the way nerves communicate information (or, as she put it, propagate 'impressions'). She observed that nerves known to perform different functions – those connected with sight, hearing, the digestive system, for example – all seemed to be made of the same material. So how was, for example, a feeling of hunger distinguished from a sound when the signal reached the brain? Not, apparently, because of the appropriate nerves producing two different chemicals. She wondered if it might be something to do with the 'movement' of the molecules in the nerves, in other words with the pattern of impulses.

It is tempting to see her groping towards something very profound here. Without the benefit of knowing about neurones or synapses, she was confronting the issue that is still taxing some of the best minds of the late twentieth century, people like the mathematician Roger Penrose and the neuroscientist Gerald Edelman. Is the brain a glorified computer, a machine for manipulating symbols that could as easily be built out of silicon chips or cast-iron cogs as flesh and blood? Or is there something special about its biology, about the way the 'nervous substance' works, that takes it beyond mere computation?

To answer such a question, she needed to find out more about the nature of the nervous substance, and this, she acknowledged, threw up 'many & great difficulties', chief among them being the design of 'practical experiments'. She needed to turn herself into 'a most skilful practical manipulator' of difficult materials – brain, blood, nerves. This would be difficult, as she had neither experience of experimentation nor access to a lab. She therefore set out once again to 'find a man' who could help her, just as she had done in her search for a mathematics tutor.

She started at the top, as only an aristocrat would. Michael Faraday was the country's leading experimental scientist. Over the past decade he had explained how electrolysis worked, revealing the laws governing the apparently alchemical power of electricity to change one chemical into another. He had also revealed the potential of electromagnetism. In 1820 the Danish physicist Hans Christian Oersted had shown how an electric current could create a magnetic field. Faraday discovered that the opposite also applied when conductive material, such as iron, moved through a magnetic field. He went on to demonstrate how this phenomenon could be used to make dynamos and motors.

Faraday's discoveries had an enormous effect on the already charged view most people had of electricity. They almost reinforced the occult quality it seemed to have as a force of nature, the 'vital spark', the quintessence. Electricity seemed to be emerging as the 'bond of union between the mind & muscular action', as Ada put it, the link between the mental and material realms that had for so long eluded scientific study.

Faraday was a fan of Ada's. He had once asked Babbage for a portrait of her, which Babbage duly supplied. He was a self-taught blacksmith's son and did not consider himself much of an intellectual or mathematician – indeed prided himself that the most complex mathematical task he had ever performed was cranking Babbage's Difference Engine. He therefore readily admitted that he did not understand Ada's Menabrea notes, but could tell from the response of others how 'great a work it is'.

Ada wrote to him, saying she would like to repeat some of his experiments under his supervision. Faraday, unfortunately, was ill – he attributed the cause to overexertion, but it seems likely to have been chemical poisoning. He had been working with a variety of toxic substances for years, including mercury, which might explain the memory lapses that he complained to Ada about. In the light of his condition, he declined her request 'most unwillingly'.

Ada was not put off. She busied herself getting hold of new research material and reading the latest books. She persuaded Robert Noel, a relative living in Germany, to send her any neurological information he could discover there. She also renewed her interested in mesmerism, despite her belief that it was the cause of the mental problems she had endured for the past three years. Her inspiration was the work of Harriet Martineau, an eager member of her mother's retinue of friends.

Martineau was the Unitarian author of numerous improving texts extolling the virtues of liberalism and laissez-faire to the middle classes, works like *Poor-Laws and Paupers Illustrated* and the better known *Illustrations of Political Economy*. She shared many interests with Annabella, including industrial education, the abolition of slavery and chronic invalidism, the latter forcing her to retreat to Tynemouth in 1839. She re-emerged from her illness five years later claiming that she had been cured by mesmerism, as practised on her by the flamboyant former Nottingham factory worker, Spenser Hall, who practised a variety of the art he called 'phrenomagnetism' because it combined phrenological manipulations of the skull with magnetic healing.

She vividly recalled her experiences in a series of *Letters on Mesmerism*, in which she described finding herself in a sort of twilight, being bathed in beautiful phosphoric light, experiencing 'strange inexplicable feelings', having sensations of 'transparency and lightness' and a complete recovery from her symptoms.

Ada read these letters, and expressed her enthusiasm to write some of her own to Miss Martineau, which she asked her mother to pass on.

Like her scientific friends, Ada remained unconvinced by the clairvoyance that some mesmerical enthusiasts, Harriet Martineau included, claimed to experience. She was particularly scathing of the 'new mesmerism' wafting in from America, which added a spiritualist twist to an already potent brew. But she remained fascinated by the way mesmerism seemed to reinforce the idea that magnetism, electricity and nervous energy were somehow linked together, elements of a single trinity of power. The question was: how? The quest to find an experimenter to help her continued, until a new and very promising name suddenly presented itself: Andrew Crosse.

*

Crosse was one of many 'electricians', electrical experimenters, at work in the late eighteenth and the early nineteenth centuries. Like the mesmerists, most were more showmen than scientists. Armed with their Leyden jars, Voltaic cells, leather straps, glass globes and brass electrodes, they would put on elaborate demonstrations for the benefit of fee-paying audiences. They would electrocute dogs and cats and then apparently bringing them back to life by making their lifeless muscles twitch, they would sling wires across rivers to carry a charge from one shore to the other, and produce enormous sparks that generated bangs louder than cannon fire.

Crosse was no showman. He preferred solitude. By day he would work alone in his laboratory, a converted ballroom. By night, he would wander the Quantock Hills that surrounded his remote Somerset house, Fyne Court, observing the wildlife, scavenging crystals from caves, making up rhymes and reciting them to himself.

He had no interest in performing demonstrations, giving lectures or publishing his results, and hated going to London. He did receive many visitors, including Wordsworth, Coleridge, Robert Southey, William Whewell, George Singer (the author of an influential book on electrical theory) and Adam Sedgwick (who spread the rumour about Ada writing *Vestiges*). But most of his work was conducted away from public scrutiny. The only signs of what he was up to were occasional flashes and bangs emanating from his laboratory, which resulted in him being dubbed the 'thunder and lightning man' by local farmworkers.

In 1837 his privacy was destroyed when he announced a sensational discovery. He had been working on an experiment to explore the effect of an electrical charge on the composition of various solutions. To do this, he slowly dripped the solution being investigated onto a large lump of pumice stone which was connected to a battery. After several days, he noticed that 'whitish excrescences or nipples' were beginning to form on the stone. A few weeks later, to his astonishment, he observed that these excrescences had turned into tiny crawling insects. He had apparently revealed that most mystical of mechanisms, spontaneous generation.

From Aristotle onwards, spontaneous generation had been postulated as the origin of life. Since the ancient Greeks, natural philosophers had believed flies came from mud, maggots from meat, mice from cheese wrapped in cloth. This idea was discredited during the Enlightenment, through the work of scientists such as William Harvey, the seventeenth-century British anatomist who proposed the biological dictum *ex ovo omnia*, 'everything comes from an egg', and the eighteenth-century Italian priest Lazzaro Spallanzani, who demonstrated that spermatozoa were necessary for the reproduction of mammals.

Crosse's discovery was therefore very much at odds with the prevailing tide of scientific opinion. He was fully aware of this, and when he announced his results to the Electrical Society endorsed the decision to have them checked by another electrician, a W. H. Weeks of Sandwich, Kent. Weeks was commissioned to try to reproduce Crosse's results independently, and report back his findings to the Society.

Meanwhile, news of his discovery broke in the *Western Gazette*, and it provoked a hailstorm of criticism. Religious leaders accused him of 'reviling' God, scientists of helping to keep old, unscientific prejudices alive. But he did have some supporters, notably Harriet Martineau and Michael Faraday. The latter denounced the treatment his fellow electrician had received, and even proposed calling the insect Crosse had discovered *Acari crossii*. But the shock of the response, of being subject to 'so much virulence and abuse, so much calumny and misrepresentation', as he described it to one friend, plunged Crosse into depression, and he withdrew back to his laboratory, where he could conduct his experiments in peace and privacy.

Thus hidden from the public gaze, assumed to be conducting secret experiments into the 'vital spark', a certain mystique began to attach itself to Crosse's name. This culminated with speculation that he was the inspiration for Mary Shelley's mad scientist, the real Dr Frankenstein. He was the 'modern Prometheus', who brought the flame of scientific knowledge, the very secret of life, from the heavens and gave it to mankind, and who now faced the wrath of the Gods.

The idea is not wholly fanciful. Mary Shelley records in her diary attending a lecture Crosse gave in London in 1814 on 'electricity, the gasses, and the phantasmagoria'. Perhaps what he described there may have come to her mind two years later, that evening in the Villa Diodati when, following a discussion about the supernatural and the principle of life, Byron challenged her, her husband Percy Bysshe and Dr Polidori to write a ghost story.

Certainly, Crosse's converted ballroom was a model of the Gothic laboratory, a veritable 'workshop of filthy creation', like Dr Frankenstein's – as Ada was imminently to discover.

In 1842, she had written to Crosse asking for more information about his *Acari* insects and the experiment that had generated them. Crosse replied with a detailed account, and directed her to the results of Weeks's own experiments, which had apparently also produced living insects. He promised to send Ada the equipment she would need to perform the experiment for herself. No record remains as to whether or not the equipment ever arrived, or whether she used it.

In 1844, Ada decided to contact Crosse again, sending him a copy of her Menabrea notes as a pretext and an outline of her plans to investigate the nervous system. It is not clear why she had not made contact with him before. Perhaps it was because she knew him to be a recluse, and anticipated a rejection. In fact, he welcomed the approach with open arms, and invited her to come and stay with him at Fyne Court.

Fyne Court was no more than twenty-five miles away from Ashley Combe. The Lovelaces had, in fact, been there in 1838, while he was away. According to a newspaper report, William and Ada were 'much gratified by their inspection of the elaborate electrical, chemical & philosophical adaptation of the premises'. Andrew had, in turn, visited the Lovelaces with his two sons Robert and John in 1843. This time Ada had been absent, and the Crosses had to be entertained by

William, who unsympathetically observed Andrew being overcome with vertigo when they went on walks along the high coastal path.

It was arranged that Ada would go to Fyne Court in November 1844. Before going, she sent a letter to Andrew warning him of her still delicate health. 'I think I may as well just give you a hint that I am subject at times to dreadful physical sufferings,' she wrote. 'If such should come over me at Broomfield, I may have to keep to my room for a time. In that case all I require is to be *let alone*. With all my wiry power and strength, I am prone at times to bodily sufferings, connected chiefly with the digestive organs, of no common degree or kind.'

Undaunted, Andrew arrived to escort her and William to Fyne Court on 20 November. Their departure was delayed until the 22nd, probably because William had gone down with the flu, and decided to stay at Ashley.

Thus Ada embarked alone with Andrew on what, from the tone of her letters, was one of the most enjoyable and stimulating visits of her adult life. The horses pulling the Lovelace carriage 'flew' along the western slopes of the Quantocks and up the great ridge leading to Broomfield, the village nearest Fyne Court. Along the way, she discussed with Andrew her plans, which he considered practical as well as exciting.

They turned into the long, sloping drive leading down to Fyne Court at half-past three in the afternoon, and spent the rest of the day reading and talking, staying up until the small hours.

Ada came down the following morning at the appointed hour to find no one around and breakfast not yet prepared. As her hosts gradually appeared, she quickly discovered that life there was lived in the midst of bewildering chaos and casualness. Even the layout of the house was haphazard, failing to observe those little delicacies that were so important to Victorian sensibilities. The lavatory, for example, led directly off the drawing room, making it quite obvious what 'errand' a person was on when he or she headed for the door. Ada was as much amused as embarrassed when she discovered on one occasion that the door had been locked from the outside, forcing her to set up a 'hue and cry' for the key.

Crosse had a daughter and two sons. She did not much care for the daughter or the younger son, Robert. But she developed a fascination for the older son, John. She referred to him as 'young' John to William, though in fact he was five years her senior. He had recently been in Germany, where he said he had been studying the latest mathematical and scientific developments. He planned a return trip to Berlin for six months the following year. This aroused Ada's keen interest, particularly when he pointed out how much more money scientific texts earned if they were published in Germany compared to England, where readers were more interested in fiction.

They talked and talked, late into each night. The convivial, permissive atmosphere, no doubt lubricated by quantities of medicinal claret, lured Ada into talking about subjects that were too dangerous to touch upon at home or among other friends – such as her father. For the first time, we hear of her speaking of him with 'rapturous admiration', of her expressing her '*passionate affection* for his memory'. The veil that she had once been forced to draw over his memory was now becoming unfurled, and she seemed to be falling in love with what she beheld.

Ada also fell in love with the informality of the Crosses, the fact that she was left to her own devices and treated as an equal (indeed, she told William that John regarded her like he would a man – though, as we shall see, she may have had an ulterior motive for suggesting this). Everyone muddled along, making no demands, having no expectations.

Such a regime was fine when it came to domestic arrangements (or the lack of them), but it extended to Andrew's scientific ones as well, which Ada found increasingly frustrating. The whole house was in the process of being haphazardly transmogrified into a laboratory, with crockery as likely to be used to store toxic chemicals as serve afternoon tea, and cellars to grow crystals and breed *Acari* as to ripen vintage wine.

Even the trees in the garden had been adapted to scientific use. A heavy cable, a third of a mile long, was suspended by elaborate glass insulators from their branches. The cable came through a window and terminated in the 'electrical room'.

The electrical room was the nerve centre of Andrew's operation. It was located in the converted ballroom to the rear of the house. The room still had its dance floor and long, elegant Georgian windows, but where there might have once been twirling couples performing elegant waltzes there was now only electrical equipment.

The most prominent item was a huge Leyden jar, an early type of battery. Crosse's was said to be the biggest in the world, with 1,600 cells capable of storing enough charge to kill an elephant. There were also several furnaces and crucibles, used in the preparation of metals and compounds.

The space Andrew used to perform his electrical experiments was up the small flight of steps that led into the hall's old organ loft. Here was to be found the apparatus he used to work with 'atmospheric electricity'. A large spherical brass conductor was connected to the end of the wire that hung from the trees outside. Near it was a brass rod, set into the floor, which was earthed to the metal pipes that drew water up from the lake. A lever varied the distance between the conductor and the discharge rod. A plaque was set next to this fearful apparatus inscribed with a warning message unhelpfully written in Latin: *Noli me tangere*, do not touch. (As might be expected, one of the servants failed to heed the warning, and got a terrible jolt when she tried to polish the brass.)

When the atmospheric conditions were right, the wire suspended over the garden would gather an enormous static electrical charge, which could be fed into the Leyden jar or used to generate great sparks between the sphere and the rod. In demonstrations, Andrew would use this charge to show how different metals would burn in the most brilliant hues, to ignite fireballs like falling stars and to create a miniature aurora borealis.

Ada was not much interested in such pyrotechnics. She wanted to find out about the effect of electricity on the nerves. However, thanks to the general level of chaos, she had few chances to investigate this phenomenon, and left a week later knowing little more about performing experiments than when she arrived.

Nevertheless, she went straight to London in a state of optimism and excitement, as charged as Crosse's Leyden jar with ideas and plans. She immediately wrote to William suggesting that John Crosse come to live at Ockham so that she could work with him on their joint projects – including one relating to *Vestiges* – while devoting future visits to Fyne Court to experimental work.

William did not detect any warning signals in this request – nor, perhaps, did she. They were now quite distant from one another, geographically as well as emotionally. When they did meet, there was often trouble. On one occasion Ada alludes to William hitting her, on another she was forced to flee after a terrible argument witnessed by some friends, leaving William with the impression that she would never return. To avoid such confrontations, they had evidently decided to live as independently of each other as possible, so it is perhaps unsurprising that William seems to have agreed to her request concerning 'young' John Crosse.

This new partnership collaborated on a number of works which combined John's expertise of German research with Ada's interests. She helped John with a review he was writing of *Vestiges*, which was to be published with a more extensive critique of Alexander von Humboldt's *Sketch of the Physical Description of the World*, the first volume of the influential German scientist's grand opus *Kosmos*. He, meanwhile, appears to have offered advice on a review she was writing of a paper by the German scientist Baron Karl von Reichenbach.

Von Reichenbach was one of the more colourful and irrepressible characters on the German intellectual scene. When he was a teenager he set off for the South Seas to set up a new Germanic state, a task that occupied him for three years before he was arrested by the French. Deported back to Germany, he started a successful manufacturing business, and dabbled in chemistry, which resulted in the discovery of those two unglamorous essentials of modern life, paraffin and creosote. He also built up what was reputed to be one of the best collections of meteorites in the world.

He was best known, however, for his views on animal magnetism, particularly his identification of a new force of nature he later dubbed 'od' (after the Norse God Odin). Od (more often rendered in English as 'odyle' and the 'odic force' because Reichenbach's term seemed . . . odd) was the animal equivalent of the electrical and magnetic forces. The paper reviewed by Ada, *Researches on Magnetism*, was written before he had developed the idea of od, but it is there in embryonic form, characterized as the 'magneto-crystalline' force.

The reason for Ada's interest in the Reichenbach paper is clear from the start. Reichenbach had reported seeing various phenomena that might indicate something about the relationship of the magneto-crystalline force and the nervous system. In particular, he had observed that patients who were particularly 'sensitive', i.e., of a nervous disposition, produced 'luminous emanations' when they came into contact with a magnet or magnetized wires. Ada considered this to be an important finding, not necessarily because she believed it to be true, but because it could be confirmed experimentally. Mesmerical phenomena were usually impossible to verify – indeed, Ada suggested in the review that the term mesmerism should be dropped from scientific discourse, as it had been so discredited by charlatans and frauds. But the sort of phenomenon reported by Reichenbach was different, as it could be objectively observed and therefore, if those observations confirmed Reichenbach's claims, scientifically analysed.

The experimental method she proposed should be used to explore this phenomenon was the still nascent technology of photography. The daguerreotype, then the most popular method of photography, was widely used in portraiture. But for Ada it was of more interest as a research tool, a means not just of proving the magneto-crystalline force's existence, but of studying and classifying the various emanations it was said to produce.

The Reichenbach review, the most substantial work of Ada's to survive other than the Menabrea notes, shows that she was not a particularly promising reviewer. But it does demonstrate how well she understood a scientific method which was still being formalized, in particular the key notion of repeatability, that experimental results only counted if they could be reproduced by others.

The review also demonstrated that her idea of a 'calculus of the nervous system' was being both broadened and refined into a new scheme: the discovery of an 'atomic law', as she called it, that united the mental with the material realm. It was at the atomic or molecular level that she saw this union developing. New ideas emerging from chemistry were beginning to reveal that it was those same electromagnetic forces which had such an influence on the nerves that held atoms together and determined their behaviour. Perhaps by studying the behaviour of these forces the mathematics that determined how they operated on particles within their field of influence, the behaviour of the nervous system, the machinations of thoughts and feelings, would be scientifically revealed.

*

John Crosse's review of *Vestiges* and Humboldt's *Sketch* was published in the *Westminster Review*, despite a fracas that provoked Crosse to complain bitterly to Ada about his treatment by the printer. Ada's Reichenbach review, however, was never published. There is no explanation for this among her surviving records. James Braid, the Manchester surgeon who invented the term 'hypnotism', published a more robust critique of Reichenbach's findings in 1846, but that was hardly pre-emptive. Several journals would publish reviews of a work acknowledged as significant as Reichenbach's.

Perhaps it was the quality of Ada's writing or analysis that was at fault. This seems unlikely simply because she did not apparently take the work beyond the stage of a first draft, as the manuscript that remains was covered with corrections, and was missing a section to be added later, once she had recovered the Reichenbach text from someone identified only as 'TC'.

Whatever the reason, it did not seem to dampen Ada's improving mood. In 1845, there are the first indications of a dramatic recovery in both her mental and physical health. 'I am very well, in very good spirits, very busy – and perhaps I might add not a little mischievous,' she told Woronzow Greig. She stopped taking laudanum and began to go out more. Her relationship with her children improved, too. She had learned a lot from her 'terrible experience', she told her mother, and now she wanted to make sure none of them became similarly afflicted.

As 1845 drew to a close, she contemplated her thirtieth birthday with a calmness that she had never shown before. She wrote to her mother, saying she no longer feared her 'influence' – indeed, quoting Hamlet, she wondered if she might not now prefer Annabella's influence to that of what she teasingly described the 'romantic school'. It was as though she had mastered her passions at last, bottled them in a Leyden jar, where they could be discharged at the pull of a lever.

As she entered the fourth and last decade of her life, she seemed to be acclaiming almost with a sense of relief the arrival of what the scientific age had always threatened and poets had for so long feared: the conquest of the spirit, the mastering of passion, the death of romance . . . except that, in her personal life, this moment also seemed to mark its birth.

CHAPTER NINE

Clinging to a Phantom

To Norman Abbey whirled the noble pair,
    An old, old monastery once, and now
Still older mansion of a rich and rare
    Mixed Gothic, such as artists all allow
Few specimens yet left us can compare
    Withal. It lies perhaps a little low,
Because the monks preferred a hill behind,
To shelter their devotion from the wind.

It stood embosomed in a happy valley,
    Crowned by high woodlands, where the Druid oak
Stood like Caractacus in act to rally
    His host with broad arms 'gainst the thunder stroke.
And from beneath his boughs were seen to sally
    The dappled foresters; as day awoke,
The branching stag swept down with all his herd
To quaff a brook, which murmured like a bird.

Don Juan, Lord Byron

THE NOBLE PAIR William King, Earl of Lovelace and his wife Ada passed through the Abbey's entrance gate on the evening of Saturday, 7 September 1850. Byron had given it the fictional name Norman Abbey in his epic poem *Don Juan*. His daughter knew it as Newstead, the ancestral seat of her father's family.

In letters to her mother, Ada represented the visit to Newstead as a mere diversion in the long tour of northern England upon which she and William had embarked a few weeks earlier. As she now approached the house, passing through the woods that had only recently recovered from the damage the Wicked Lord had done in his efforts to despoil the estate, she began a descent into the deepest anxiety and trepidation.

They rounded the corner leading up to the entrance, and saw the 'lucid lake' . . .

> Broad as transparent, deep, and freshly fed . . .

On its opposite shores stood the mock castles that according to local legend were the venues for orgies, and the pontoons from which nearly lifesize model frigates had once embarked to re-enact naval battles.

And then, in the gloom of an autumn evening, looming over the lake and mirrored in its waters, Newstead itself: a building split into two, one half, the newly renovated baronial hall, built of stones plundered from the other, the remnants of the Abbey. Thus Ada would have beheld the home of her ancestors for the first time, virtue next to vice, angels neighbouring demons, the house of God joined to a den of iniquity, one in ruins, the other restored . . .

> A glorious remnant of the Gothic pile
>   (While yet the church was Rome's) stood half apart
> In a grand arch, which once screened many an aisle.
>   These last had disappeared, a loss to art.
> The first yet frowned superbly o'er the soil
>   And kindled feelings in the roughest heart,
> Which mourned the power of time's or tempest's march
> In gazing on that venerable arch.

Ada and William drew up to the entrance to be welcomed by their host, Colonel Wildman, a former classmate of Ada's father who had bought the Abbey in 1817, within a few days of Ada's second birthday. Though she had never seen it before, she would have heard endlessly about it, because in restoring the mansion Wildman had revived its infamous reputation.

In the final years he had owned the Abbey, Byron had not been able to afford to maintain even the fabric of the building. A contemporary account records how the roof had all but collapsed by the time Wildman got there. The rain had seeped in for years, until the 'paper had rotted on the walls, and fell, in comfortless sheets, upon the glowing carpets and canopies, upon beds of crimson and gold, clogging the wings of glittering eagles, and destroying gorgeous coronets.'

Wildman had bought it for £94,500, saving both it and its former owner from total ruination.

His renovations had led to many of the features being changed. For example, he removed from the forecourt the fountain that had so prominently featured in *Don Juan*:

> Amidst the court a Gothic fountain played,
>   Symmetrical but decked with carvings quaint –
> Strange faces like to men in masquerade,
>   And here perhaps a monster, there a saint.
> The spring gushed through grim mouths of granite made,
>   And sparkled into basins, where it spent
> Its little torrents in a thousand bubbles,
> Like man's vain glory and his vainer troubles . . .

The Gothic fountain and its gargoyles now played in a more solitary position, in the middle of the Abbey's inner cloister.

But everyone agreed that, for all the changes, the atmosphere of the Abbey had been preserved – indeed, enhanced. Through his judicious use of Byronic relics and portraits, the place still resonated with the spirit of its former owner, and as a result it had become established throughout the country, indeed the world, as his most important shrine.

It had also become a shrine to the spirit of Gothic romanticism and horror. This was the home not just of Byron, not even of Childe Harold and Don Juan, but of the ghosts and monsters that came to haunt the Victorian mind. If the images of the Promethean scientist and his grotesque creations that were to saturate popular literature and later films came from the laboratories of electricians like Andrew Crosse, the images of the haunted mansions they inhabited surely came from places like Newstead.

One of the men who brought its architecture and history into the popular imagination was the American writer Washington Irving. In 1835 he had visited Newstead as part of a tour of the area, and quickly fell under its spell. He portrayed a place still enchanted with the myths of the Middle Ages: the legend of Robin Hood and his outlaw band of merry men who once lurked in the surrounding Sherwood Forest; the ghost of a 'goblin friar' who paced the cloister by night, portending some terrible fate that would befall the mansion's master; the Morris dancers and mummers who were even now to be found performing ancient rites and frolics in the servants' hall; the wizened old maid discovered living in a cave in a nearby quarry; the woodland that recalled to mind German fairy tales of knights and water nymphs; the mysterious lady dressed in white who roamed the Abbey grounds from dawn until dusk, refusing to speak to anyone who encountered her.

Most suggestive of all was Irving's description of the 'Rook Cell'. This was the name given to Byron's old bedroom, because its window overlooked a rookery. Everyone wanted to sleep in this room, hoping to pick up the reverberations of Byronic passions before time dissipated them.

'Never was a traveller in quest of the romantic in greater luck,' Irving wrote,

> for in this chamber Lord Byron declared he had more than once been harassed at midnight by mysterious visitors. A black shapeless form would sit cowering upon his bed, and, after gazing at him for a time with glaring eyes, would roll off and disappear. The same uncouth apparition is said to have disturbed the slumbers of a newly married couple, that once passed their honeymoon in this apartment.

It was into this mansion, this once holy, now unholy shrine still full of spirits and ghosts that Ada now stepped.

*

Wildman had evidently been looking forward to the visit. He had even boned up on the latest scientific topics so he would have something to talk about. However, he was to find all his preparations were useless. As he led Ada and William into the Abbey's great hall, she was almost mute, responding to eager questions with reluctant monosyllables.

Still, there was plenty to talk to William about. Both of them had just renovated large houses. William, whose own taste in decorations was described by many as dubious, gallantly admired the quality and refinement of Wildman's work. Though Newstead's great hall did not feature the steam-pressed beams he had installed at East Horsley, it did have the same profusion of medieval motifs: the heraldic devices and armour, the panelled walls, the oak screen, the minstrels' gallery. William had to admit the hall was both larger than that at East Horsley and more comfortably appointed, but then it did not have to serve as an entrance to the house.

After a brief tour of the main chambers, they were shown to their rooms, which were above the old cloister, overlooking the Gothic fountain. William thought them gloomy, which at least meant they matched Ada's deteriorating mood.

The following morning both Ada and William felt obliged to write to Annabella, who must have been anxious for news about her daughter's response to this immersion in the Byronic legend. Annabella knew for herself how the place could bring him alive in the imagination of those who went there – she had discovered as much when she made a secret visit to the Abbey in 1818.

She left an account of her experience in a journal, which was published by E. C. Mayne in her official 1928 biography of Annabella:

Just come from Newstead. The sunshine, the blue lakes, the reappearing foliage of the remaining woods, the yellow gorse over the wild wastes gave a cheerful effect to the surrounding scenes. My feelings were altogether those of gratification. In becoming familiarised with the scene I seemed to contemplate the portrait of a friend.

I entered the hall – and saw the Dog [one of Byron's beloved Newfoundlands]; then walked into the dining-room – not used by Ld. B. as such. He was wont to exercise there. His fencing-sword and single-sticks – beneath the table on which they stood, a stone coffin, containing the four skulls which he used to have set before him, till (as he told me himself) he fancied them animated.

I saw the old flags which he used to hang up on the 'Castle walls' on his birthday. The apartments which he inhabited were in every respect the same – he might have walked in. They looked not deserted. The Woman who has lived in his service regretted that the property was transferred. He should have lived there, particularly after he was married but his Lady had never come there, and 'she, poor thing! is not likely to come there now'. She said that he was 'very fond of Mrs L [Augusta], very loving to her indeed,' as if this were the only part of his character on which she could dwell with commendation, for she drew a very unfavourable comparison between him and G. B. [George Anson Byron, Byron's successor] in regard to charity . . .

The parapet & steps where he sat – the halls where he walked. His room – where I was rooted having involuntarily returned.

The experience inspired Annabella to pen a poem:

> . . . I remember when beside the bed
> Which pillowed last that too reposeless head,
> I stood – so undeserted look'd the scene
> As there at eve its habitants had been.
> Struck by that thought, and rooted to the ground,
> Instinctively I listen'd, look'd around,
> Whilst banish'd passion rush'd to claim again
> Its throne, all vacant in my breast till then;
> And pardon'd be the wish, when thus deceiv'd,
> To perish, ere of hope again bereav'd!

These lines were typical of the sort Annabella liked to compose, little demonstrations of spontaneous feeling designed to show that she amounted to more than 'methodised feelings'. Yet this, like her other poetic offerings, is unconvincing. This isn't an eruption of hot lava, but the careful sculpting of something supposed to look like it.

In Ada, however, unruly passion, banished by the reign of terror that she had been subjected to during her childhood, really might seize the throne. What then? Would all Annabella's work be undone? Would Byron finally win the heart of the daughter whose mind had been claimed by Annabella?

Ada's first letter home was reassuring. There was no hint of a passionate reawakening. Wildman had done an excellent job on the place, she said, better than any Byron could have afforded. She had not yet seen her father's rooms, but she had heard the well-told stories about his great-uncle, the Wicked Lord. She also learned that the huge fortune that had once belonged to her forebears had been seized by the Roundheads during the Civil War, and only a part of it restored afterwards. That explained why the estate and its owners had remained in such a state of dilapidation since.

However, for all the expense that he had lavished on the place, she felt Wildman had failed to bring it back to life. Indeed, it depressed her, feeling like the 'Mausoleum of my race', where history and life had become petrified. She could not wait to get away before she, too, was turned into stone.

William was more matter-of-fact, giving his considered assessment as a fellow renovator of stately homes of Wildman's work. He admired the interior decoration, though he found it fussy, and criticized the number of trees that had been planted in the park, which he thought excessive as they would block the views.

Ada did not write again for a week, but when she did her tone had changed completely. There was nothing too troubling in the opening paragraphs. She felt some affection for the place, she said, and wrote admiringly of Wildman. He had saved the house for posterity, and was devoted to her family's memory.

Having marked a more positive frame of mind, she now became exultant, each word providing Annabella with yet more disturbing evidence that something dramatic had happened to Ada. Ada reported an old myth that the Byrons would one day return and reclaim their inheritance, and it was in Ada's generation that this would be achieved. Ada proclaimed the visit as marking a turning point in her life, a 'resurrection'. 'I do love the venerable old place & all my *wicked forefathers*', she proclaimed.

This transformation in attitude came as a complete shock to Annabella. It had occurred on 10 September, three days after Ada's arrival. That morning, Ada had decided to go for a walk in the gardens. Like the house, these too were haunted with memories of her father. There was, for example, Devil's Wood, a thick, dark grove of trees overlooking one of the fishponds. It had been planted by the Wicked Lord and populated with statues of satyrs and fauns. Its trees were the only ones to survive the Wicked Lord's subsequent destruction of the estate, and it had become one of Byron's favourite haunts.

In the centre of the wood, by now dark and overgrown, there was an elm with two trunks springing from the same root. The trunks grew up side by side, and the branches that grew from them had become intertwined to form a single crown. It was into the bark of this tree that Byron and Augusta, during a last melancholy visit to Newstead before its sale, carved their names. This scintillating memento of the still publicly unacknowledged incest allegations had become so famous that P. T. Barnum, the American showman whose freak show featuring General Tom Thumb had just made a tour of England, offered Wildman £500 for the tree (the offer was rejected).

While wandering alone in the midst of this evocative landscape, Ada was approached by the colonel. He was by now so concerned and perhaps impatient with her taciturn manner that he had decided to confront her. When he did so, the rabble of unruly passions that she had been trying to suppress broke free and ran amok. She had found herself overwhelmed with feelings of sadness and loss among these powerful memorials to her father, she told him, and oppressed by the need to keep her real feelings to herself.

Having thus unburdened herself, she now felt able to talk to Wildman about her father without restraint, and the colonel found himself captivated by a lady he later said was more agreeable and cultivated than any he had met. She now wanted to know all about Byron, beginning with what the colonel remembered of him from their days together at Harrow.

Nothing is known about what Wildman told her, but it would have been favourable, probably adulatory. After they left school, the two had very little to do with one another, as Wildman had gone immediately into military service. But he had since become an authority on the poet's time at Newstead, and would have spent the latter days of her stay regaling her with the tales that he told all his visitors.

Thus, for the first time, Ada would have discovered the little things she had in common with her father, for example their love of animals. This was something that decisively separated them from Annabella. When Ada had once asked her mother why animals had to suffer, Annabella responded with a verse:

> You ask me why the Brutes should be
> As liable to pain as we,

Since they have not a heavenly rest,
Nor suffer here to rise more blest . . .
If victims of an early blow
They dream not it will lay them low.

Byron's attitude was very different. He had doted on his Newfoundlands just as Ada doted on her dogs Sprite and Nelson. While in a despairing mood he was reputed to have rowed out into the middle of the lake and thrown himself overboard, allowing his faithful dogs to pull him back to the shore. As Boatswain, his most beloved dog, suffered the final convulsions of rabies, he had cradled the huge shaggy creature in his arms, wiping the infected saliva from its mouth, and after it had died, he laid it to rest in a magnificent tomb he had built over the site of the Abbey's old high altar. Ada would also have discovered how, just as she had her little menagerie which she allowed to roam the corridors and rooms at Ashley Combe, he had his more exotic collection – a bear, a wolf, some tortoises and a hedgehog – wandering the galleries and halls of Newstead.

Inspecting the plunge-bath that Byron had adapted out of a passage the monks had once used to reach the monastic burial grounds, she would have heard of their shared love of bathing, and their efforts to control their fluctuating weight, she with her regime of galvanism and swimming, he with his of mud-baths and cold plunges, and regular visits to the local miller to get himself weighed.

She must also have been told the more sensational stories about life at Newstead – Byron quaffing from the flagon fashioned by a Nottinghamshire jeweller out of the cranium of a dead monk, the parlour games with the parlour maids, the drunken parties at which the guests would dress up as monks and take turns to lie in the stone coffin resting in the great hall.

The interaction of poignant details, outrageous anecdotes, all that she knew from her mother, his poetry and the Medora affair inevitably produced an overwhelming effect on Ada. She had reached, she believed, an 'epoch' in her life. She had come home.

Annabella responded to Ada's moment of revelation by writing a letter saturated with resentment and vehement self-justification.

Suddenly this mountain of cold reserve erupted with the lava of true feeling, and Ada found herself about to be engulfed.

She rebuked Ada for falling for the idea that she had somehow abandoned Byron, an idea spread by the 'partisans' of her former husband, among whose number Wildman was obviously now to be included. In fact, she had helped him even after the Separation, when she had stopped him from destroying his character by (she implied) preventing his planned elopement with Augusta.

Searching for a weapon with which to threaten her daughter, Annabella seized upon Ada's children. She would have nothing further to do with them if they were allowed to believe their grandmother to be the cold, calculating woman of Byronic mythology.

Annabella concluded by claiming that she hated having to justify herself, as she found any attempt at self-justification repugnant, but had been forced to do so for the sake of others (presumably meaning her grandchildren).

Ada was not the first to receive this astonishing epistle. It was enclosed with a note to William, and he held it back for a while to consider ways of softening the very powerful blow it was intended to inflict upon his wife, and perhaps to consider how it might alter his own devoted feelings to its author. Even if he accepted Annabella's fantastic delusion that she did not indulge in self-justification, the threat against his family would have shaken him. Annabella had helped enormously to raise not just his children but his own public profile, by providing loans of large capital sums and introductions to important national figures. She was also now publicly revered (thanks to all that self-justification) as a model of Victorian virtue. If any rift were allowed to form, it was obvious who society would blame.

To his credit, William did not try to save the situation by forcing Ada into any sort of retraction or apology. He gave the letter to her, and left her to reply for herself, which she did with a robust defence – perhaps enlivened by feelings of guilt – that implicitly accused Annabella of overreacting. Annabella's letter had been addressed not to her but to a figment of Annabella's imagination, Ada pointed out. Newstead was no hotbed of anti-Annabella or pro-Augusta agitation.

Wildman had not tried to sway Ada in favour of her father. He was no enemy of the cause.

To prove her point, Ada added a denunciation of her father. His conduct towards his own daughter, she announced, had been 'unjust & vindictive' (what she meant by this is unclear; perhaps it was a reference to him using her in his poetry).

Denunciations were not enough. Having failed to get what she was after by threatening to cut herself off from the grandchildren, Annabella resorted to saying that the incident had made her ill. Annabella had used this tactic before – claiming that Ada's behaviour had put her at risk of debilitation, even death. However, on this occasion the threat did not work. If her health had suffered, Ada bluntly told Annabella, it was her own delusions, not her daughter's actions, that were responsible.

The Newstead incident revealed a fundamental change in Ada. It was as though she had spent most of her life subject to her mother's mesmerical powers, but had now broken free of them. Newstead had shaken her out of a life-long trance.

*

After a week, William and Ada left Newstead to continue their tour. They stopped over at Bolsover Castle and whisked around Hardwicke Hall before reaching Castleton. This was another location with Byronic connections. Fifty years earlier, Byron had visited it with his cousin and first love Mary Chaworth.

Byron and Mary had taken a memorable candle-lit boat trip through Castleton's deep caves and subterranean lakes. When the tunnel ceiling dipped down low, the ferryman would wade through the water, pushing the boat ahead of him, while Byron and Mary lay flat side by side in the bottom of the boat. Being pressed up against the first love of his life provoked feelings so powerful in Byron that even twenty years later he felt unable to describe them.

It is unlikely Ada and William would have re-enacted such a romantic escapade, and after they left Castleton they went their separate ways. William took a tour of Lincolnshire, no doubt to examine the latest agricultural techniques and methods of animal husbandry, which remained two of his principal interests and were now the subject of his own scientific writings. Ada, meanwhile, went off to the races at Doncaster.

Horse racing had been a recurring theme during their tour, with one of its most eagerly awaited stops being the home of the Zetlands at Aske in Yorkshire, which they first visited before their arrival at Newstead. The second Earl of Zetland, Thomas Dundas, and his wife Sophia had become friends of Ada through their racing interests. They owned her favourite horse, one of the most celebrated of the time: Voltigeur.

Ada's enthusiasm for the horses had developed in the late 1840s. By the spring of 1850 it had reached such an intensity that she even told her mother about it, despite Annabella's strong disapproval. Gambling was a Byronic trait. It had been the ruin of Byron's father, and though Byron himself renounced it, he sought out the company of gamblers, intrigued by their thirst for risk, excited by the way that 'every turn of the card, and cast of the dice, keeps the Gamester alive', as he put it.

Gambling by now seemed to be keeping Ada alive, too. Her stakes were becoming ever higher, the risks ever greater. By 1850 she was already deeply in debt, owing £500 to a London banker called Henry Currie, and several hundred more to her mother, which she had borrowed on the pretext of needing a 'travelling fund'. She had made attempts to cut down on her expenditure to meet her growing obligations and retain at least a semblance of financial independence. She stopped having a personal maid, and kept detailed accounts of her domestic expenses (in which she would try to convince herself of her natural parsimony by including the most trifling expenditure, such as the cost of ice, which amounted to one penny per month). These measures failed to produce the desired effect, and by 1849 she was writing to Woronzow pleading with him not to let her mother know how desperate the state of her finances had become.

Annabella knew Ada had debts, but not how large they had become. At first, Ada explained them away as arising from incidental expenses on music and books. It was not a very convincing explanation, and Annabella gradually wheedled out another. By May of 1850, the day after the Derby at Epsom, Ada was openly admitting to running up gambling losses. However, she kept Annabella in the dark as to their size.

That year the Derby was won by her favourite, the Zetlands' Voltigeur, but she had evidently backed several other, less successful horses, and finished the day well down as a result. She backed Voltigeur during her visit to the Doncaster races, and was rewarded by a surprise victory over the much fancied Flying Dutchman in the Cup.

From Doncaster she met up with William for the final leg of the northern tour, a trip to the Lake District. At the first opportunity, Ada was off into the mountains, climbing Helvellyn and Skiddaw, shrouded in a heavy mist, and trekking across Borrowdale and Buttermere, where a rainstorm set off torrents that cascaded down the mountain side.

Ada seemed to be as happy there as she had been anywhere, as though she had liberated herself of some impediment. She returned to Ashley Combe alone in October feeling fit and energized.

The first thing she did when she got back was make contact with her friends, a small group of men who shared her attitudes and her sense of humour and who she could trust to be discreet with her confidences. That circle had become the focus of a whole new life for her, one that she had embarked upon soon after meeting John Crosse. This life still involved science, but a host of other interests, ones that have remained to this day obscured by mists of subterfuge and intrigue. 'The great object of life is Sensation,' her father had once written to her mother, 'to feel that we exist, even though in pain . . .' That now appeared to be her guiding philosophy.

Among the first people she wrote to was Charles Babbage. This was one of many letters she sent him during this period which contained cryptic references. In this case, it was to an unspecified 'invalid'.

The reference may well have been an innocent one – perhaps it was one of her beloved pets. She had already called upon Babbage to help her when one of her favourite birds died in November 1848. Babbage wrote to her to say he had organized for it to be sent to a taxidermist for stuffing – an appropriate end for a creature that, according to a bird-fancying friend of his, had died of overeating.

Perhaps, though, he was writing in code on that occasion. A letter he sent soon after it contains a strangely detailed post-mortem on the cause of the unhappy bird's death, and ends with a report about visiting a 'starling', which he intended to invite home despite the risk that it would arouse the jealousy of 'Polly', who was presumably a parrot. It is an odd missive, dwelling more than one would expect on such a trivial matter, and made odder by throwaway references at the beginning and end to much weightier matters, such as his 'Engine' and a debate at the Royal Society.

Ada did own a parrot, and kept starlings. Maybe Babbage was exploiting this to provide an innocent vocabulary for encoding secret messages. On the other hand, he may have simply been indulging her ornithological interests.

Alternatively, the invalid may just have been a person. On one occasion Babbage refers directly to 'her' being seen by his friend Sir James South, an eminent astronomer and physician – surely too eminent a man to attend to a poorly parrot. Indeed, this and a further letter point to an entirely different solution to the puzzle. Babbage later mentions Sir James giving him a book to be sent on to Ada's maid. The maid in question was probably Mary Wilson, recruited from Babbage's own household when Ada finally gave up trying to do without one. Perhaps she was the invalid. But then Mary was also becoming Ada's intermediary in her gambling activities, and was used on numerous occasions to deliver messages to secretive associates, so it may not have been her invalidism that was the source of their concern.

With suspicions aroused by these suggestive allusions, other mysteries begin to insinuate themselves into Ada's correspondence. In other letters to Babbage, for example, there are several references to a 'book'. Unlike the many other volumes mentioned in their letters, neither correspondent ever identifies this 'book' by its title, indicating it was in some way special. In a note she sent to Babbage in 1849, Ada wrote: 'I was so hurried & bothered the evening you came (not expecting you in the least), that I could scarcely speak to you. Yet I was particularly glad to see you, even in that uncomfortable way, & it was a very good thing as regards the book.'

They regularly referred to and passed this precious book between them. Indeed, it was so regularly used that Ada became concerned about the state of its binding.

One suggestion is that the book was the mathematical scrapbook Ada had first mentioned in 1840. This seems unlikely, as in other correspondence she simply referred to her scrapbook as a scrapbook. Why would she suddenly adopt the enigmatic language she and Babbage now used?

A more enticing possibility is that it was the manuscript for Ada's second intellectual 'child', the sequel to her Menabrea notes. Cornelia Crosse, Andrew's second wife, mentioned that Babbage had promised to show her some 'interesting papers respecting Lady Lovelace's mathematical studies' when she visited him in the 1860s. Could these have included an unpublished manuscript?

If they did, the subject was almost certainly related to mathematics and music (a theme that would become a preoccupation of some of the modern intellectual pioneers of computing). In a letter to her mother in the autumn of 1851, she mentions working on 'certain productions' related to music and numbers. The subject first appeared in the Menabrea notes, when she considered how an Analytical Machine might be used to compose music: 'Supposing, for instance that the fundamental relations of pitched sounds . . . were susceptible of such expression and adaptations [as the Analytical Engine could perform on numbers], the engine might compose elaborate and scientific pieces of music of any degree of complexity and extent'. Perhaps she had come to think that the 'fundamental relations' underlying musical harmony might somehow reflect fundamental relations underlying the operations of the nervous system and brain, which might explain why humans have such an instinctive appreciation of musical harmony. Rather than being the organic magnet imagined by mesmerists, perhaps Ada started to consider the human body as some sort of tuning fork, vibrating in sympathy with the music of the spheres. The mathematics of music might therefore provide the basis of the 'Calculus of the Nervous System' she was now searching for.

The same search also took her into the more speculative realms of other scientific studies, beyond electricity and magnetism into ideas about the 'molecular' structure of the universe. 'I want to see you again, to ask you some scientific questions,' she one day announced to Babbage in her characteristically bossy fashion. 'I can hint to you their nature now. I want to know what are the *German* scientific & mathematical authors & works which especially bear *my* subjects, viz: molecular forces & theories. I believe the Germans are *ahead* of us in some of these matters, if one can but get at them.' Her interest in these subjects no doubt related to their promise of linking the physics of the planets with the biology of the body.

She kept in close contact with another man to help her 'get at' the Germans, John Crosse, who had visited the country just before she met him at Fyne Court and would do so again soon after. However, unlike 'dear old Bab', as she now referred to Babbage, she sought the company of 'young Crosse' to stimulate more than her interest in German science.

*

From the moment he enters her life, John Crosse's presence is a looming one. He was an elusive character who preferred to cover his tracks. He was forced to destroy nearly all his correspondence with Ada – including eighteen letters thought to be particularly incriminating – as part of a deal brokered by Woronzow Greig to protect Ada's name following her death, and hardly a trace of his existence is to be found among the usual biographical sources.

But the little that does remain is enough to reveal a man quite unlike any other she knew. He was a passionate, forceful character with an explosive temper and caustic wit. That much is obvious in the one substantial letter he wrote to her which survives, which concerns his review of *Vestiges* and Humboldt's *Sketch* for the *Westminster Review*. It is preserved because Ada forwarded it on to Babbage, among whose papers it was to remain safely preserved:

> I have just glanced through a portion of the article in the Westminster – the printer seems to be some sarcastic dog who rejoices in the most mal à propos distortions & displacements. Thus he makes me (or rather Humboldt who is far from being so original) say that

the time of the earth's revolution around *the sun* (instead of her own revolution) depends on her volume! Proper names of eminent individuals are defaced without pity – Commencements of paragraphs are made to end them. *Condensations* of the original are given as *quotations* – Greek words are transmuted into forms that would puzzle Bopp, Pott, Grimm or any other monosyllabic philosopher (the very names of these fellows are most truculent roots) – Mr Babbage's note, which belonged to the Oceanic part of the argument (where I requested its insertion) is placed in solitary dignity amongst all sorts of things that don't belong to it – And to crown all, the notice of 'the Vestiges' . . . comes in as a weary straggler instead of being dovetailed into its proper place.

Another scanty but telling piece of evidence attests to the intensity of feelings this man aroused in Ada. When he died in a state of apparently respectable gentility at Fyne Court in 1880 he bequeathed to his children two artefacts. Four generations on, they are still owned by the family. They are the bloodstone signet ring and lock of hair that Byron had sent to Ada along with his final letter to Annabella, the only heirlooms the father personally bequeathed to his daughter.

To Ada, the ring and the lock of hair must have been like Ariadne's thread, strands laced through the labyrinth of her life and family history, linking the outside world to the dark interior where roamed the mythic Minotaur – half grotesque beast, half beautiful man – that in the popular Victorian imagination her father had become. Visiting Newstead was like visiting the threshold of that labyrinth. She could smell her family's past emerge from it like hot, pungent breath. By giving these artefacts to Crosse, she had made him her Theseus, someone to follow into the labyrinth's convoluted depths.

It was by taking this step that she suddenly became exposed to the world her mother's influence had kept beyond her reach. She began to meet a more colourful and altogether less respectable class of people. An impecunious member of the Augusta Leigh's family (it could have been any of them – the entire tribe was in a state of almost perpetual financial crisis) emerged from the shadows to offer her a rifle and a set of pistols that he claimed had been owned by her father.

She developed a strong liking for another shadowy character,

Babbage's friend Fortunato Prandi. Prandi is straight out of a Wilkie Collins story, a member of the community of spies and exiles that gathered in London in the 1840s and 1850s to hatch their plans for revolution back home. He was an associate of the great Italian patriot Giuseppe Mazzini, who agitated tirelessly for republican revolt in Italy from his London base.

Mazzini ignited a famous and important debate about the relationship between state and individual by proving that the British government had been opening his (and no doubt Prandi's) private mail. This is perhaps why, at some time in the late 1840s or early 1850s, Ada chose to avoid the postal system in communicating an extraordinary note to Prandi (though pressing urgency seems to be a more likely reason).

The note was sent anonymously to two addresses, one being the Athenaeum Club in London's Pall Mall, in the hope of catching him for some secret mission:

> Dear Prandi. I have a more *important service* to ask of you, which *only you* can perform. I had hoped to find you; – but I leave this note here, trusting that you may come in in an hour or two . . .
>
> I can *in writing* explain nothing excepting that you must *come to me at 6 o'clock*, & be prepared to be at my disposal till midnight. You must be *nicely* yet not *too showily* dressed. You may have occasion for both *activity* & *presence of mind*. Nothing but *urgent* necessity would induce me thus to apply to you; – but you may be the means of *salvation*. I will not *sign*. I am the lady you went with to hear *Jenny Lind* [the singer]. I expect you at 6. –

There is no way of knowing what 'service' Ada wanted Prandi to perform, and why only he could perform it. From her instructions about his dress, it presumably involved appearing in public, perhaps at a social engagement or theatrical event, where he was to blend in.

Whatever she was up to, Ada seemed to delight in it. As the Victorian elite became ever more bourgeois in its manners and attitudes, so Ada became ever more outrageous. This was reflected in her views as well as her actions, in the growing extremity of her 'heresies', as Hester referred to her sister-in-law's sometimes shocking religious opinions. In a brief but characteristically intimate correspondence with Edward Bulwer-Lytton, the novelist and politician who, like her father, notoriously separated from his first wife, she revealed herself to have become all but a full-blown materialist. 'Futurity', the word then often used to refer to a belief in the afterlife, was nothing more than a physical instinct, a reflex, like hunger, she told Lytton. She did not believe it had any spiritual dimension at all. 'I am too honest with myself to cling to a phantom . . .' In this, as now in so much, she was expressing the same views as her father. 'I will have nothing to do with your immortality,' he once wrote to Francis Hodgson, a clergyman and Cambridge tutor who saw it his purpose to save Byron's soul. 'We are miserable enough in this life, without the absurdity of speculating upon another.'

Another man who seemed to bring out the Byronic streak in her was Richard Ford, son of a magistrate (who was also a government spy, and may have recruited Wordsworth as an informant). Ford was a travel writer, also, as his letters to Ada reveal, a fervent anti-Catholic, and a mischievous correspondent. He addressed her as 'My Senora & Duena mia', an outrageous familiarity in polite society but one which Ada, preferring the unpolite, probably encouraged. He poked gentle fun at her husband William, feigning an admiration for the 'ductile irrigation' systems he had installed as part of his ceaseless renovation of Ashley Combe.

Ford also took a central role in coordinating Ada's circle of friends, which seemed increasingly to be taking the more sinister form of a ring of confederates engaged in a high-risk mission relating to Ada's latest interest, indeed passion: gambling.

*

A particularly imaginative theory about the mysterious 'book' referred to in Ada's correspondence with Babbage is that the word was being used in the gambling sense, to refer to a record of bets. According to this idea, the book was based on some sort of mathematically based betting system she and Babbage had jointly developed.

In his memoirs Babbage mentioned that a 'considerable time after the translation of Menabrea's memoir' he started to consider whether it would be possible to devise an automaton based on the Analytical Engine that played simple games of skill such as tick-tack-toe. He thought such an application of his technology would attract more public and therefore political interest than the less glamorous promise of calculating logarithms and nautical tables. In another context, he also wrote a paper about games of chance, in which he tried to formalize ways of calculating odds.

Ada knew of this work, indeed encouraged it. 'Sometimes I think . . . of all the Games, & notations for them,' she wrote to him on one occasion. 'If any good idea should accidentally strike me, I will take care to mention it to you.' 'I want you to complete something,' she scolded him on another, when she detected his interest in the tick-tack-toe machine flagging, 'especially if the something is likely to produce silver & golden somethings.' It is conceivable that the 'something' was a machine that calculated odds, which applied to gambling might produce those 'silver & golden somethings' she so desperately needed.

It seems unlikely that a man as actuarial as Babbage would have participated in developing any sort of 'machine' to calculate odds, or participated in any other gambling scheme. However there is no question that, by 1851, Ada was using mathematics in her gambling activities. She was also busily turning her circle of closest friends into a syndicate, which evidently had the job of raising money to finance her scientifically selected bets. Its membership comprised Richard Ford, John Crosse, someone called Nightingale (almost certainly William, father of the nursing pioneer Florence, who Ada had visited during her tour of northern England), and two mystery figures identified only by their surnames, Fleming and Malcolm. It was to this group of men that, as the 1851 racing season approached, she proposed making a book – taking, rather than making bets.

It was a daring proposal, and some of the syndicate members balked at the idea. Nightingale seemed poised to pull out. Richard Ford admired her audacity, but was nervous. Ada, however, was confident. She was sure that she was on the threshold of winning a fortune.

But why did she want a fortune? What was it for? This is one of the most puzzling mysteries surrounding the whole episode. She undoubtedly needed some money to pay off gambling losses; and more to buy the sorts of gifts she had begun to lavish on Crosse, such as a suite of furniture for his home in Reigate, Surrey. But it seems unlikely that even these demands would require the thousands of pounds she so desperately sought.

An intriguing possibility was that she was making one final bid to raise the money Babbage needed to build his Analytical Engine. She had suggested she would help him do as much when she had completed the Notes. References in subsequent letters show that she remained interested in his efforts to raise the finances, and sympathized with his frustration at the lack of interest shown by the public and government. Could this explain why she was prepared to stake so much on her gambling activities, why she set such store on the mathematical formulae – 'wonderful combinations', as a bewildered Ford called them – that she was now convinced would lead her to gambling success? As she had shown in her Notes, the Analytical Engine promised to put the Promethean power of mathematics into human hands. Did she believe that the same power, applied to the speculative world of gambling, would actually help bring the Engine into being?

If so, then it was not just her bank balance but the future of technology that rested on the outcome of the opening fixtures of the 1851 racing season. For these were to provide her with the first chance to test out her new mathematical system. If it succeeded, she stood to win thousands of pounds – enough to help Babbage build at least a demonstration model of the Analytical Engine, and thereby perhaps ignite the steam-driven computer revolution imagined by William Gibson and Bruce Sterling a hundred and forty years later.

*

By the early spring of 1851, the groundwork for the first deployment of the new betting system had been completed. In particular, Ada, apparently encouraged by Crosse, had persuaded William to write a letter giving her permission to gamble. It was essential to her scheme, as bookies and punters would not have otherwise dealt with her, not least because an aristocratic wife, despite her class, owned very little independently of her husband, and therefore could be taken for very little. Having secured the necessary permit from the trusting William, she handed it over to Crosse, no doubt justifying her actions on the grounds that it was her money the letter enabled her to use, as William had acquired it as part of the wedding settlement.

The other important figure in the preparations was Malcolm. He was now Ada's point man, a professional punter who went to the courses on her behalf on race days. It seems that, with his help, she took bets from members of the syndicate, and, piling risk upon perilous risk, used that money to finance further bets placed with professional bookies.

The first test of the system Ada had developed probably came with the York Spring Meeting. Ada's favourite, Voltigeur, was running, and she evidently was putting most of her money on him.

Voltigeur lost, and so did Ada, badly. It was her 'Doomsday', she told her mother, hoping that jocularity would disguise her desperation. With her losses now multiplied, all her hopes rested on the Derby, held on 21 May. It is unclear whether she backed Voltigeur on that day; it seems more likely that her fortunes rested on those of several horses, and the continuing support of her loyal associates.

She lost again. A note from a tipster called H. L. Browne still among her papers reminded her in a 'told-you-so' tone that he had advised her against the horse she had backed – though his own tip, too, had obviously fared no better, as he points out that it would have undoubtedly won had it not been for the weather.

A few days after the Derby, she met with the other members of the syndicate to settle accounts. The full horror of their situation then emerged.

It was revealed that the losses amounted to an astronomical £3,200. Malcolm had himself lost £1,800. Like Ada, he had nothing like the means to discharge such a debt. His annual income was the same as hers, and this had to cover alimony for his wife, from whom he was separated, as well as his own living expenses.

Ada managed to raise the money to cover her own losses through loans from Ford, Crosse and Fleming. That still left Malcolm's debts. He was evidently prepared to resort to extortion to get himself out of trouble, and recognized that Ada was a vulnerable target. She had mentioned to her mother a year before a report in one of the newspapers about a countess's gambling activities leading to her husband's suicide. If news of her own antics were ever publicized, it would inevitably revive the whole issue about Byron's moral legacy, possibly the Separation debate too, and the human cost of such exposure might be just as tragic.

She had no choice but to turn to William. His reaction is unrecorded. All that is known is that he wrote to the Zetlands for advice. Through their racing connections they would have knowledge of people like Malcolm, and would know how to deal with them. There is no record of their reply, but they apparently suggested he settle, since that is what he did, despite it being a course of action that a man as proud as William would have found hard to follow. He offered to lend Malcolm the money he needed on the condition that the man sign an undertaking never to reveal what Ada had done.

Soon after the Derby incident had run its course, William received another piece of bad news concerning Ada. Two medical opinions had been sought about her health, which was increasingly prone to relapses. Having considered the symptoms, the physicians, a Dr Lee and Sir James Clark, came to the same view, which they reported to William: she had cancer of the womb.

Charles Locock, the Lovelaces' family doctor, was called upon to perform the internal examination needed to confirm the diagnosis. It was a delicate business, but if anyone should do it, he was regarded as the best person. He was Ada's most trusted physician, and had the specialist knowledge required.

Locock was one of the Victorian era's most respected gynaecological experts. In 1840 he had been appointed the Queen's accoucheur, and attended most of her births. In 1854 he also performed an examination on Ruskin's wife Effie, to confirm that her marriage had never been consummated. It was Locock's evidence that led to the marriage being annulled.

Despite his experience and reputation, Locock's diagnostic skills appear, in retrospect, to have been questionable. Whenever Ada was struck down by illness, he only seemed to have the vaguest idea as to the cause, and prescribed from a very limited range of treatments, which invariably included laudanum and hot baths. On this occasion, however, there could be no doubt of the problem. Locock found Ada's cervix to be riddled with cancers. He told William what he had found, but they decided not to tell Ada.

In the mid-1840s, Ada seems to have experienced a period of comparatively rude health which also produced a marked improvement in her mental condition. She often felt energetic and this translated into a new-found optimism about the future. There were no further violent oscillations in her mood, no more moments of wild exaltation.

Then in 1848 things started to get worse. She suffered a series of what she called heart and rheumatic attacks, which were treated with wine or laudanum. She also tried chloroform, which had first been used as an anaesthetic during an operation the year before.

She suffered one of her worst attacks at the offices of John Murray, the son of Byron's publisher, and himself the publisher of texts which, in their own way, were to prove as explosive as Byron's, such as Darwin's *Origin of Species* and Lyell's *Principles of Geology*. Whether she was there for professional reasons or purely social ones goes unrecorded, as the whole visit became overshadowed by her suffering a series of 'spasms'. Murray himself had to resuscitate her.

She was struck down once more at the Zetlands', when she visited them with William during their tour of the north. On that occasion she was attended by the Zetlands' own physician, a Dr Malcolm. He prescribed plenty of rest and an immediate stop to her writing (he did not specify what writing, but since she continued with her correspondence, it was presumably the mysterious book on which she was working at the time). The regimen apparently worked – she certainly felt so. She suffered no further attacks during the rest of the tour, and felt strong enough to set off on perilous expeditions into the mist-swathed mountains during their visit to the Lakes.

But in 1851 her condition worsened quickly, and she experienced a series of heavy periods. These no doubt triggered Dr Locock's intervention, and even though she was not told the full truth, she certainly suspected that she was running out of time.

Her response to this realization was not self-pity but a new determination to make the best of the years she had left. 'I'd rather have 10 or 5 what I call *real* years of life, than 20 or 30 such as I see people usually dawdling on, without any spirit,' she wrote to a friend.

William, meanwhile, had to consider how to recover from the two dreadful blows he had received in such rapid succession, and how to cope with the complex feelings they must have aroused – anger at her irresponsibility, pity for her suffering.

The decision he came to was a fateful one, its consequences more extreme than he could have guessed, not just for Ada, but for himself, and for his descendants for years to come: he would tell Annabella everything.

CHAPTER TEN

Beyond the Shallow Senses

19 June 1851. Annabella was staying at Leamington Spa, undergoing one of her endless series of health cures, when William came to see her. He arrived unannounced at 11 p.m., when she was preparing for bed. The disturbance annoyed and upset her, and she felt no better when she discovered who the visitor was.

He had brought a letter from Locock reporting on the grave nature of Ada's illness. He had also decided to tell her everything about Ada's gambling activities.

He would no doubt have expected Annabella to be angry at her daughter. He may even have considered apologizing for her. What he could not have anticipated was Annabella turning the full force of her fury upon him. He had revealed the worst flaw a man could have, weakness. Showing no interest in the reports about Ada's health, she accused William of failing to care for Ada's moral welfare; he had allowed her to drift into immorality and abandoned her there. His decision to let her go alone to Doncaster during their tour of the north was proof enough of this. If only Ada had turned to *me*, Annabella bitterly asserted, I could have saved her.

In one of her more breathtaking examples of self-delusion, one that would surely have provoked William into a counterattack had he not been so in awe of his mother-in-law, she claimed never to have interfered in his or Ada's lives, but now, discovering what a mess they had made of them, felt compelled to give her judgement.

William had invested a huge amount to win his mother-in-law's approval. The school he had built at Ockham according to her educational principles was virtually a monument to her. She had acted as a substitute mother to him. It was to her he turned for advice, approbation, succour.

He returned home with this treasured relationship in tatters. And if he had harboured any hopes that her feelings would soften with time, they were quickly destroyed. Every bewildered, submissive letter he sent to her in subsequent weeks provoked even fiercer condemnations and curt dismissals. The old implacability was revived, and no appeal to reason, to justice, to sentiment would shift her from the conviction that William had betrayed a trust, and was now only worthy to be considered an obstacle between her and her daughter's urgent need of redemption.

Making a last but precarious stand on the crumbling remnants of his dignity, William responded by gently but firmly refusing Annabella access to Ada on the grounds that she was too ill. He was not simply being vindictive; Ada did not want to see her mother either.

Ada's relationship with Annabella had been troubled for some time. It wasn't just particular incidents, such as the Newstead visit. It was the way Annabella had surrounded herself with those dreadful Furies. They were the principal cause of many of her grievances, spying on her, spreading lies about her (or rather truths she did not want others to hear). And, worse, Ada had recently discovered that Sophia De Morgan, a woman who was supposed to be an ally, was one of them, deceitfully exploiting Ada's confidence. Who could she trust?

The position deteriorated when Annabella developed a passion for a charismatic cleric called Frederick Robertson, whom Ada instinctively disliked – a sentiment she suspected to be mutual.

The particular brand of radicalism and piety which Robertson preached attracted a huge and devoted following, and Annabella became one of his most devoted fans. 'I love him,' she wrote simply to one friend. His moral earnestness was absolute, and he was as liable to preach it at dinner engagements as from the pulpit of his Brighton church.

Himself a former fan of Byron, Robertson was intrigued by Annabella, and she obliged by enacting an epistolary dance of the seven veils for him, each letter giving a broader hint of the 'crime' which her former husband had committed until it was finally revealed to Robertson in its full horror, like a monstrous deformity at a freak show.

All Ada's prejudices against Robertson – a lurking suspicion, perhaps, that he was another moral voyeur enjoying the chance to rummage through her family's dirty linen – were confirmed when he accompanied Annabella on an expedition to have a final confrontation with the woman who had once been Annabella's treasured sister and was now her sworn enemy.

Having isolated her for years, Annabella had agreed to meet Augusta. Her purpose was not to seek a reconciliation but to extract a confession. She wanted to hear from Augusta's own lips, and witnessed by the man to whom she had told so much, that it was Augusta who had 'fed' the animosity that Byron had felt towards Annabella following their Separation – who, indeed, by dripping the poison of slander into his ear had made it impossible for them to reunite.

The two elderly ladies met in late March 1851, at the White Hart Inn at Reigate, coincidentally within a few streets of where John Crosse was comfortably ensconced in the furniture supplied to him by Ada. Annabella, now fifty-nine, was as robust as ever in her frailty, and supported by her young male friend. Augusta, sixty-six, was ill, impoverished and unaccompanied, and discomfited by having just stepped off a train, only the second she had ever taken in her life.

Annabella immediately noticed how sunken and old Augusta looked. This did not arouse any feelings of pity, however, only a realization that this was probably the last opportunity she would have to get the confession she wanted, the vindication she required for the years of enmity to which Augusta had been subjected.

No such confession was forthcoming, however, only denial followed by denial followed by denial . . . followed by an expression of gratitude for the kindness Annabella had once shown towards herself and her family. This latter gesture – so unexpected, so generous, so utterly pitiable – threatened, perhaps for the first time since Byron broke her heart, to penetrate Annabella's armour of self-righteousness. At the point of tears, on the edge of uncontrollable rage, Annabella could do nothing but escape Augusta's presence. The meeting was over.

A few days later, Augusta wrote imploringly to Annabella in a further bid to explain herself, but by now Annabella had recovered herself, and resorted to her old tactic of returning the letters unopened, a job she in fact delegated Robertson to perform for her. Augusta then addressed a plea to Robertson himself. He was unmoved. In a reply marked with the complacent cruelty that comes when all shades of grey are banished from a monochrome mind, he told her that since she was about to meet her maker, nothing any mortal might say would reconcile her with her conscience.

Augusta met her maker six months later.

*

Ada knew that Annabella's meeting with Augusta was going to take place, but Annabella's letter to her reporting the outcome came while her gambling activities were at their height, and it took her a week to respond. When she eventually did so, she adopted a tone she had rarely if ever used with her mother before. She wanted to condemn Annabella's decision to go ahead with the meeting – only her sense of duty as a daughter prevented her doing so, she wrote. That duty, she added tersely, now hung around her neck like a 'millstone'.

Such an open demonstration of irritation suggested that Ada was on the threshold of a decisive change in her attitude towards her mother. She was almost bridling at Annabella's attempt to keep old wounds open, even allowing a scintilla of suspicion that Augusta was not the only guilty party in the tragic saga of her parents' separation. These were dangerous thoughts, more dangerous than any scientific theory about evolution or materialism, and, egged on by the rediscovery of her Byronic legacy, she for the first time in her life seemed ready to embrace them.

In letters following the revelations about her gambling activities, Ada became if anything bolder, showing increasing impatience at Annabella's hypochondria, refusing meekly to accept her mother's criticisms about her behaviour.

But just as she seemed ready to stage this last stand against her mother, to declare an independence she had for the first time her life savoured, she was cut down by her cancer. She started to haemorrhage huge quantities of blood. Every five days, she would suffer a painful discharge, each one leaving her weaker than before.

All Annabella could do was rap at the closed door of her daughter's life. The reckoning she demanded was constantly postponed, the desire to regain control of Ada's destiny denied. But she did not give up. She stayed at the door, provisioned by Robertson's reassurances of moral righteousness, and waited.

On 13 August 1851 Ada was finally told of the full seriousness of her condition. She was by now on doses of laudanum so powerful that her letters became saturated with hallucinatory references to 'magic crystals' and 'prismatic drops'. She almost tauntingly wrote to her mother about her experiments with cannabis, supplied to her by the explorer Sir John Gardner Wilkinson, and referred to the relief she got from her 'old friend Opium'. It was yet more proof to Annabella that William had let Ada yield to her Byronic impulses.

There is certainly evidence that Ada was now an addict. At one point she was prescribed strong camphor pills to overcome the effects – the 'jumps', she called them – of withdrawal. Knowledge of this must have taken Annabella back to a few days before her flight from Piccadilly with her infant daughter, when she surreptitiously searched Byron's room and found a vial of laudanum, together with an illicit copy of the Marquis de Sade's *Justine*.

Despite the pain, the constant haemorrhaging, the drugs, Ada's thirst for life did not diminish. Throughout 1851 she not only pursued her gambling and continued her writing, but she kept up her connections with the scientific world. In October, she wrote to her son Byron, who in exasperation at his lack of ambition and initiative William had put into the navy when he was just thirteen. Her subject was the Great Exhibition. She had shown a close interest in it since 1850, when she met one of the members of the committee responsible for setting it up. Babbage, whose continuing bad relations with officialdom had ensured his exclusion from the same committee, had shown her round the site of the Crystal Palace when it was being built at Hyde Park. He had instructed her to dress up in thick stockings and cork-soled shoes for the occasion, and they had together walked round the rising structure of glass and cast iron with Sir James South, the physician who had advised them on the health of their mysterious 'invalid' in 1848.

Though it coincided with the period when her gambling reached its height, Ada would surely have attended the opening ceremony for the Great Exhibition, when Prince Albert gave a speech in which he exultantly declared the event marking a period 'of most wonderful transition, which tends rapidly to accomplish that great end, to which, indeed, all history points – the realization of the unity of mankind'. She certainly went with William to the magnificent gala ball hosted by Queen Victoria at Buckingham Palace, an occasion for members of the new Victorian elite – more than two thousand of them – to indulge in a little un-Victorian excess, with dinner and dancing stretching on until dawn.

The period also saw a new relationship emerging between Ada and her children. She missed young Byron dreadfully, using every means, including contacts Babbage had with the Admiralty, to keep in touch with him as his naval duties in HMS *Daphne* took him across the breadth of the Empire. She wrote to him regularly, and relished the replies that managed to make their way through a capricious international postal system.

Ralph, the youngest of her three children, was identified by her as the most Byronic. Her initial response was to prescribe the same dose of science she had received to dampen down the spirits, but by September 1851 she was openly celebrating the fact that he was so like her father, a 'gush of warm rays from the rising sun'.

Her relationship with Annabella was the most complex and least successful. Indeed, her dealings with her only daughter revealed a streak of malice in Ada. When Annabella had just turned eight years old, all she could say about the girl was that her behaviour was coarse and her gait too masculine. Two years later the child was being criticized for impudence and insolence. But by 1851, she, too, had become an object of affection and admiration, elected 'my vice-Queen in everything'. Ada, now more or less permanently confined in London, wrote regular letters to her, promising to keep her informed about the latest animals to be displayed at London Zoo, reporting news about young Byron's ship getting caught in a terrible storm, encouraging her to drink wine for her health and providing her with bulletins about Ralph, who had been struck down with scarlet fever while in Brighton, and whose progress Ada was tracking via reports sent over the electric telegraph.

At the beginning of 1852, Ada's illness took a turn for the worse. Until then, the pain had been controllable; she even considered it milder than when she had suffered heart spasms three years earlier. Now it was at times beyond endurance. She was prescribed even heavier doses of laudanum – ten drops every four hours, together with draughts of effervescent ammonia and warm wine.

Despite the deterioration, Ada continued to show fortitude and optimism. William was astonished, and feelings that had been trampled by recent events flourished again. He watched in silent admiration as she continued with her scientific interests. 'Babbage was a constant intellectual companion,' William noted, 'their . . . philosophical discussions begetting only an increased esteem & mutual liking'.

If he had hoped to be left to rebuild his relationship with his wife in what everyone now acknowledged to be the last months of her life, he was to be bitterly disappointed. Annabella was on the warpath, and now she was properly armed.

*

Annabella had learned from Woronzow Greig about the letter William had written giving Ada permission to gamble. This was precisely the sort of hard evidence she needed to vindicate her now openly belligerent position. Despite pleas from Woronzow begging clemency for William, she immediately sounded the battle cry, and her army of lawyers came running.

Dr Stephen Lushington was chosen to undertake the first sortie. It was he who had so brilliantly managed to defy legal gravity and extract Annabella from her marriage to Byron. He was now to be deployed as the instrument by which Annabella would regain access to her daughter.

Lushington was dispatched to visit Ada in London. William could hardly stand in the way of such an overture, not least because Lushington was a tenant of his, having taken over Ockham Park after the Lovelaces had moved to East Horsley Towers. In any case, Lushington would no doubt have made his intrusion innocuous to the point of unnoticeable. Before anyone had realized, he was established as a soothing presence next to Ada's sickbed, where he would sit and listen to her agonize over her future and her relationship with her mother.

His first report back to Annabella shows that he was clearly shocked by what he saw. She was thin and obviously in great pain. Dr Locock had 'happened' to be there at the same time, and, ever the cheery optimist, was still hopeful of a recovery. But Lushington feared the worst.

He had, as required by his mission, talked to her for some time about her 'alienation' from her mother. Ada's feelings on the subject were obviously ambivalent, perhaps still hostile, and Lushington felt unable to communicate them via a letter. He added that he had left Ada when she was about to be taken out in her bath chair, which, typically, he observed to be fitted with the latest innovation in transport technology, India-rubber wheels.

During subsequent visits, Lushington gently began to apply the leverage that Annabella needed to open up a gap for her own intervention. Couching his request with reassurances that her mother wanted to help, he asked Ada for a list of her debts, promising to deal appropriately with 'circumstances of delicacy'. He got his list, but the amounts involved were much smaller than Annabella had expected – just a few hundred pounds.

It is a indication of the power of Ada's charm that, even as she was in the terminal stages of cancer, she could hoodwink such a wily old lawyer as Lushington – the man who in another context had been worldly enough to stop Byron asserting his conjugal rights over Annabella by claiming that she would otherwise face the risk of catching venereal disease.

Annabella probably knew better than to accept Ada's list as comprehensive, but it did not matter. It served her purpose. She was now effectively in charge of Ada's finances, and there was nothing that William could do about it. The London home in which they were now based, at Great Cumberland Place near Hyde Park, was owned by her. The very walls around him had been refurbished, the very rooms in which he sat furnished, the very sickbed upon which Ada lay supplied by Annabella, offered and eagerly taken during happier times.

Within three weeks of his first visit, Lushington could exultantly report to Annabella that Ada was beginning to feel closer to her mother that she had before. Her foot was in the door. To William's evident exasperation, Annabella began to issue advice on Ada's treatment, and in particular to voice her disapproval of the increasing use of drugs. To mollify her, mesmerists were called, a Mrs Cooper and a Mr Symes. They passed hands over her tortured body, tried to entrance her, manipulate her magnetic field, but to no avail. They were soon after sent away.

Her mother drew remorselessly closer. Ada now wrote directly and openly to Annabella of her ambivalence towards their estrangement. Each letter edged nearer to accepting the inevitable reckoning. At first she both dreads and desires a confrontation, then she anticipates the need for one – 'the volcano is better than the glacier' – then she fears the consequences – 'What is once said never can be unsaid' – finally . . . 'I hope I shall have courage to let you walk quite in'.

She did walk in, and immediately set about trying to extract the full story of what had happened to her daughter while Ada had been beyond her reach, frolicking with the forces of evil. It did not take Annabella long to get at the truth.

Within a few days, she had learned that Ada had handed the family jewels to John Crosse to pawn – necklaces, brooches, tiaras made of diamonds and pearls. They were worth far more than the £800 he got for them from Mr Vaughan, a jeweller on the Strand, who had additionally charged £100 for a replacement set to be made from paste so the absence of the originals would not be noticed.

Ada slept well the night after she had made her confession. She compared the experience to giving birth to a monster, and felt relieved to have finally evacuated it from her body. But another monster still grew inside her, the cancer of her womb.

Annabella immediately instructed her bankers to retrieve the jewels, which they did. Ada had asked her not to say anything to William about what she had done. Annabella was happy to comply – she was not speaking to him anyway. Instead she continued her visits to her daughter's bedside. Mr Symes was recalled to attempt more mesmerical ministrations, but without any noticeable effect, and William admitted to Woronzow that he no longer dared ask Dr Locock for his prognoses.

At the end of July, a new symptom appeared, a hard swelling of the uterus. A further internal examination was performed. 'They then told

me it must end fatally sooner or later for my beloved wife,' a distraught
William told Woronzow.

As if that, and his mother-in-law, and the debts, and the round-the-
clock care were not enough for the poor man, a further blow was to
befall William, and stretch to the limit the love he now so strongly felt
for Ada.

*

On 25 July 1852 William went for a walk with his old school friend
Woronzow Greig in the grounds of East Horsley. A thunderstorm
broke, and they were forced to take shelter in a shed. There they began
to talk about recent events.

In passing, Woronzow mentioned John Crosse's wife. What wife?
asked an astonished William. He had met Crosse several times, the
man had even advised him on his writings. Never had a family been
mentioned. He had presented himself as a bachelor, for only as a
bachelor could he have been welcomed into the household on such an
informal basis.

Woronzow told William that Crosse had a wife and two children.
He had learned of their existence from Mr Hart, the Clerk of Petty
Sessions at Reigate in Surrey. Mr Hart had come to Woronzow's office
on some other business just before Christmas, and mentioned they had
a mutual acquaintance in Crosse. When Woronzow confirmed this,
Hart mentioned meeting Crosse and his family at Reigate. At first
Woronzow assumed it must therefore be a different Crosse, the name
being a fairly common one. But Hart confirmed that the man he had
met had the same distinctive scar on his jaw that John Crosse had. Just
to stir the pot a little, he added that he had four horses and liked to
entertain. Hart himself had recently been invited to a Christmas Tree
party at the man's family home.

This picturesque image of settled domesticity was worse than
William realized. A wife? Horses? Christmas Tree parties? Not only
did Crosse, Ada's 'intimate friend' as Woronzow candidly described
him, appear to be keeping a secret family, but he had introduced it to
society, in the very same county of which William was Lord Lieutenant.

William decided he must confront the man directly to get at the bottom of the matter. Crosse was a regular visitor to the Lovelaces' house in Great Cumberland Place, his father's London residence being just a short walk away in Manchester Square. Indeed, he had been there to see Ada a few days before William heard the revelations about him from Woronzow. He had learned that the family jewels that he had pawned had been recovered. Now he was back to pawn them again, to take a second bite at the encrusted cherry.

The last time he had done this, Ada had obviously been terrified of discovery. That explains why she spent so much money on getting fake substitutes made, and felt such relief when her mother finally retrieved the originals. So it seems extraordinary that she would expose herself to the same risk a second time. Why would she? She knew her days were numbered, and that her financial affairs would soon be sorted out one way or another without her intervention. She knew there was a fair chance that her mother might find out what happened, and on this occasion decide to tell William. Yet she directed Crosse to where the recovered jewels could be found, and he carried them away once more to Mr Vaughan in the Strand.

In an effort to be charitable towards Ada, one might try to argue that she did this because Crosse somehow forced her, perhaps with threats of going public about her gambling. This is possible. Perhaps their relationship turned sour following the Derby Day fiasco. Crosse was capable of extortion. Following Ada's death, he forced Greig to make him the beneficiary of Ada's life insurance policy in return for burning her letters to him.

But as her gambling activities demonstrated, Ada was no innocent party in her subterfuges with Crosse. Furthermore, she was eager to see him even in the final days of her life. It also appears that she urged William to go to East Horsley, despite the feelings of loneliness he now endured when he was parted from her, so she could see Crosse alone at Great Cumberland Place – he made at least two visits in the period William was away.

In other words, allowing the man to pawn the family jewels a second time seems to have been a gesture of wilful, desperate, selfish recklessness – a gesture of love.

Love is absent from all Ada's surviving letters, and from all the memoirs and biographies relating to her. We can see her being flirtatious, affectionate, impulsive, but never in love. But here, in the last moments of her life, perhaps we finally glimpse its inimitable signature, the flourish of romance, the human act that cannot be explained by any science or calculated using any calculus, the ghost playing havoc within the machine.

On the very day that Woronzow had sheltered with William in that shed while thunder played around the Gothic towers of East Horsley, Crosse once again insinuated himself into Ada's sickroom, perhaps to tell her how much money he had raised on the jewels, perhaps to discuss alternative means of extracting yet more money from her depleted estate before death snatched it from them and passed it into the hands of testamentary law.

Soon after, William cornered him alone and, still unaware of what the man had just done with his heirlooms, demanded an explanation concerning his family. Like a little boy caught with his hand in the sweet tin, Crosse came out with a series of increasingly improbable excuses for his incriminating position. They were a cousin and his wife, who happened to be sharing the house, he suggested. William refused to accept this. Crosse tried again: his uncle Colonel Hamilton had taken possession of the place and installed his mistress and illegitimate children there, and the mistress coincidentally was also called … Crosse … No, no, the uncle had installed his mistress but the mistress wasn't really called Crosse, but John – yes, this is it – being grateful to the uncle for taking on the expense of the house, had let her pretend to be his wife, and he had been at that Christmas Tree party mentioned by Mr Hart because he had been invited as her guest …

Crosse tried a variety of unconvincing permutations of uncles and cousins and women coincidentally called Crosse before eventually arriving at the explanation William had been waiting for: the woman was his mistress and they were his children, and he had pretended he was married so as not to cause embarrassment.

Even this seems almost certainly to have been a lie. According to the 1851 census records for Reigate, John Crosse, landholder and proprietor, of Springfield Villas, Reigate was married to a Susannah Crosse, and father of two legitimate children, John junior and Mary.

He lived as husband and wife with this same Susannah when he inherited Fyne Court from his father Andrew in 1855. So, either he was married, or married at a later date the woman he had passed off as his wife, or continued the subterfuge for the rest of his life to the complete ignorance of his family and his heirs.

William had no time to pose such questions. He was only too eager to accept Crosse's final story – a man who concealed an illegitimate family was socially much more acceptable than a man who concealed his marital status. He exchanged some 'sad words' with Ada on the subject, and suggested that Crosse should be henceforth excluded from the house.

This provoked panic in Ada. She pleaded with Dr Locock to intervene, which he obediently did by drafting a letter to William claiming that Ada should continue to see Crosse on medical grounds. 'I think in her present deplorable state it would be both cruel and mischievous to debar her from what has been such a source of comfort & happiness,' he wrote.

The letter was too late. Annabella had now installed herself at Ada's bedside. It would be she, not William, who would decide who Ada saw, and within days exclusion orders had been issued against not just Crosse, but all Ada's friends and confidants. One by one, connections were severed, finally leaving her with just one independent link with the outside world: Charles Babbage.

Annabella had always had her suspicions about Babbage, and these were by now developing into outright hostility. On 12 August Ada had managed to put a pencil-written note into his hands asking him to be her executor, using money that she assumed he could get from her mother following her death. Babbage wrote back to her to say that the letter did not give him the authority to act in the way she wanted. Before she could do anything about this, the door was closed on him, too. A few days later Mary Wilson, who had once worked for Babbage's family and was therefore tainted, was dismissed, and paid a £100 severance fee on condition that she return a letter written to her by Ada specifically to put on record how loyal she had been.

*

Ada was too preoccupied with making arrangements for her own death to notice much of what was going on around her. 'She busied herself with papers – making arrangements & giving directions, remembrances, etc,' wrote William in a journal he was now keeping for Annabella, as a sort of peace offering. 'She finds sitting at the piano for a few minutes and playing over on it some of the airs (that used to enchant all who heard them from her in old times) a great relief. The pain and malaise are constant but not always intense – & in this way the nervous energy finds escape.' As she sat there, Henry Phillips, son of the Thomas Phillips who painted the heroic picture of Byron in Albanian costume, worked on a portrait of her. The result depicted a thin, melancholy woman, pale to the point of luminescence, her eyes not looking at the delicate fingers poised over the piano keys, but staring as though she were in an hypnotic trance, or surveying the contours of a distant landscape in another world.

Her preparations were spiritual as well as practical. She underwent a series of religious transformations that showed the growing influence of her mother. 'She spoke freely of the future state – & how necessary a sequence it was to this world, how incomplete all here was – how pervading the mind of the Deity and yet how inscrutable His design,' William noted in his journal, probably exaggerating the extent of her devoutness for the benefit of Annabella, his intended reader.

> She considered how all lives had in the view of their creator their use and mission – that they ended when that was over – how hers might be in that predicament. It put her in mind how often in our rambles among the hills I had observed her eyes gazing wistfully into space as though ready to float off into the future. She smiled assent with a melancholy pleasure.

On 15 August she revealed to William she wanted to be buried next to her father in the Byron vault at Hucknall Church. The following month she also specified that the epitaph on her monument should be a quotation from the letter of St James, v: 6: 'You have condemned, you have killed the righteous man; he does not resist you.'

These requests were not the result of some impulsive desire. It emerged that she had discussed her burial arrangements with Colonel

Wildman two years earlier, during the visit to Newstead. She now asked William to draft a letter confirming the arrangements she and the colonel had secretly discussed then.

No gesture could more emphatically declare her Byronic conversion. At the moment of her death, she wanted to be taken to the man who had left her at the moment of her birth. Annabella had little choice but to accept the decision in a spirit of public magnanimity, and tried to make the most of it by suggesting to a friend that it had been her idea.

A few days later, Wildman replied, saying he would hold the fulfilment of her wish to be a sacred duty.

She now felt she had little time, and was desperate to see her two boys. Both were away when her illness had been diagnosed as terminal, Byron at sea, Ralph in Switzerland, where Annabella had sent him to be educated according to the stern and sombre principles she had once tried to apply to Ada. Byron was the first to reach her bedside, on 6 August, and his arrival brought an instant improvement in her condition. A few days later he joined his father at East Horsley, presumably having been sent away by his mother for the same reason she had dispatched William, so she could have what she probably imagined to be her final tryst with John Crosse. He returned with his father and Annabella, and together they all now attended at her bedside morning and night.

On 21 August, she told William she thought the end was near, and asked repeatedly for Ralph, who was still in Switzerland. 'She walked about the room on my arm for a time – speaking almost with satisfaction of the posthumous arrangements,' William wrote in his journal.

> She told me she felt all was fast ending in this life . . . Then as to being laid by the side of her father, whether at Hucknall or Newstead – indifferent so that it were by him. Very anxious not for life but for a day or two in full possession of her faculties . . . Pleased at having her temples and hands bathed. We talked again of her strange destiny . . . She did not know why so much suffering was necessary – & yet she bowed in submission to its infliction.

Then she asked to see Charles Dickens.

Ada had known Dickens since 1843, when he asked her to back a charitable scheme he was helping to organize. At the time, when her ambitions to make a name for herself were at their most intense, she suspected him of approaching her simply as the daughter of Byron. Dickens obviously managed to persuade her otherwise, as their friendship subsequently flourished, with him later gallantly comparing himself opening a note from her with Aladdin opening the cave.

He received the summons to her sickbed after returning from a trip, and came immediately. They spent an hour together, during which time he read to her from her favourite book, *Dombey and Son*. In particular, she asked to hear the chapter entitled 'What the Waves were Always Saying', which describes little Paul Dombey's life draining away like water from a river into the sea.

In one passage, Paul notices something at the end of the bed. 'What *is* that?' he asks his sister Floy.

'Where, dearest?'

'There! at the bottom of the bed.'

'There's nothing there, except Papa!'

The figure lifted up its head, and rose, and coming to the bedside said: 'My own boy! Don't you know me?'

Paul looked it in the face, and thought, was this his father? But the face so altered to his thinking, thrilled while he gazed, as if it were in pain; and before he could reach out both hands to take it between them, and draw it towards him, the figure turned away quickly from the little bed, and went out at the door . . .

Like little Paul, Ada now seemed to be approaching her final hours in a state of semiconsciousness, only partly aware of who was around her, who was talking to her. There were periods of calm and tranquillity, when William and her two children, Annabella and Byron, would gently sponge her. As the water ran down her body, it momentarily took her back to the cascades running down the sides of the mountains at Buttermere.

Then there were attacks of agony and delirium. The day after Dickens's visit she had to endure one of the worst so far. She could not lie in her bed, and started to pace the room, 'restrained rather than supported by her mother & me', as William put it. Maddened by the continuing spasms of pain, she would slip their grip and throw herself against the furniture or down onto the floor. Eventually the room had to be lined with mattresses to prevent her causing any further injury to herself. The pain would relent and she would fall into a fretful sleep on the floor. Then it would resume and she would writhe around in agony.

Most of this was witnessed by William and her mother, whose fragile armistice inevitably came under considerable strain. Much of it was seen by Ada's traumatized children too, young Byron squatting down on the floor where she lay to sponge her face during her ever shorter respites from the pain. But still there was no Ralph. Ada became obsessed with seeing him before she went, repeatedly asking for the time as though she knew how much she had left. She also developed a terror of being buried alive.

On 26 August, an exhausted Ralph arrived after a frantic two-day dash from Switzerland and there was a moment's calm. It did not last for long. The next day she suffered a seizure, and the day after that recurrent convulsions. She only had hours now, said the doctors.

Don't let me be buried alive, don't let me be buried alive . . . the incantation was ceaseless. Where was the funeral to be? What time is it? Who is that standing outside? Who is that at the door? Who is that standing at the end of the bed? William was there, reassuring her there was no one at the door. But there was, someone, someone out to get her.

Her father – he had sent her this disease, she told her mother. It was a cruel God who made her suffer so.

She started to scribble notes in an almost indecipherable hand, many of which were transcribed by her mother. She quoted a few words from her father's poem *Cain* – 'Believe – and sink not! Doubt – and perish!' Perhaps Annabella took this to be a sign of a deathbed conversion, forgetting that the line was spoken by Lucifer, mimicking the platitudes of the Godly . . .

> . . . miserable things,
> Which, knowing nought beyond their shallow senses,
> Worship the word which strikes their ear, and deem

> Evil or good what is proclaim'd to them
> In their abasement. I will have none such . . .

In other notes she mentions Hester and Charlotte, her two beloved sisters-in-law, not there and for some reason not called for by William or Annabella . . . then she wants a nursing institution to be set up in her name . . . then there is a bequest to her mother: '*malgré tout*', in spite of everything . . . Psalm 17, verse 8:

> Keep me as the apple of the eye;
> Hide me in the shadow of thy wings,
> From the wicked who despoil me,
> My deadly enemies who surround me . . .

. . . lines of poetry, phrases in German, lists of book titles, it all came pouring out.

In the midst of all this, a mysterious initial appears: 'T'. It is next to the title of a translation of Homer jotted down on a piece of paper, and again next to the line 'One thought of thee, & I am calm . . .'. There is a letter she evidently tried and failed to send to Babbage which had inside a note addressed 'To T'. T was to have treasured mementoes from her literary life, the note instructed – her gold pencil case, which she wants him to use habitually, and any twelve books that will remind him of the hours they spent together.

T, clearly, was a close collaborator. But which one? In her review of Reichenbach's *Researches into Magnetism*, she added a marginal note referring to a 'T.C.' having some papers she needed to finish a section of the paper. There are also notes added to the manuscript in another hand – perhaps that of the same T.C. It seems reasonable to assume that 'T' and 'T.C.' were the same person.

Since John Crosse was so closely involved with the work she did at the time she wrote the Reichenbach review, one suggestion is that 'T' is the initial of Crosse's petname. It is a credible theory, but not decisive. First, in the final days, she actually had more access to Crosse than to Babbage, which makes it puzzling that she would use the latter as an intermediary. Also, given the rather cerebral and literary nature of the bequest, one could argue that these were not the tokens of love you would expect in a final bequest.

So there is room for a little more speculation, and with Crosse removed from the list of suspects, the field is left temptingly open. Perhaps T.C. is the husband of Ada's sister-in-law Charlotte, a Greek curate called Calliphonas. There are only a few tangential references to him among all the Lovelace papers . . . and none that mentions his first name. However, it does emerge from the journal of the educational reformer and besotted admirer of Ada Dr Kay that Calliphonas took part in the mesmerical experiments in which Ada participated in 1841.

Perhaps T is not the same person as T.C. Ada did correspond with an artist called Field Talfourd, brother of the more famous lawyer and author Thomas Noon Talfourd. Could he be T? He wrote several letters to Ada in the autumn and winter of 1851, one of which turned up among Woronzow Greig's papers, which would indicate that it was treated, perhaps following her death, as legally sensitive. In it he writes in a manner that recalled William Carpenter's letters, describing a note from her as being like a 'chemical process . . . precipitated into words', setting off a 'triumphant sensation' in him. 'How deeply I value your confidence,' he added, 'a confidence growing more valuable as the quality of the mind which bestows it is unfolded to me.'

The problem with T being Talfourd is that Ada only appears to have met him in 1850, and the few specimens of their correspondence that remain do not suggest anything like the sort of long-term familiarity implied by the bequest.

Of course, rather than attempt to apply scientific, rigorously academic standards of evidence to identify T, one could deploy a more intuitive, artistic, adventurous approach. Then another name enticingly presents itself: that of Thomas Carlyle. There is barely any evidence linking the two of them. In 1822, Carlyle's wife to be, Jane Baillie Welsh, wrote to him suggesting he compose a poem about Ada following the death of Byron, which he attempted to do. Carlyle also corresponded briefly with William in 1846 about the former's historical work on Oliver Cromwell, and he met Ada on at least one occasion, when they both attended a dinner held by Charles Dickens.

They shared a number of acquaintances in common besides Dickens, for example Sir David Brewster, to whose *Edinburgh Encyclopaedia* Carlyle contributed a number of articles, and who visited Ada on several occasions while he was in London in 1851. Carlyle also knew Babbage, the intended intermediary for Ada's bequest – indeed the two shared a mutual dislike, one that Babbage would no doubt have overcome to perform such an important service on Ada's behalf. Carlyle was also a keen and distinguished Germanist, just the sort of person Ada needed in the late 1840s, when her interests turned to the philosophical and scientific developments then taking place in that country.

To propose Carlyle as Ada's literary collaborator is both perverse – as all the best unprovable intuitions should be – and somehow fitting. He was so utterly attuned to and at odds with the scientific and mechanical world of which Ada was a part. It is him we have to thank for capturing the flavour of the modern, industrial age by dubbing economics the 'dismal science', the media the 'fourth estate', and plutocrats the 'captains of industry'. It was he who flippantly described Napoleon's brutal attempts at crowd control as a 'whiff of grapeshot', who characterized the public, that newly enfranchised entity in a democratizing age, as an old maundering mumbling woman. Even as Karl Marx scribbled away in the British Museum library, it was Carlyle who expressed a growing distrust of the ideological and dogmatic, who thought moral rather than political reform to be the only sort possible, who unlike Babbage and his ilk thought the universe far too big to comprehend, who lionized great heroes and despised democracy. After a lifetime of her mother's cool rationalism, Ada would have found a shot of his hot-blooded, high-handed humanism scintillating, enough to convince her that life was to be lived rather than mastered, the guiding philosophy of her final years.

*

Now, as she struggled to communicate her dying wishes, there was barely any life left in her. On 28 August she began to suffer a series of fits. At four o'clock in the morning of 29 August her pulse stopped. The moment had at last apparently arrived . . .

But she revived. On the 30th she regained consciousness. An exhausted William noticed a strange new behaviour. She started to play with a handkerchief, repeatedly spreading it out, measuring it, waving it about, as though she had become a machine. Annabella saw these strange movements too, and considered them 'idiotic'.

That evening, William went to see how she was and found her 'not quite insensible & yet not conscious' – she reminded him of Ophelia floating down the stream. The following evening she regained consciousness once more. He kissed her hand, almost as though to welcome her back, as though she were a guest in the land of the living who must imminently take her leave. 'She then silently put her arms around my neck & drew me to kiss her face, but with only a sort of instinct, as she spoke nothing.' A little later, she asked William if he would forgive her. Annabella watched him give her his tender assurances. It was for this moment Ada had been kept alive, she thought.

If only Ada had died then, in her husband's arms, the object of both her husband's love and her mother's approbation. If only . . .

But she did not die. Death continued to dally cruelly at the doorstep of Great Cumberland Place.

Ada had her own idea as to why she was being kept alive. It was, she seemed to realize, in order to make her confession to poor William. She had asked his forgiveness, now she must tell him why she had sought it.

He had been hovering at the door, placed flowers just outside so she could see them from her bed. She called him in several times, then sent him away, then finally invited him to stay and close the door.

It is not known exactly what she told him, but it was probably of her adultery with John Crosse. The shock for William was unbearable. Annabella was not in the room, but evidently waited outside. She saw him come out, and was almost gleeful to witness his wretched, miserable state. He admitted he had lost his temper with her, shouted to a poor, penitent, dying wife that he could only hope that God would have mercy on her soul – further proof, as if Annabella needed it, of what a weak man he was. He locked himself away in his own room, and would not come out.

Ada was now Annabella's. As William brooded resentfully next door, Annabella sat in vigil at Ada's bedside. In the brief intervals of consciousness between her daughter's continuing convulsions and comas, Annabella patiently, relentlessly, sometimes, as she herself later admitted, too roughly coaxed out of Ada admissions of guilt and promises to redeem.

Yes, Ada eventually confessed, she had on innumerable occasions behaved badly, tried things out for herself rather than heeded the voice of experience, taken scientific experimentation too far – experimented with the precious, God-given gift of life as though it were so much electricity or molecular matter.

And yes, she had allowed herself to fall prey to the idolaters of Byron – John Crosse now being regarded as one of the worst examples of that benighted species – been lured by them into worshipping not just the poet, but the man, a crucial distinction for Annabella, who even after all she had gone through could not quell the shivers of excitement his seductive verses summoned. Ada had let his terrible vices go disregarded and – it was undeniable, though awful to admit – in the process allowed her mother to be cast into an unfavourable light – a silent but sinister calumny that so cruelly turned victim into villain.

Ada could no longer deny any of it. She submitted utterly to her mother, and Annabella immediately set about organizing her redemption. This was achieved by calling upon Ada to perform numerous little demonstrations of loyalty and like-mindedness – prayers of supplication, further confessions of wrongdoing (including to the second pawning of the peripatetic family jewels, which were once more recovered), acceptance that her mother should have complete control over her papers and affairs following her death, and pledges of affection for Annabella's friends. An example of the latter was a scrawled note nominating Mary Millicent Montgomery as the woman Ada most wished had been her best friend, in acknowledgement of which she left a bequest of a precious brooch, a portrait and a selection of books. Miss Montgomery had been one of the fearsome Furies who had harried Ada through her adolescence at Fordhook, one of Annabella's coterie of spinsters that Ada bitterly told Woronzow Greig 'hate me like poison'.

By September, Annabella's work was done. Ada had repudiated her friends, her conduct, her life. Her illness, which Annabella and one or two of her friends believed to be providential, had served its purpose, 'weaning her from temptation, & turning her thoughts to higher and better things'. Now it was time for Ada to die.

But, defiant to the last, Ada did not die. Night and day her retinue of attendants and doctors laboured to keep her comfortable in the expectation of her imminent demise. September and October passed, but she remained, now withdrawn into a private world of pain, narcotic hallucinations, swirling memories, and, no doubt, flashes of cold, sparkling clarity when she could at last have thoughts that were entirely her own, and never to be anyone else's.

People were now slowly falling away from her: her friends had gone, her husband too, and now her children – Annabella, Ralph, Byron. It was decided that Byron, who was to return to naval duties, would not say goodbye, as the pain of parting would be more than she could bear. On 20 October he came to the door of her room, which had been left ajar. As Ada lay there, awake but unaware of his presence, he took one last look at his mother and left.

He walked out of Great Cumberland Place, posted off his midshipman's uniform to his father in a carpetbag, and disappeared. Some later interpreted this act as an example of extreme selfishness on the part of the boy, but it can also be seen as a sort of tribute to his mother, a demonstration that the rebellious spirit crushed out of her now lived on in him. Whatever the intention, poor William was forced to dispatch a relative to try and find his fugitive heir. An advertisement was placed in *The Times*, offering a

> reward for the discovery of a youth . . . nearly 17 years of age, 5 foot 6 inches high, broad shouldered, well-knit active frame, slouching seaman-like gait, sunburnt complexion, dark expressive eyes and eyebrows, thick black wavy hair, hands long and slightly tattooed with a red cross and other small black marks . . . nails bitten, deep voice, slow articulation . . .

He was eventually traced by a police detective to an inn in Liverpool, where he was trying to get passage to America. He was forced back into the navy by his father.

Young Byron's subsequent story is a sad, romantic one. After further years with the navy, he pleaded with his father to be discharged. He eventually left on his own initiative. How he achieved this goes unrecorded, though it seems unlikely he deserted.

Having won his freedom, he found himself marooned in a port on the Black Sea, where he managed to get passage back to England on a small trading ship. He arrived in Hull many months later, sick and bedraggled, and made himself known to friends of his grandmother's who lived in Yorkshire. From there he was received back into Annabella's care. She placed him under the supervision of one Lieutenant Arnold, son of the formidable and famous headmaster of Rugby School. Forced once again to endure the sort of military regime he had just escaped, he absconded, and ended up working in a colliery in Sunderland. Finding himself suited to a labourer's life, he went on to work in a shipyard run by Brunel on the Isle of Dogs under the assumed name of Jack Okey. He was briefly engaged to a humble girl, who ultimately rejected him because of suspicions about his identity, and in 1860 he became the very reluctant twelfth Baron Wentworth, following the efforts of his alienated father to force upon him a title he inherited after the death of his grandmother. He died two years later of consumption, aged twenty-six.

In November 1852 his mother, too, finally approached death. On the 27th Annabella, still at Ada's bedside, observed her suffering a series of faintings and convulsions that went on for several hours. She did not record her daughter's last words, only her own, a reminder to her to surrender herself to the Almighty. At nine thirty in the evening Ada suddenly became still, but continued breathing. A little later, unnoticed by those around her, she finally slipped away.

*

Byron's body had been conveyed from London to Nottingham in a hearse drawn by six black steeds, the route lined with crowds and strewn with flowers. Ada was delivered to the same destination by the Midland Railway Company.

The funeral was a week after she died, and a week before her thirty-seventh birthday. The hearse arrived from Nottingham at ten o'clock, followed by the procession of mourners: William, George Anson Byron, Stephen Lushington, Colonel Wildman, Woronzow Greig. Her sister-in-law Hester and her husband, Sir George Craufurd, were there, but her other sister-in-law, Charlotte and her husband, Calliphonas, appear not to have been.

Another, more notable figure was also absent: Ada's mother.

A large crowd had gathered in the square around Hucknall Torkard church in expectation of the opening of Byron's vault, which was to receive Ada's body. William had put on quite a show for Ada – too much of a show, loyalists later reassured Annabella. The coffin he had chosen for her was covered in violet velvet and studded with silver coronets and an escutcheon bearing the Lovelace arms.

In the final lines of the third canto of *Childe Harold*, which he considered the best he ever composed, Byron addressed these lines to his daughter:

> Yet, though dull Hate as duty should be taught,
> I know that thou wilt love me; though my name,
> Should be shut from thee, as a spell still fraught
> With desolation, – and a broken claim:
> Though the grave closed between us, – 'twere the same,
> I know that thou wilt love me; though to drain
> *My* blood from out thy being were an aim,
> And an attainment, – all would be in vain, –
> Still thou would'st love me, still that more than life retain.

After the service was over, Ada's coffin was carried down the short flight of steps leading to the Byron vault beneath the nave. The grave that had closed between them was now opened, and there she was laid to rest next to the father she had in life been forced to renounce, and in death could embrace.

Epilogue

'ONE THING certainly contemplated in the scheme of creation,' wrote the eminent physician and long-standing member of Annabella's medical retinue Dr Herbert Mayo, 'is a progression from barbarism to civilization.' Thanks to the gift of reason bestowed upon humanity by Providence, Mayo had not a shadow of doubt in his mind that primitivism would proceed to perfection just as surely as one proceeds to two.

In the year that Ada died, the American writer Nathaniel Hawthorne published a novel called *The Blithedale Romance*, which in terms of its themes and approach was an interesting precursor of what we would now call science fiction. It concerned the setting up of an experimental commune that simulated the sort of utopia Mayo argued progress would one day create. And there, even in this state of rational, scientific purity, Hawthorne argued that human weakness, folly and selfishness would lurk, like the worm in the bud.

The worm was busy following Ada's death. It gnawed away at Annabella and William, until their relationship finally collapsed, bequeathing yet more rancour and recrimination to Ada's children, which resulted in both Ralph and young Annabella eventually becoming estranged from their father. It devoured the residue of Ada's reputation, with a variety of grotesque characters arising from the depths the moment she died, hands outstretched to a bewildered and very busy Woronzow Greig, demanding portions of the estate he had been charged to settle.

And the worm feasted on the foetid remains of her love life. Right up to the end, Crosse had tried to see Ada, but had been repulsed by Annabella. Cut out of her life, he went for her life insurance policy instead, and eventually forced Woronzow to make him a beneficiary in return for destroying incriminating letters he had received from Ada – 108 in all, Woronzow estimated, none of which now remain.

The worm kept turning, not because of Ada but the event that so influenced her life: the Separation. Harriet Beecher Stowe, the author of *Uncle Tom's Cabin*, made Annabella's incest allegations against Byron public for the first time in an article for the *Atlantic Monthly* and a subsequent book, *Lady Byron Vindicated*, published in 1870, a decade after Annabella's death. Though Stowe was, as the title of her book made clear, attempting to defend Annabella, her indiscretion angered Ada's son Ralph, who started to collate all the documents he could find relating to his grandparents with an aim of setting the record straight once and for all. When William died in 1893 and he became Earl of Lovelace, he took possession of the full collection of family papers, including key correspondence between Annabella, Augusta and Byron. This he used as the basis of *Astarte*, a book he completed in 1905 and circulated privately among friends.

The title for the book came from Byron's dramatic poem *Manfred*. Astarte was the woman loved by the eponymous antihero, who was so 'like me in lineaments'.

> She had the same lone thoughts and wanderings,
> The quest of hidden knowledge, and a mind
> To comprehend the universe: nor these
> Alone, but with them gentler powers than mine,
> Pity, and smiles, and tears – which I had not;
> And tenderness – but that I had for her;
> Humility – and that I never had.
> Her faults were mine – her virtues were her own –
> I loved her, and destroy'd her!

In the climactic scene of the drama the intensity of Manfred's remorse summons up the ghost of Astarte. Ralph's book had a similar effect. It was quickly procured 'at extortionate prices, and, one is told, by subterranean methods' by journalists and its contents made public. Thus, as the *Tribune* newspaper reported on 30 August 1906, soon after Ralph's death, 'the rotting bones of the sordid story were clothed again in flesh and blood. The dead ashes of controversy were revivified, and the poor ghosts of Byron and Augusta pilloried once more in public shame.'

In 1929 Ethel Colburn Mayne published the first official biography of Annabella, which sympathetically fleshed out Stowe's portrait of a moral exemplar. In 1962 Malcolm Elwin produced *Lord Byron's Wife*, which took very much the opposite view (as is only too obvious from the index: Annabella's entry includes subentries for 'complacency', 'egoism', 'jealousy', 'lack of loyalty', 'pedantry', 'self-absorption' and 'smugness').

What all this seemed to show was that the reconciliation Ada had tried to enact in her life, the utopian state she had attempted to create in which the mind and the heart, science and art, truth and beauty would be united, was unachievable. Her efforts to find that 'poetical science' were doomed.

But as the Age of Industry passes into what has been portentously hailed the Age of Information, perhaps such a reconciliation will become possible. There is a new cosmology, we are told, a 'New Physics' which seems to embrace both the subjective and objective, the conformist and the creative. Quantum mechanics has given us the Uncertainty Principle, chaos science unpredictability in the midst of determinism, neo-Darwinism the selfish gene and the evolution of social behaviour. Could this be what Ada anticipated? When she described herself as being a prophetess in 1844, was this what she prophesied?

An alternative view is that her life foreshadowed the very reverse, the view that so-called progress leads only to greater polarization. As we have more of one thing, we can only have less of another – technology diminishes our humanity, freedom of expression corrupts our culture, liberty threatens anarchy, order produces constraint. Indeed, these antagonisms are the very stuff of life, the reason why we live neither as selfish savages nor as soulless robots, but poised precariously somewhere in between, like Manfred on the mountain edge . . .

> How beautiful is all this visible world!
> How glorious in its action and itself!
> But we, who name ourselves its sovereigns, we,

Half dust, half deity, alike unfit
To sink or soar, with our mix'd essence make
A conflict of its elements, and breathe
The breath of degradation and of pride
Contending with low wants and lofty will,
Till our mortality predominates . . .

Notes and Further Reading

This is by no means the first full-length biography of Ada. That honour goes to Doris Langley Moore, the formidable and devoted Byronist. Her book, *Ada: Countess of Lovelace*, first published in 1977, sheds very little light on Ada's intellectual pursuits, but does provide a vivid picture of Ada's relationship with her mother.

Dorothy Stein's biography, *Ada, a Life and a Legacy* (discussed above, see page 276), published in 1986, complemented Moore's, as it concentrated on Ada's mathematical work. However, it would be wrong to think that Stein did not contribute to our understanding of Ada's personal life. It was she who first discovered the short biographical sketch of Ada written by Woronzow Greig, an important document, as it contained the revelation of Ada's elopement. Stein also produced the first proper assessment of Ada's notes on the Analytical Engine. Finally, she managed to decipher many of the letters written by Ada's husband William, whose handwriting was, as Ada diplomatically put it, 'very pretty' but never 'very legible'. William's hand often defeated Langley Moore, and almost always me.

In 1992, Betty Alexander Toole published *Ada, the Enchantress of Numbers*, a comprehensive selection of Ada's letters. Dr Toole also added a 'narration' which provided many biographical details about Ada.

All the above, like this work, draw on four main collections of primary sources. The most important is the Lovelace Byron collection held at the Bodleian Library in Oxford. This contains many of the original manuscripts of Byron's poems and letters, as well as a large amount of correspondence between Ada, Annabella and William. The Somerville Papers, also at the Bodleian, contain Ada's letters to Mary Somerville and Woronzow, as well as Woronzow Greig's biography of Ada (reference Dep b 206/MSIF-40).

Ada's letters to Babbage are to be found among his papers, held at the British Library (Add. MSS 37190 to 37194). This extensive collection has yet to be published, which is a pity, as it amounts to one of the most important and interesting sources of primary material on the history of Victorian science and technology. Also at the British Library is the Wentworth Bequest, which contains some further papers relating to Ada, including William's journal of the final days of her life (Add. MSS 54089).

Other letters are scattered elsewhere, and I have managed to find a few that have never before been published (for example, Ada's letters to the novelist Edward Bulwer-Lytton – a frustratingly incomplete record of what appears to have been a close friendship). I found a particularly interesting stash in the Pforzheimer Collection (full title: 'The Carl H. Pforzheimer Collection of Shelley & His Circle, The New York Public Library, Astor, Lenox and Tilden Foundation').

For details about Byron, I have relied heavily on his voluminous correspondence and diaries. The entire collection has been edited and indexed by Leslie A. Marchand in his magnificent twelve-volume *Byron's Letters and Journals*.

For other biographical details, I have been spoilt for choice. Byron is surely the subject of more biographies than just about any other poet, and every aspect of his life has been documented. I have drawn mostly on the prolific Professor Marchand, in particular his three-volume *Byron: A Biography*, which appeared in 1957, and his updated *Byron: A Portrait*, published in 1971. I cannot claim to have read all the other available works, but found several to be useful, including the recent *Byron: The Flawed Angel* by Phyllis Grosskurth (which drew heavily on the Lovelace Byron papers).

Accessibility and intelligibility rather than protocol or scholarly precision have been my guiding principles when it comes to the treatment of quotations, names and aristocratic titles. Where there was no risk of changing the meaning, I have corrected and updated spellings and punctuation. I have avoided titles, and adopted the informality of first names for the main characters in the story: Ada, Annabella, William and Woronzow (Byron and Babbage, however, are referred to by their surnames, because it is by those names that they are almost universally known).

The notes that follow provide brief commentaries on points arising out of the text. I have given references for the Byron poems quoted, in the hope that readers who are unfamiliar with his work will use them as convenient ways into it.

INTRODUCTION

page 1 – 'Tis strange, but true ... *Don Juan*, Canto XIV, verse 101. As far as I know, this is the origin of the phrase that now has such popular currency.

Information regarding the history of the computer language Ada can be found in *Defence Science*, March 1984. In 1997, the US Department of Defense softened its policy on the mandatory use of a standardized language, preferring to adopt a less purist, more 'engineering'-oriented approach.

page 2 – Some have argued ... that for historical purposes the nineteenth century began not in 1800 but in 1789, and did not end until 1914. This was the period covered by Eric Hobsbawm in his great trilogy *Age of Revolution*, *Age of Capital* and *Age of Empire*. See his work on the 'short' twentieth century, *The Age of Extremes*, p. 6.

page 3 – These changes do not amount to a coordinated artistic or conceptual movement, more a 'mood', as one observer put it. See *The Naked Heart*, p. 38. This book is volume four of Peter Gay's huge survey of nineteenth-century culture, *The Bourgeois Experience: Victoria to Freud*.

page 5 – Science had paved 'the common road to all departments of knowledge', [Coleridge] once wrote ... Coleridge's views on science are scattered through his works, though the most comprehensive statement is contained in *Hints Towards the Formation of a more comprehensive Theory of Life*, published posthumously in 1848.

page 6 – Technology meant that science was not just interpreting the world, but changing it. In this passage I have treated science and technology as in some way necessarily connected but, as many historians of science (and scientists) have pointed out, this is not really the case. Many technological developments, for example steam locomotion, occurred without their inventors needing to know anything about the scientific principles underpinning them. Nevertheless, it is the discovery of those principles, particularly a belief in uniform laws of nature, that, I believe, provided the necessary intellectual setting for technology's rapid spread from the eighteenth century onwards.

CHAPTER ONE

page 10 – **For the first time in her life Ada, the daughter of Lord Byron, beheld in its full romantic glory the life-size face of her father.** No record remains of Ada's reaction to seeing the portrait of her father, but from the lack of any anxious communications between Dr King and Ada's mother it can be assumed she kept her feelings to herself, as any display of them would have been immediately reported and minutely dissected. The unveiling episode has been reconstructed, with a few embellishments, from a letter from Annabella to Ada which accompanied the present, and various pieces of correspondence between Ada, Annabella and Dr King. For an account of the veiling of the portrait at Kirkby, see *The Late Lord Byron*, Doris Langley Moore, p. 429. Annabella's mother Judith specifically stated in her will that the picture was to remain 'inclosed' until Ada reached twenty-one or Annabella decided she was ready to have it; see Elwin, *Lord Byron's Family*, p. 228. The portrait now hangs in the British Embassy in Athens. A copy is on display at the National Portrait Gallery in London.

The same year she saw her father's portrait, Ada was painted by the society portraitist Margaret Carpenter. Ada complained that Carpenter had made her jaw look so big the word 'Mathematics' could have been written across it (Ada to Annabella, 29 October 1835). The picture now hangs in 11 Downing Street, the official residence of the Chancellor of the Exchequer.

page 12 – **It appeared when the population of Britain was a fifth its current size and when less than half that population could read or write, and that rate of sale puts Byron's work on a par with *Diana: Her True Story in Her Own Words*.** Comparisons are obviously difficult over such an expanse of time. Nevertheless, according to Rita Alam of Whitaker Booktrack, which compiles sales data for the British book trade, Byron is, in sales terms, easily in the same league as Andrew Morton, whose Diana book sold forty-one thousand copies in the first week of publication. Typically, a book now needs to sell between six and eight thousand copies in the first week to reach the top of the best-seller charts. The statistics for population and literacy are taken from *Industry and Empire*, E. J. Hobsbawm.

page 13 – **And this hero, this Childe Harold was – everyone knew it, though he then denied it – himself.** On 31 October 1811 Byron wrote to his literary agent/adviser, Robert Dallas: 'I by no means intend to identify myself with *Harold*, but to *deny* all connection with him . . .' By the time he came to write the fourth and last canto of the poem, he had given up on denials altogether. In an open letter to his friend John Hobhouse, published as an introduction to the canto, he wrote: 'the fact is, that I had become weary of drawing a line which everyone seemed determined not to perceive'. Thomas Moore, Byron's biographer and friend, wrote: 'In one so imaginative as Lord Byron, who, while he infused so much of his life into his poetry, mingled also not a little of poetry with his life, it is difficult, in unravelling the texture of his feelings, to distinguish at all times between the fanciful and the real.' Annabella later observed that the source of all his main characters was 'self-inspection, and that his feelings & sympathy are only kindled by what he can identify with himself.' She was being caustic, but not entirely inaccurate.

page 13 – **'Ah, happy she!'** *Childe Harold's Pilgrimage*, Canto I, 5.

page 14 – **Lady Melbourne was a powerful, clever, beautiful Regency socialite and courtesan**. Contemplating Lady Melbourne's portrait in his old age, her son, who had by then become Prime Minister, said: 'A remarkable woman, a devoted mother, an excellent wife – but not chaste, not chaste.' (Lord David Cecil, *The Young Melbourne*, 1939, pp. 214–15.)

page 14 – **'Then dress, then dinner . . .'** *Don Juan*, Canto XI, 67.

page 16 – **It was his presentation copy of – received from the author that Annabella had originally read.** Harriet Beecher Stowe, the author of *Uncle Tom's Cabin* and a later champion of Lady Byron, did not pull her punches in describing the power of *Childe Harold*, to which she was obviously not immune:

> I remember well the time when this poetry, so resounding in its music, so mournful, so apparently generous, filled my heart with a vague anguish of sorrow for the sufferer, and of indignation at the cold insensibility that had maddened him. Thousands have felt the power of this great poem, which stands, and must stand to all time, a monument of what sacred and solemn powers God gave this wicked man, and how vilely he abused this power as a weapon to slay the innocent.

See Stowe's *Lady Byron Vindicated*, p. 29.

page 20 – **Caroline did exactly that, sending on the entire letter except for the last page, which she retained.** This is the conclusion drawn by Malcolm Elwin from the fact that the letter from Byron that Caroline forwarded to Annabella is to be found among the Lovelace Papers, and is missing the last page. The full version was published in *Letters and Journals*, ed. Rowland E. Prothero.

page 21 – **Was this another example of her reverse psychology, another attempt to pique Annabella's interest, rouse her redemptive instincts into action?** Not according to most biographers, who have settled on the simpler explanation that Caroline wrote to Annabella about the dangers of becoming entangled with Byron's world simply to discourage her. This may be correct, but ignores the role Caroline played in keeping Annabella's connection with Byron alive. It seems very likely that had not Caroline been there to act as an intermediary between the two of them, their relationship would have fizzled out.

page 23 – **Translating from her almost technical language . . .** Here is a typical example of Annabella's style: '[Byron's passions] have often enveloped [his intellect] in the obscurity of temporal delusion unenlightened by the *Faith* of an immortal existence.'

page 24 – **And it was all a part of his 'genius' . . .** The meaning of the word genius has changed over the last two centuries, and Annabella's use of it combines both older and more modern senses. She is here using the word as we now use it, to refer to some quality which is innate rather than acquired. But there is also an echo of the word's older meaning, the idea of a genius being a personal god or spirit allotted to each one of us at birth, and responsible for shaping our destiny thereafter.

page 33 – **'He lived – he breathed – he moved – he felt . . .'** *The Bride of Abydos*, Canto I, 12–13.

page 36 – **'That man of loneliness and mystery . . .'** *The Corsair*, Canto I, 9–10.

CHAPTER TWO

page 46 – **One evening, when they were alone together, she asked him . . . if there was anything that had changed his feelings about their marriage . . .** Only Annabella's account of this 'scene' survives, in one of the many narratives she wrote about her times with Byron, and in a conversation with Harriet Beecher Stowe reported in *Lady Byron Vindicated*. Both versions were recorded decades after the event. Byron mentioned the incident in passing in a letter to Lady Melbourne written a few days after it happened, but made light of it.

page 51 – **Annabella walked in accompanied by her childhood nanny, Mrs Clermont.** Mrs Clermont was never actually called a nanny by the Milbankes, but it is probably the most appropriate term to describe her relationship with Annabella.

page 52 – **Within the hour she was dressed in her travelling clothes – a slate-grey cape – and was escorted to a carriage waiting outside.** Ethel Colburn Mayne, Annabella's official (and sympathetic) biographer, saw the wedding dress and cape preserved at Ockham Park, and disputed Hobhouse's description, emphasizing instead the intricacy of the embroidery and the softness of the colours.

page 54 – **Whatever the truth of the [honeymoon] ... all the accounts of them that remain ... portray a couple sinking into a mixture of Gothic romance and horror ...** Most of what is known about the honeymoon comes from a series of narratives drawn up by Annabella after she separated from Byron, which are now to be found among the Lovelace papers. They were written to support her case for permanent separation and are therefore inevitably biased. But even Byron admitted his behaviour during this period was bad.

page 58 – **'For the Angel of Death spread his wings ...'** *The Destruction of Sennacherib*, III–IV.

page 59 – **[The lines to Thyrza] were not for Lady Falkland, nor for any woman; they were addressed to a youth, John Edleston ...** Many biographers have accepted at face value Byron's description of his love for Edleston as 'violent, though *pure*', with the notable exception of Louis Compton, whose *Byron and Greek Love: Homophobia in Nineteenth Century England* provides a detailed examination of Byron's ambiguous sexuality.

page 60 – **'There's not a joy the world can give ...'** The poem forms part of *Stanzas for Music*, along with lines Annabella thought might be about her: 'I speak not, I trace not, I breathe not thy name / There is grief in the sound, there is guilt in the fame ...'

page 61 – **In her account of the final days of their stay at Six Mile Bottom, Annabella resorted to a shorthand system to disguise several allusions to sexual issues.** The shorthand was transcribed by Ada's son Ralph when he collated Annabella's papers for his own work about the Byron marriage.

page 67 – '. . . **the lean dogs beneath the wall . . .**' *The Siege of Corinth*, XVI.

page 67 – '**All that of living or dead remain . . .**' *The Siege of Corinth*, XXXIII.

page 72 – **In a 'fantasy' newspaper interview with Byron . . .** The interview with Byron by Paul Johnson was published under the headline 'when I woke up next to my wife on our wedding night I thought that I was in Hell', and appeared in the *Daily Mail* on 27 December 1997.

page 73 – **Annabella learned of Le Mann's report on Byron's health: there was nothing wrong with him other than a 'liver complaint', the doctor announced, for which he had prescribed a course of calomel pills.** Le Mann's choice of medication is suggestive, calomel being a mercury-based preparation that was mainly used for the treatment of syphilis as well as liver complaints. Though Le Mann reportedly diagnosed liver disease as the cause of Byron's illness, it seems possible that there was a suspicion (entertained perhaps by Byron himself who was complaining of pains in his 'loins') that he was suffering from syphilis, the terminal stages of which are a form of madness which manifests itself as sexual incontinence and, intriguingly, an inability to do mathematical sums. During this period, many men who slept around became paranoiac about infection from the 'pox'– a form of hypochondria so common it was dubbed *syphilis imaginaria*. Syphilis was certainly a disease that, after a few often mild initial symptoms, would make itself known in a slow, surreptitious fashion, lying dormant for years before producing a variety of effects.

To prevent Byron from exercising his conjugal rights over Annabella following the separation, the lawyer Stephen Lushington argued that Byron's promiscuous behaviour meant he was likely to have become infected, and that he therefore threatened to pass the disease on to Annabella.

Some may argue that if Byron really was infected he surely would have passed it on to Annabella before her departure, and certainly to his later lover, Countess Teresa Guiccioli. This, of course, is entirely possible. Not everyone who is infected shows symptoms, and those that appear can often be hidden.

It is impossible to prove any of this, as venereal disease was not a subject written about openly, and any references to it may well have been removed from the archive by Annabella's descendants.

page 80 – **In a gesture of magnificent defiance, he proposed being carried off to the Continent in a replica of Napoleon's carriage, which he had commissioned at enormous cost from a London coachmaker.** The poor coachmaker was never to recover his costs; all he got from Byron were complaints about its uncomfortable ride.

CHAPTER THREE

page 98 – **Over 170 years later, a few days after the burial of Princess Diana, the similarities between Byron's life and death and hers were noted by at least one newspaper.** 'The princess and the people's poet', David Crane, *Daily Telegraph*, 9 September 1997.

page 112 – **'My very chains and I grew friends . . .'** *The Prisoner of Chillon*, XIV.

page 116 – Dr Herbert Mayo's views on food are to be found in his *Management of the Organs of Digestion*, a work that captures perfectly prevailing views of food, and offers an interesting contrast to our own.

page 120 – Ada's views on the 'Furies' and the account of her subsequent attempt to elope are contained in the manuscript of Woronzow Greig's 'biography' of Ada.

page 120 – **[Ada's lover's] identity remains a mystery, but he may have been William Turner . . . taken on to teach Ada shorthand around this time.** This suggestion is made by Betty Alexandra Toole in her collection of Ada's letters, *Ada, the Enchantress of Numbers*, p. 45. If it was the shorthand tutor, then the elopement must have taken place before the late spring of 1833, as I found among the Lovelace Papers a letter (in shorthand) dated 4 June 1833 from a 'James B', who identified himself as Ada's shorthand secretary. He had evidently been helping her with her shorthand since before her trips to London to meet Mary Somerville and Charles Babbage.

CHAPTER FOUR

page 125 – 'The sun went down, the smoke rose up . . .' *Don Juan*, X, 81–82.

page 125 – He or she would behold the glittering, gross Regency city glimpsed in Byron's other great epic, *Childe Harold*. See *Child Harold's Pilgrimage*, I, 69–70.

page 126 – In 1834, the Poor Law Amendment Act was passed, which brought new scientific thinking into the heart of public policy by enshrining in law Thomas Malthus's theory of population growth. Byron enjoyed poking fun at Malthus's heartless prescriptions for growing populations: see for example *Don Juan*, XII, 14.

page 127 – In Bloomsbury, Jeremy Bentham's University College . . . had just opened its doors to non-Anglicans. It was founded in 1826 as the University of London, and changed its name to University College in 1836, the year it received its Royal Charter.

page 130 – His audience may have seen complex clocks, perhaps the most complex of all, the great chronometers built by John Harrison . . . In *Don Juan*, Byron compared the mind of Donna Inez, the woman clearly modelled on Annabella, to 'the best timepiece made by Harrison', see I, 17.

page 131 – The Ada who came to see Babbage's marvellous invention was a changed woman . . . The chronology of Ada's attempts to reform herself is hard to piece together, as there are no dates for certain key episodes, notably the elopement. Going by the internal evidence from Ada's letters, it seems that she eloped in the early spring of 1833, and embarked on her programme of rehabilitation in April or May of that year, in preparation for her presentation at Court. The picture is confused by a letter to her mother dated 19 May 1833. This has been interpreted as an outburst of petulance in reaction to her frustrated elopement, but I have assumed that it refers to the incident with Charles Knight Murray.

Information on Charles Knight Murray was drawn from *The Work and Records of the Ecclesiastical Commission* by Elizabeth Finn, and supplied by the Church of England Record Centre.

page 136 – **Providence really did seem to be smiling on Ada at that moment . . . because her mathematical studies led to her being introduced to . . . Mary Somerville.** Other than the Somerville Papers themselves, much of the information about Mary Somerville comes from Elizabeth C. Patterson, author of *Mary Somerville and the Cultivation of Science, 1815–1840*, Boston, 1983, and 'Mary Somerville', *British Journal of the History of Science*, vol. 4, pp. 311–37.

page 141 – **'The hills are shadows, and they flow . . .'**, Tennyson, *In Memoriam*, vii.

page 144 – **'. . . *he* has his £20,000 snug!'** *The Collected Letters of Thomas Carlyle and Jane Welsh Carlyle*, ed. Charles Richard Sanders *et al.*, vol. 11, p. 11. To be fair on Carlyle's London Library, which was regularly used in the preparation of this book, it has probably one of the best collections of Victorian scientific books in the country, including most of Babbage's works.

page 150 – **'I conceived all of a sudden,' [De Prony] wrote in 1824, 'the idea of applying the same method . . . to manufacture logarithms as one manufactures pins.'** Translated and quoted in I. Grattan-Guinness, 'Work for the Hairdressers', *Annals of the History of Computing*, vol. 12, p. 179.

page 156 – **A particular draw was the tarantula spider made of steel . . .** See Simon Schaffer, 'Babbage's Dancer', in *Cultural Babbage*, ed. Francis Spufford and Jenny Uglow, p. 54. *Cultural Babbage* contains an excellent selection of papers relating to technology of this period. Of particular interest besides contributions by Schaffer and Doron Swade (see following note) is Tom Paulin's wonderfully titled article 'Serbonian Bog and Wild Gas: A Note and a Pamphlet', which relates in colourful detail some of the 'cultural wars' that erupted around science in the late eighteenth century.

page 157 – **Babbage would crank the handle once more and, without him intervening in any way, the number would suddenly leap to a new value, say 117.** Annabella gives her own account of Babbage's demonstration of the Difference Engine in a letter to Dr King dated 21 June 1833. However, her version is rather confused, so I have instead given a version here based on the example given in the essay 'It will not slice a pineapple', written by Doron Swade, the curator of computing at the London Science Museum, who oversaw the building of Babbage's Difference Engine

No. 2 at the Museum in 1971. His essay appears in *Cultural Babbage* (see previous note).

CHAPTER FIVE

page 179 – **This was the King estate, Ada's new rural domain. From here Ashley Combe could be reached via a path that disappeared into a thick wood . . .** There is an alternative, less dramatic road to Ashley Combe, via Porlock Weir, which they may have taken, but I suspect that William would have wanted to take the most picturesque route.

page 179 – **'. . . In the leafy month of June . . .'** Coleridge, 'The Rime of the Ancient Mariner', Part V.

page 180 – **And among the volumes, a collection of works by Byron, including *Mazeppa* and *Childe Harold*.** The list of books in the Ashley Combe library is contained in an undated memorandum book written in William's scrawled and largely illegible hand in the 1840s. It can now be found at the Somerset Record Office, DD/CCH 3/3.

page 181 – **'My daughter! with thy name this song begun . . .'** *Childe Harold*, Canto III, 115.

page 182 – **A geographer sent to map Somerset's wilder terrain had stood on the same spot in the 1780s.** The geographer was Edmund Rack, whose observations appeared in *The History and Antiquities of the Country of Somerset*, by the Reverend John Collinson, 1791. Rack died before the volume was completed. His notes on Culbone were among many contributions that were omitted from the published book. They are now held at the Somerset Record Office, A/AQP 8/6.

page 182 – **. . . this was one of the birthplaces of . . . English Romanticism.** The basis of this claim is the importance of the *Lyrical Ballads*. For example see Peter Gay, *The Naked Heart: The Bourgeois Experience*, vol. IV, p. 41, and Tom Mayberry, *Coleridge and Wordsworth in the West Country*, p. 123. The poetical innovations in *Lyrical Ballads* brought them a poor critical reception when they were first published in 1798, but Thomas De Quincy perceived them to be the 'ray of a new morning'.

page 183 – **He reached Culbone and decided to rest at Withycombe**

Farm . . . It is not certain that Coleridge stopped at Withycombe. He much later fancifully recalled the farm's name as Brimstone (there is no farm of that name, or any similar, near by). Two other possible contenders are Ash and Silicombe. Withycombe is the choice of Tom Mayberry, a local historian who has written a detailed account of Wordsworth's and Coleridge's time in Somerset. Silicombe is favoured by Michael Grevis in his 'Notes on the place of the composition of "Kubla Khan" by S. T. Coleridge' (*Charles Lamb Bulletin*, June 1991). All three farms were on William's estate, and were inhabited by his tenants. Ada would have known the tenants, and passed the houses on her frequent expeditions. I have tried to retrace Coleridge's steps by walking the coastal path that runs from Porlock to Culbone. It seems likely the route has changed little since then.

page 183 – **'The hanging Woods, that touch'd by Autumn seem'd . . .'** 'Poem to Alhadra, moorish woman in *Osorio*'; see *Coleridge's Letters*, ed. Earl Leslie Griggs, vol. I, p. 350. Quoted in Tom Mayberry, *Coleridge and Wordsworth in the West Country*, p. 103.

page 185 – **This was an idea explored by the physician John Elliotson in 1837 . . .** See John Elliotson, 'Clinical Lecture on Animal Magnetism', the *Lancet*, 9 September 1837, pp. 866–73.

page 187 – **'He kneels at morn, and noon and eve . . .'** 'Rime of the Ancient Mariner', part VII.

page 187 – **'And now the storm-blast came . . .'** 'Rime of the Ancient Mariner', part I.

page 196 – **A veritable epidemic of nervous diseases was, according to the physician George Cheyne, released . . .** Cheyne's book *The English Malady or a Treatise of Nervous Diseases of All Kinds* (1773) provides a fascinating glimpse into changing attitudes to mental illness. He seems to have been one of the earliest scientific writers to ascribe psychological disorders to social change.

page 198 – **. . . sexuality was allowed to 'flood into all areas of life', as one historian has put it.** See Michael Shortland, 'Courting the Cerebellum', *British Journal for the History of Science*, vol. 20, 1987, p. 195.

page 198 – **The phrenologist was one Dr Deville . . .** See Elizabeth S. Ridgway, 'John Elliotson (1791–1868): a bitter enemy of legitimate medicine?

Part I: Earlier years and the introduction to mesmerism', *Journal of Medical Biography*, 1993, vol. 1, p. 193.

CHAPTER SIX

page 207 – **According to Medora's self-vindicating account of her life** ... The account was first published in *Medora Leigh, a History and an Autobiography*, ed. C. Mackay, London, 1869. Doris Langley Moore published an unedited version in her biography of Ada, pp. 104ff.

page 210 – **This ingenious transaction was made possible by the Reversionary Interest Society.** The Byron and Leigh family finances were so complex they would probably defeat even a modern-day corporate tax accountant. They did not, however, daunt Doris Langley Moore, who devoted a whole book to them, *Lord Byron: Accounts Rendered*. She wrote it after a discussion in 1960 with John Murray, the descendant of Byron's publisher, on whether there was any aspect of Byron's life left to be written about. He suggested a book on Byron's finances might be interesting and (astonishingly) he was right. See Langley Moore, *Ada*, p. 117 for an explanation of the intricacies of reversionary interest.

page 211 – **She began to spot biographical details hidden in the words** ... The examples of suggestive references in Byron's poetry to his relationship with Augusta are to be found in the following poems: *Manfred*, II, ii, 105–7, *The Giaour*, lines 1048/9, *The Bride of Abydos*, Canto I, 12, *The Corsair*, Canto I, 12.

page 215 – '. . . **Godlike crime . . .**' *Prometheus*, III.

page 219 – '. . . **Deformity is daring . . .**' *The Deformed Transformed*, I, i, 313–21.

page 219 – **But mesmerism's influence runs even deeper than that.** See Adam Crabtree, *From Mesmer to Freud: Magnetic Sleep and the Roots of Psychological Healing*, London, 1993 for a full discussion of the relationship of mesmerism and psychoanalysis. I have taken the text for the secret annexe to the report Benjamin Franklin's committee wrote on mesmerism from pp. 92–3.

page 224 – **The historian Fred Kaplan attributes its popularity to** the way it 'anticipated the coming psychological crisis of Western man . . .' See Fred Kaplan, 'The Mesmeric Mania', *Journal of the History of Ideas*, 1974, 35, p. 693.

page 226 – Elizabeth Okey, then about fourteen, and her sister Jane, twelve, were almost wild . . . In contemporary reports, Okey is sometimes spelled O'Key. Accounts of the experiments Elliotson undertook with the girls can be found in 'Experiments performed on Elizabeth and Jane Okey at the house of Mr Wakley, Bedford Square', *The Lancet*, 1838, ii, 805–14. The pamphlet sensationalizing Elliotson's work, 'A Full Discovery of the Strange Practices of Dr. Elliotson', is held at the Wellcome Institute library.

page 231 – . . . she half-jokingly referred to her stomach as the seat of *"original sin"* '. See Pforzheimer Collection, B.ana 535 Ada to Mr Fonblanque, 13 Nov. [1837].

page 233 – When he met Annabella, Kay was living in Battersea . . . Kay had a street named in his honour in Battersea called Shuttleworth Road. It is very close to where this book was written. The manuscript for Kay's journal is held among the Kay–Shuttleworth papers at the John Rylands Library, Manchester University, ref. 219, 1/25.

page 237 – Annabella had now pledged to help Medora get hold of this document . . . Technically speaking, Medora would have been the beneficiary of the deed drawn up by Augusta, but its main purpose was to provide for Marie. Being illegitimate, the girl could count on no inheritance to help her in the event of Medora's death, which, due to chronic illness, was thought to be imminent.

CHAPTER SEVEN

page 248 – There could be found such innovations as a 'reading machine', an educational device she had apparently designed for Dr Kay . . . The machine is mentioned in Kay's journal: 'Instructed cabinet maker how to construct one of the "Reading Machines" of Lady L's . . . Saw Mr Thurgau respecting the preparation of the work on the Phonic method which is to accompany the Reading Machine.' Kay–Shuttleworth papers, John Rylands Library, Manchester University, ref. 219, 1/25.

page 258 – **One baffled professor asked how it could do such a thing**
... The professor was Ottavino Fabrizio Mosotti; Babbage relates the story
in *Passages*, p. 98.

page 262 – **By the mid-nineteenth century, fairies were everywhere**
... For an excellent account of the way fairies fluttered into nineteenth-
century culture, see Nicola Brown, *Victorian Fairy Painting – An Introduction to the
Exhibition*. The exhibition was held at the Royal Academy of Arts in November
1997.

page 274 – **Even time as we know it now was then a novelty.** For a
general discussion about the evolution of the modern view of time, see G. J.
Whitrow, *Time in History*, 1988.

page 281 – **The following morning she sent some more of her revisions to him, together with a brief note.** This note, never published
before, throws an important new light on Ada's state of mind at the time, and
demonstrates that she was not as delusional about her mental state as some
have suggested. The original is to be found in the Pforzheimer Collection, ref.
B.ana 90 Ada to Babbage, Tuesday Morning 15 August [1843].

CHAPTER EIGHT

page 285 – **The Dr Carpenter who arrived at Ashley Combe for his
first encounter with Ada was a tall, wiry, earnest man** ... I have
made certain assumptions about the chronology during this period, as the
dating of much of the correspondence between the parties involved is
inaccurate or unclear. The key events all took place between mid-December
1843 and mid-January 1844, but the exact order is hard to establish. In
particular, a meeting Carpenter had with Ada at her home in St James's
Square, London may have occurred before or after Christmas; Doris Langley
Moore suggests the former, I have chosen the latter.

page 289 – **The *Edinburgh Medical and Surgical Journal* attacked him
for reducing the miracle of Creation to a machine** ... See 'Carpenter's
Principles of General and Comparative Physiology', *Edinburgh Medical and
Surgical Journal*, 1840, vol. 53, pp. 213–28. His defence appeared in 'Remarks
on some Passages of the "Review of Principles of General and Comparative

Physiology" in the *Edinburgh Medical and Surgical Journal*, January 1840, by William B. Carpenter', *British and Foreign Medical Review*, vol. 9, appendix.

page 290 – **Having agreed to keep everything she told him secret, even from her mother, he stood open-mouthed as she began to unravel herself before him.** Ada's letters to Carpenter, which contained her confessions and revelations, do not remain, so their contents can only be surmised from his (very long and detailed) replies, which are all to be found among the Lovelace Papers (box 169).

page 290 – **'For all was blank, and bleak, and grey . . .'** *The Prisoner of Chillon*, IX.

page 297 – **William loved tunnels and towers.** See Stephen Turner, 'William, Earl of Lovelace, 1805–1893', *Surrey Archaeological Collections*, vol. 70, pp. 108–9. East Horsley Towers, as they became known, were taken over by the Central Electricity Generating Board and then by a company which turned them into a management training centre. The tunnels and tower are still there. Most of the tunnels leading to Ashley Combe have been blocked up. However, one, just a few yards in length, remains open. The coastal public path from Porlock to Culbone passes through it.

page 303 – **Sedgwick was one of those who suspected Ada to be the author of [*Vestiges*], mentioning his suspicions in a letter to Mcvey Napier, editor of the *Edinburgh Review*.** See British Library Add. MSS 34625, Segwick to Napier undated.

page 304 – **When he was asked years later what reasons he had for not putting his name to the work, Chambers pointed to his house, in which lived his eleven children, and said, 'I have eleven reasons.'** Quoted in James A. Secord, 'Behind the Veil: Robert Chambers and *Vestiges*', *History, Humanity and Evolution*, ed. James R. Moore, p. 186.

CHAPTER NINE

page 321 – **'To Norman Abbey whirled the noble pair . . .'** *Don Juan*, XIII, 55.

page 322 – **A contemporary account records how the roof had all**

but collapsed . . . See *The Mirror of Literature, Amusement and Instruction*, vol. 3, no. LXVII, 24 January, 1824. A copy is on display at Newstead.

page 327 – **In the centre of the wood, by now dark and overgrown, there was an elm with two trunks springing from the same root.** The segment of trunk upon which Byron carved his name alongside Augusta's is preserved in the Newstead collection, and is on public display. Thanks to a policy of low-key intervention adopted by its owner, Nottingham City Council, the Abbey has kept much of its character. The atmosphere is mercifully preserved from interactive exhibits or theme park rides, and will thrill anyone with even a passing interest in Byron or, indeed, Romanticism.

page 328 – **While wandering alone in the midst of this evocative landscape, Ada was approached by the colonel.** The account for this version of events appears in *Notes in England and Italy*, ed. Julian Hawthorne (New York, 1869). It is based on a diary kept by Sophia Amelia Hawthorne when she visited England in 1857, the story being related to her by a local innkeeper. Though there are inaccuracies, Sophia's account is consistent enough with Ada's letters to make it appear reasonably reliable.

page 332 – **[The Zetlands] owned her favourite horse, one of the most celebrated of the time: Voltigeur.** A race run at York is named in Voltigeur's honour.

page 334 – **Perhaps [Mary Wilson] was the invalid.** The evidence that the 'invalid' mentioned in Ada and Babbage's correspondence was Mary is rather convoluted. It is contained in a letter found among the Lovelace papers, sent by Babbage to Ada on 13 January 1851. This mentions that he was forwarding something given to him by Sir James South which the latter thought Mary might find useful. That something appears to have been some sort of book, though Babbage's writing is difficult to decipher.

page 336 – **[John Crosse] was forced to destroy nearly all his correspondence with Ada . . .** There is one letter among the Lovelace Papers from him to her reporting his health following an operation, and another perhaps written by him bearing scrawled initials that appear to read 'JC'.

page 338 – **Ada chose to avoid the postal system in communicating an extraordinary note to Prandi . . .** This intriguing and, because of its lack of details, frustrating letter, never before published, can be found in the Pforzheimer Collection, Ada to Prandi, ref B.ana 70, undated.

page 339 – **In a brief but characteristically intimate correspondence with Edward Bulwer-Lytton . . .** The correspondence is to be found among the Bulwer-Lytton papers, Hertfordshire Record Office, ref. D/EK C3/15.

page 339 – **Another man who seemed to bring out the Byronic streak in her was Richard Ford, son of a magistrate . . .** Intriguing material recently uncovered concerning Richard Ford's father and Wordsworth can be found in Kenneth R. Johnston, *The Hidden Wordsworth*, 1998.

page 339 – **A particularly imaginative theory about the mysterious 'book' referred to in Ada's correspondence with Babbage was that the word was being used in the gambling sense, to refer to a record of bets.** This theory was first proposed by Maboth Mosley in her 1964 biography *Irascible Genius: A Life of Charles Babbage, Inventor*.

page 340 – **[The gambling syndicate's] membership comprised . . . someone called Nightingale (almost certainly William, father of the nursing pioneer Florence, who Ada had visited during her tour of northern England) . . .** It is not certain that the Nightingale who formed part of Ada's circle was William; it may have been another male member of the family, or indeed another Nightingale altogether. However, William was of the right age, and he would have been her host when she visited the Nightingale home at Lea Hurst in Derbyshire during her tour of the north.

page 343 – **Through their racing connections they would have knowledge of people like Malcolm, and would know how to deal with them.** There is a letter from William to Lord Zetland dated 24 May [1851] (held at the North Riding Record Office, ref. ZNKX8) in which he thanks Zetland's wife for the 'shrewdness of her advice' regarding the 'whole affair', and expresses his appreciation for the 'sympathy' the Zetlands have shown. Given the date, it seems likely that the 'affair' was Ada's disastrous Derby Day adventure.

page 343 – **Locock was one of the Victorian era's most respected gynaecological experts.** Biographical information about Locock is sparse. I have drawn mostly on William Munk, *Roll of the Royal College of Physicians*, 1878, vol. 3. The episode concerning Locock and Ruskin is documented in Ronald Pearsall, *The Worm in the Bud*, 1969.

page 344 – **'I'd rather have 10 or 5 what I call *real* years of life, than**

20 or 30 such as I see people usually dawdling on, without any spirit,' she wrote to a friend. The friend was the chemist H. Bence Jones. The letter is held at the Cambridge University Library, ref. Add. 8546/I/144.

CHAPTER TEN

page 349 – **Annabella was staying at Leamington Spa, undergoing one of her endless series of health cures when William came to see her.** The only account of this meeting that remains is one written by Annabella and given to her solicitors, dated 1 July 1851, which is among the Lovelace Papers. The text appears in Doris Langley Moore, *Ada*, p. 288.

page 367 – **Then another name enticingly presents itself: that of Thomas Carlyle.** For references to Ada in Carlyle's correspondence, see *The Collected Letters of Thomas Carlyle and Jane Welsh Carlyle*, ed. Charles Richard Sanders *et al.*, vol. 2, 13 June 1822; vol. 16, 24 March 1846; vol. 20, 17 May 1849.

EPILOGUE

page 375 – **'One thing certainly contemplated in the scheme of creation,' wrote . . . Dr Herbert Mayo, 'is a progression from barbarism to civilization.'** See Herbert Mayo, *The Nervous System and its Functions*, London, 1842, p. 157.

page 376 – **'like me in lineaments.'** *Manfred*, II, ii, 105.

page 377 – **Quantum mechanics has given us the Uncertainty Principle, chaos science unpredictability in the midst of determinism** . . . The cosmological implications of such developments are not always as far reaching as they sound. Scientists have given words like 'uncertainty' and 'chaos' technical meanings that are not quite the same, nor quite as suggestive, as their normal ones.

Select Bibliography

Anderson, Ian G., *History of Esher*, 1948.

Arnold, Rev. Frederick, *Robertson of Brighton with Some Notices of His Times and Contemporaries*, 1886.

Babbage, Charles, *The Ninth Bridgewater Treatise, A Fragment*, 1837; *Passages in the Life of a Philosopher*, 1864.

Barber, Rev. Canon T. G., *Byron and Where he is Buried*, 1939.

Bishop, Morchard (ed.), *Recollections of the Table-Talk of Samuel Rogers*, 1952.

Bowra, Maurice, *The Romantic Imagination*, 1950.

Byron, Anne Isabella Noel, *Remarks occasioned by Mr. Moore's Notices of Lord Byron's Life*, 1830.

Campbell-Kelly, M. (ed.), *The Works of Charles Babbage*, 11 vols, 1989.

Carlyle, Thomas, *Sartor Resartus*, 2nd edn, 1841, in *Selected Writings*, Alan Shelston (ed.), 1971.

Carpenter, William Benjamin, 'Carpenter's Principles of General and Comparative Physiology', *Edinburgh Medical and Surgical Journal*, 1840, vol. 53, pp. 213–28; 'Remarks on some Passages of the "Review of Principles of General and Comparative Physiology" in the *Edinburgh Medical and Surgical Journal*, January 1840, by William B. Carpenter', *British and Foreign Medical Review*, vol. 9, appendix.

Cecil, Lord David Gascoyne, *The Young Melbourne*, 1939.

Cheyne, George, *The English Malady or a Treatise of Nervous Diseases of All Kinds*, 1773.

Coleridge, Samuel Taylor, *Hints Towards the Formation of a more comprehensive Theory of Life*, 1848.

Collinson, Rev. John, *The History and Antiquities of the Country of Somerset*, 1791.

Combe, George, *A System of Phrenology*, 5th edn, 1843.

Crabtree, Adam, *From Mesmer to Freud: Magnetic Sleep and the Roots of Psychological Healing*, 1993.

Crompton, Louis, *Byron and Greek Love: Homophobia in 19th-century England*, 1985.

Crosse, Cornelia, *Red Letter Days of My Life*, 2 vols, 1892.

Dallas, R. C., *Correspondence of Lord Byron, with a friend*, 1825.

De Morgan, Sophia Elizabeth, *Memoir of Augustus De Morgan*, 1882; *Threescore Years and Ten*, 1895.

Desmond, Adrian, and Moore, James, *Darwin*, 1991.

Eadie, John *et al.* (eds), *Imperial Dictionary of Universal Biography, c.* 1845.

Elwin, Malcolm, *Lord Byron's Wife*, 1962; *The Noels and the Milbankes*, 1967; *Lord Byron's Family*, 1975.

Foster, Vere (ed.), *The Two Duchesses: Georgiana, Duchess of Devonshire, Elizabeth, Duchess of Devonshire*, 1898.

Gay, Peter, *The Bourgeois Experience, Victoria to Freud*, 5 vols, 1984–1998.

Gibson, William and Sterling, Bruce, *The Difference Engine*, 1990.

Golby, J. M. (ed.), *Culture and Society in Britain, 1850–1890*, 1986.

Grattan-Guinness, I., 'Work for the Hairdressers', *Annals of the History of Computing*, 1990, vol. 12, pp. 177–185.

Grosskurth, Phyllis, *Byron: The Flawed Angel*, 1997.

Hall, Marshall, *Lectures on the Nervous System and its Diseases*, 1836.

Hamilton, James, *Turner and the Scientists*, 1998.

Hawthorne, Julian (ed.), *Notes in England and Italy*, 1869.

Himmelfarb, Gertrude, *Marriage and Morals among the Victorians*, 1986.

Hobhouse, John Cam, *Contemporary Account of the Separation of Lord and Lady Byron*, 1870; *Recollections of a Long Life*, 6 vols., 1909–1911.

Houghton, W. E., *The Victorian Frame of Mind, 1830–1870*, 1957.

Irving, Washington, *Abbotsford and Newstead Abbey*, 1835.

Jalland, Pat, *Death in the Victorian Family*, 1996.

James, George Payne Rainsford, *Morley Ernstein*, 1846.

Johnston, Kenneth R., *The Hidden Wordsworth*, 1998.

Kaplan, Fred, '"The Mesmeric Mania": The Early Victorians and Animal Magnetism', *Journal of the History of Ideas*, 1974, vol. 35, pp. 691–702.

Knight, Frida, *University Rebel: The Life of William Frend*, 1971.

Laplace, Pierre-Simon, *Philosophical Essay on Probabilities*, trans Andrew I. Dale, 1995.

Lightman, Bernard (ed.), *Victorian Science in Context*, 1997.

Lovelace, Ada, 'Sketch of the Analytical Engine invented by Charles Babbage Esq., by L. F. Menabrea, of Turin, Officer of the Military Engineers, translated with notes by A. A. L.', *Taylor's Scientific Memoirs*, 1843, vol. 3, pp. 666–731.

Lovelace, Mary, Countess of, *Ralph, Earl of Lovelace, A Memoir*, 1920.

Mackay, Charles, *Medora Leigh: A History and an Autobiography*, 1869.

Marchand, Leslie A., *Byron: A Biography*, 3 vols, 1957; *Byron: A Portrait*, 1971; (ed.), *Byron's letters and journals*, 1973–1994.

Martineau, Harriet, *Letters on Mesmerism*, 1845.

Mason, Michael, *The Making of Victorian Sexual Attitudes*, 1994.

Mayberry, Tom, *Coleridge and Wordsworth in the West Country*, 1992.

Mayne, Ethel Colburn, *The Life and Letters of Annabella Lady Byron*, 1929.

Mayo, Herbert, *Management of the organs of digestion in health and in disease*, 1837; *A Treatise on Siphilis*, 1840.

Milbanke, Ralph Gordon Noel, *Astarte*, 1905.

Moore, Doris Langley, *The Late Lord Byron*, 1961; *Lord Byron: Accounts Rendered*, 1974; *Ada, Countess of Lovelace*, 1977.

Moore, Thomas, *Letters and Journals of Lord Byron*, 1830.

Moseley, Maboth, *Irascible Genius: The Life of Charles Babbage, Inventor*, 1964.

Newsome, David, *The Victorian World Picture*, 1997.

Nichol, John Pringle, *Views of the Architecture of the Heavens, in a Series of Letters to a Lady*, 1837.

Nicolson, Harold, *Byron, the Last Journey*, 1924.

Patterson, Elizabeth Chambers, 'Mary Somerville', *British Journal of the History of Science*, vol. 4, pp. 311–35; *Mary Somerville and the Cultivation of Science*, 1983.

Pearsall, Ronald, *The Worm in the Bud: The World of Victorian Sexuality*, 1969.

Quennell, Peter, *Byron: A Self-portrait*, 1990.

Ridgway, Elizabeth S., 'John Elliotson (1791–1868): a bitter enemy of legitimate medicine? Part I: Earlier years and the introduction to mesmerism', *Journal of Medical Biography*, 1993, vol. 1, pp. 191–8; 'Part II: The mesmeric scandal and later years', 1994, vol. 2, pp. 1–7.

Robinson, Henry Crabb, *Diary, Reminiscences and Correspondence*, ed. Thomas Sadlier, 1869.

Rowell, Geoffrey, *Hell and the Victorians: A Study of the nineteenth-century theological controversies concerning eternal punishment and the future life*, 1974.

Sanders, Charles Richard *et al.* (eds), *The Collected Letters of Thomas Carlyle and Jane Welsh Carlyle*, 1970– .

Secord, James A., 'Behind the Veil: Robert Chambers and *Vestiges*', *History, Humanity and Evolution*, James R. Moore (ed.), 1989, pp. 165–94.

Selleck, R. J. W., *James Kay-Shuttleworth: Journey of an Outsider*, 1994.

Shortland, Michael, 'Courting the Cerebellum', *British Journal for the History of Science*, 1987, vol. 20, pp. 173–99.

Somerville, Mary, *The Connexion of the Physical Sciences*, 1834.

Spufford, Francis and Uglow, Jenny (eds), *Cultural Babbage: Technology, Time and Invention*, 1996.

Stein, Dorothy, *Ada, A Life and a Legacy*, 1985.

Stowe, Harriet Elizabeth Beecher, *Lady Byron Vindicated*, 1870.

Swade, Doron, *Charles Babbage and his Calculating Engines*, 1991.

Thomas, Clara Eileen McCandless, *Love and Work Enough: the life of Anna Jameson*, 1967.

Toole, Betty Alexandra, *Ada, the Enchantress of Numbers*, 1992.

Trinder, Barrie, *The Making of the Industrial Landscape*, 1982.

Turner, Stephen, 'William, Earl of Lovelace, 1805–1893', *Surrey Archaeological Collections*, vol. 70, pp. 101–29.

Veith, Ilza, *Hysteria: The History of a Disease*, 1970.

Warter, John Wood (ed.), *Southey's Common-Place Book*, 1876.

Yeo, Richard, 'Science and Intellectual Authority in mid-19th century Britain: Robert Chambers and *Vestiges of the Natural History of Creation*', *Victorian Studies*, 1984, vol. 28, pp. 5–31.

Index

www.panmacmillan.com